W0106917

Techniques and Concepts of High-Energy Physics II

NATO Advanced Science Institutes Series

A series of edited volumes comprising multifaceted studies of contemporary scientific issues by some of the best scientific minds in the world, assembled in cooperation with NATO Scientific Affairs Division.

This series is published by an international board of publishers in conjunction with NATO Scientific Affairs Division

A Life Sciences **B Physics**	Plenum Publishing Corporation New York and London
C Mathematical and Physical Sciences	D. Reidel Publishing Company Dordrecht, Boston, and London
D Behavioral and Social Sciences **E Applied Sciences**	Martinus Nijhoff Publishers The Hague, Boston, and London
F Computer and Systems Sciences **G Ecological Sciences**	Springer Verlag Heidelberg, Berlin, and New York

Recent Volumes in Series B: Physics

Volume 95 —Advances in Laser Spectroscopy
edited by F. T. Arecchi, F. Strumia, and H. Walther

Volume 96 —Atomic Physics of Highly Ionized Atoms
edited by Richard Marrus

Volume 97 —Mass Transport in Solids
edited by F. Bénière and C. R. A. Catlow

Volume 98 —Quantum Metrology and Fundamental Physical Constants
edited by Paul H. Cutler and Amand A. Lucas

Volume 99 —Techniques and Concepts in High-Energy Physics II
edited by Thomas Ferbel

Volume 100—Advances in Superconductivity
edited by B. Deaver and John Ruvalds

Volume 101—Atomic and Molecular Physics of Controlled Thermonuclear Fusion
edited by Charles J. Joachain and Douglass E. Post

Volume 102—Magnetic Monopoles
edited by Richard A. Carrigan, Jr., and W. Peter Trower

Techniques and Concepts of High-Energy Physics II

Edited by

Thomas Ferbel

University of Rochester
Rochester, New York

Plenum Press
New York and London
Published in cooperation with NATO Scientific Affairs Division

Proceedings of the Second NATO Advanced Study Institute on Techniques and Concepts of High-Energy Physics, held July 1–12, 1982, at Lake George, New York

Library of Congress Cataloging in Publication Data

NATO Advanced Study Institute on Techniques and Concepts of High-Energy Physics (2nd: 1982: Lake George, N.Y.)
Techniques and concepts of high-energy physics II.

(NATO advanced science institutes series. Series B, Physics; v. 99)
"Published in cooperation with NATO Scientific Affairs Division."
"Proceedings of the Second NATO Advanced Study Institute on Techniques and Concepts of High-Energy Physics, held July 1–12, 1982, at Lake George, New York"—Verso of t.p.
Includes bibliographical references and index.
1. Particles (Nuclear physics)—Congresses. I. Ferbel, Thomas. II. North Atlantic Treaty Organization. Scientific Affairs Division. III. Title. IV. Series.
QC793.N38 1982 539.7 83-11073
ISBN-13:978-1-4613-3747-8 e-ISBN-13:978-1-4613-3745-4
DOI: 10.1007/978-1-4613-3745-4

Softcover reprint of the hardcover 1st edition 1983

A Division of Plenum Publishing Corporation
233 Spring Street, New York, N.Y. 10013

PREFACE

The second Advanced Study Institute (ASI) on Techniques and Concepts of High Energy Physics was held at the Roaring Brook Resort at Lake George, New York. As in the case of the first ASI, our aim was to bring together a small group of promising young experimenters and several outstanding senior scholars in high energy physics in order to learn about the latest trends in the field and develop stronger contacts among scientists from different countries and different backgrounds. The setting at Roaring Brook was particularly congenial and the staff, under the direction of George Green, was both friendly and efficient.

The ASI was supported mainly through funds provided by the Scientific Affairs Division of NATO. It was cosponsored by the U. S. Department of Energy, Fermilab, the National Science Foundation and the University of Rochester. A special grant from the Oliver S. and Jennie R. Donaldson Charitable Trust provided a valuable degree of flexibility for supporting worthy students.

The scientific program of this ASI was, once again, designed primarily for advanced graduate students and recent Ph.D recipients in experimental particle physics. I believe, however, that the contents of the present volume will prove useful to an even wider audience of physicists.

It is a pleasure to acknowledge the encouragement and support I have received from many of my colleagues and friends in organizing this meeting. I am indebted to the members of my Advisory Committee, particularly to Maurice Jacob and Chris Quigg, for their patience and good advice. I am grateful to the superb lecturers for their participation in the ASI. I thank Earle Fowler, Bernie Hildebrand and Bill Wallenmeyer for the support of the Department of Energy, and Dave Berley for the assistance from the National Science Foundation. I thank Leon Lederman for providing me the excellent services of Angela Gonzales and Jackie Coleman at Fermilab. At Rochester, I am indebted to Mike Anthony for his assistance with the financial planning, and I am grateful to Edna Hughes, Judy Mack and Connie Murdoch for sundry help and expert typing. I wish to acknowledge the generosity of Chris Lirakis and of Mrs. Marjorie Atwood of the Donaldson Trust. Finally, I thank Drs. Craig Sinclair and Mario di Lullo of NATO for their cooperation and support of the ASI.

T. Ferbel
Rochester, N.Y.
March 1983

CONTENTS

Grand Unification.................................... 1
C. H. Llewellyn Smith

Gauge Theories and Monopoles........................... 47
N. Cabibbo

Lectures on Neutrino Physics........................... 63
F. Sciulli

Topics in Electron-Positron Interactions............. 97
P. Söding

Probability, Statistics, and Associated Computing Techniques.. 189
F. James

A Thousand TeV in the Center of Mass: Introduction to High Energy Storage Rings............ 233
J.D. Bjorken

New Developments in Gaseous Detectors................ 301
F. Sauli

Land of QCD.. 351

Participants... 353

Committee.. 355

Index.. 357

GRAND UNIFICATION

C.H. Llewellyn Smith
Department of Theoretical Physics
University of Oxford
1 Keble Road, Oxford OX1 3NP, England

1. INTRODUCTION

The purpose of these lectures is to provide a critical introduction to Grand Unified Theories (GUTs) in which technicalities are minimized but the issues are not simplified. Although only the standard SU(3)xSU(2)xU(1) model of low energy interactions and the minimal SU(5) GUT are considered in any detail, I refer briefly to many alternatives. However, I have not attempted to present a systematic review of all models nor do I give careful references to the original literature, which may be found in the comprehensive review of GUTs by Langacker[1] and the excellent lectures on the phenomenology of gauge theories by Ellis[2]. A set of problems is provided at the end of the lectures, which are intended to help readers check and amplify their understanding.

In order to study GUTs it is necessary to believe that there is a reasonable chance that both the SU(2)xU(1) model of electroweak interactions (or some alternative gauge theory) and QCD are correct and to understand these models. I therefore begin by reviewing the essential features of the standard SU(3)xSU(2)xU(1) model and the reasons for taking it seriously. After discussing the shortcomings of the standard model, I briefly consider attempts to make a truly unified electro-weak theory out of SU(2)xU(1). This leads into a discussion of GUTs. I describe the structure of the minimal SU(5) model in some detail. Following a digression on the general question of baryon conservation and cosmology, expectations for nucleon decay in SU(5) are reviewed. Next I summarize the shortcomings of SU(5) before briefly considering attempts to make better GUTs based on larger groups. This is followed by a more general phenomenological discussion of baryon number violation and neutrino masses, whose validity transcends GUTs. Finally I summarize the Grand Problems unanswered by GUTs.

2. WHY BELIEVE IN SU(2)xU(1)?

According to SU(2)xU(1) the couplings of vector bosons to fermions are described by[3]

$$L_{int} = \sum_i \bar{\psi}^i_L \gamma_\mu (g\vec{T}.\vec{W}_\mu + g'\frac{Y}{2}B_\mu)\psi^i_L + \sum_j \bar{\psi}^j_R \gamma_\mu g'\frac{Y}{2}B_\mu \psi^j_R \tag{1}$$

where $\vec{T}$ are 2x2 isospin matrices, $\vec{W}_\mu$ the isotriplet of SU(2) gauge bosons and B_μ the U(1) gauge boson. For the first "generation" of fermions the fields are

$$\psi_L = \begin{pmatrix} \nu_e \\ e \end{pmatrix}_L, \begin{pmatrix} u \\ d \end{pmatrix}_L ; \qquad \psi_R = e_R, u_R, d_R,$$

(neglecting Cabibbo/Kobayashi-Maskawa mixing), where

$$u_{L,R} = \left(\frac{1\mp\gamma_5}{2}\right) u \text{ etc.}$$

The neutral mass eigenstates are defined to be

$$\begin{aligned} A_\mu &= W^3_\mu \sin\theta + B_\mu \cos\theta \\ Z_\mu &= W^3_\mu \cos\theta - B_\mu \sin\theta. \end{aligned} \tag{2}$$

Examining the couplings of the photon (A_μ) we find

$$e = g \sin\theta = g'\cos\theta, \quad Q = T_3 + \frac{Y}{2}. \tag{3}$$

The $(\text{mass})^2$ matrix for W^3_μ and B_μ necessarily has the form

$$\begin{pmatrix} \hat{M}^2_W & \pm\hat{M}_W M_B \\ \pm\hat{M}_W M_B & M^2_B \end{pmatrix} \tag{4}$$

where the fact that $M_\gamma = 0$ requires zero determinant and fixes the off diagonal elements. The simplest possible assumption is that $\hat{M}_W$ in (4) is equal to M_W, the mass of the charged W. This gives $M_Z = M_W \sec\theta$, using (2).

In the limit of zero momentum transfer, the Born amplitudes for single $W^\pm$ and Z exchange are given by the following "effective" Lagrangian":

$$L_{eff} = \frac{4G_F}{\sqrt{2}} [(J^c_\mu)^+ J^c_\mu + J^N_\mu J^N_\mu] \tag{5}$$

where

$$J_\mu^c = \sum_i \bar{\psi}_L^i \gamma_\mu T^+ \psi_L^i$$

$$J_\mu^N = \sum_i \bar{\psi}_L^i \gamma_\mu (T_3 - Q\sin^2\theta)\psi_L^i + \sum_j \bar{\psi}_R^j \gamma_\mu (-Q\sin^2\theta)\ \psi_R^j \qquad (6)$$

Eqn. 5 successfully describes all existing leptonic and semi-leptonic data. Possible deviations are less than or equal to 10% in amplitude. For example:

1. If we add a charged current term with $(e,\nu,u,d)_L \rightarrow (e,\nu,u,d)_R$ and $G \rightarrow \eta G$, the data[4] require $\eta < 0.1$.

2. If the neutral current term is multiplied by a parameter ρ, the data give[5] $\rho = 0.992 \pm .017$. With $\hat{M}_W \neq M_W$ in eqn. 4, we would have obtained $\rho = \frac{\sec^2\theta \hat{M}_W^2}{M_Z^2} \neq 1$ so the data confirm the simple assumption $\hat{M}_W = M_W$ in the context of SU(2)xU(1).

3. If we add a term

   $$C \frac{4G}{\sqrt{2}} (J_\mu^{em})^2$$

   $\bar{e}e \rightarrow \bar{\mu}\mu$ data require[6] $C < 0.020$ (95% CL).

It is not clear whether eqn. (5) correctly describes non-leptonic interactions. Measured $\Delta I = 1/2[3/2,5/2]$ amplitudes are bigger [smaller] than naive guesses based on (5) by a factor of about 4. Given our ignorance of how to calculate non-leptonic matrix elements[7], the data are certainly not inconsistent with (5) but there is plenty of scope for non-standard contributions.

Thus apart from non-leptonic processes, eqn. (5) is rather well established. However, there is no evidence for the W and Z, let alone for the W^+W^-Z vertex predicted by gauge theories or the Higgs mechanism. Why then should we believe in SU(2)xU(1)? The elegance of local gauge invariance is often invoked but, when the Higgs interaction is included, the SU(2)xU(1) Langrangian is complex and ugly compared to L_{eff} (eqn 5). Indeed L_{eff} has an elegant global SU(2) symmetry in the limit $\theta=0$, which might be generated in other ways[8] e.g. by coupling the standard weak isospin current $\vec{J}_\mu$ to heavy fermions (F) whose interactions respect an SU(2) invariance; $\theta \neq 0$ would be generated by mixing with electromagnetism, which inevitably occurs if $Q_F \neq 0$.

The reasons for believing in SU(2)xU(1) are connected with unitarity and renormalizability. Breaking the argument into three steps[9]

1. L_{eff} (eqn. 5) cannot be a true Lagrangian. If used beyond Born approximation it leads to hopelessly divergent results:

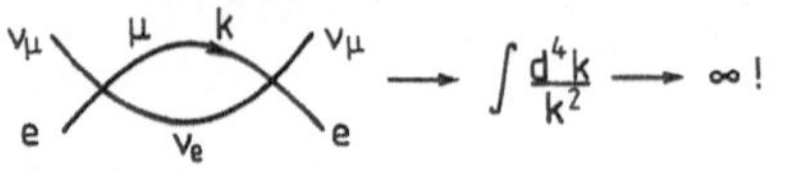

$$\to \int \frac{d^4k}{k^2} \to \infty!$$

However the Born approximation breaks down at high energies, giving s and p wave cross-sections which grow like s

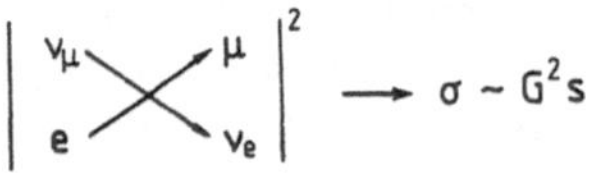

$$\to \sigma \sim G^2 s$$

and violate the unitarity bound $\sigma < c/s$ for $\sqrt{s} \sim 300$ GeV. These difficulties occur because the Fermi constant G has dimensions M^{-2}, which must be compensated by powers of k[$\sqrt{s}$] in integrals over virtual momenta [cross-sections]. The obvious remedy is to introduce the W and Z which, like the photon, have a dimensionless coupling constant g_W (with $\hbar$=c=1 as usual):

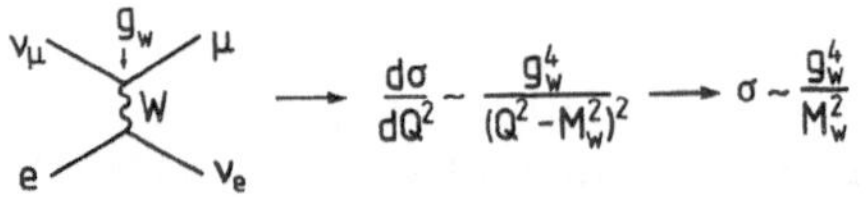

$$\to \frac{d\sigma}{dQ^2} \sim \frac{g_W^4}{(Q^2-M_W^2)^2} \to \sigma \sim \frac{g_W^4}{M_W^2}$$

(Alternatively the exchange of multiparticle states, as in the heavy fermion (F) model referred to above, could provide damping at large Q^2).

2. The mere existence of the W/Z does not cure all problems. Before the discovery of neutral currents, $\nu\bar{\nu} \to W^+W^-$ would have been described by

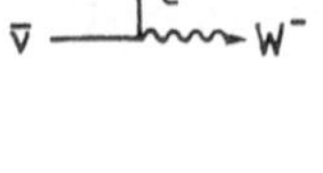

The W's polarization states are described by the three four-vectors ε_μ:

$$\begin{matrix} t \\ x \\ y \\ z \end{matrix} \quad \begin{pmatrix} 0 \\ 1 \\ 0 \\ 0 \end{pmatrix} \begin{pmatrix} 0 \\ 0 \\ 1 \\ 0 \end{pmatrix} \begin{pmatrix} 0 \\ 0 \\ 0 \\ 1 \end{pmatrix}$$

in its rest frame. The two transverse vectors (ε_μ^T) are invariant under a boost along the Z axis but the third (longitudinal) vector behaves as

$$\varepsilon_\mu^L \to \left(\frac{|\vec{k}|}{M_W}, 0, 0, \frac{k_0}{M_W} \right) \overset{k\to\infty}{\simeq} \frac{k_\mu}{M_W}$$

This introduces a factor M_W^{-1} into the amplitude for each longitudinal W and although

$$\sigma(\nu\bar{\nu} \to W_T W_T) \sim g_W^4/s$$

is well behaved, we find

$$\sigma(\nu\bar{\nu} \to W_L W_L) \sim \frac{g_W^4 s}{M_W^4}$$

which violates unitarity for $\sqrt{s} \sim 1$ TeV. The cure is to arrange that the amplitude vanishes when ε_μ is replaced by k_μ. This is equivalent to requiring local gauge invariance. In the case of $\nu\bar{\nu} \to WW$, good high energy behaviour is achieved by a cancellation with the s channel Z exchange diagram in SU(2)xU(1). Generally, if we consider fermion + antifermion → WW:

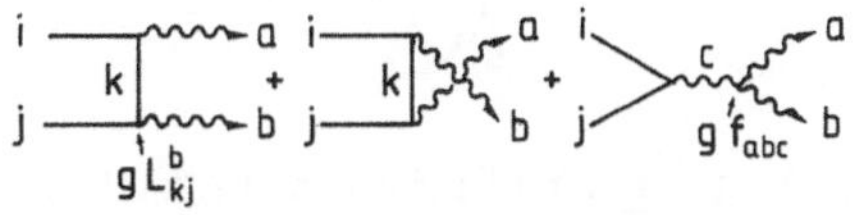

where gL and gf are coupling constants, the coefficient of the worst part of the amplitude, which grows like s at high energy, turns out to be

$$[L^a, L^b]_{ij} - if_{abc}L^c_{ij}.$$

This is zero in gauge theories because the L's form a Lie algebra with structure constants f. Likewise the worst terms ($\sim s^3$ and s^2) in the amplitude for WW → WW vanish because of the (Jacobi) identities satisfied by the f's and the gauge structure of the 4W vertex. In fact gauge theory vertices are necessary as well as sufficient for cancellation of the worst terms and may be derived by insisting on cancellation. (Alternatively if W and Z were $F\bar{F}$ bound states their composite structure could become important at a few hundred GeV and cure the unitarity crisis.)

3. There is still trouble due to residual terms in $f\bar{f} \to WW$ and $W_LW_L \to W_LW_L$ which behave as $g_W^2 \frac{m_f\sqrt{s}}{M_W^2}$ and $g_W^2 \frac{s}{M_W^2}$ respectively even with gauge theory couplings. These contributions can only be cancelled by introducing scalar (Higgs) fields. (Alternatively if W_L is a bound state, as in technicolour theories, its structure could damp out this bad behaviour.) With an exchange

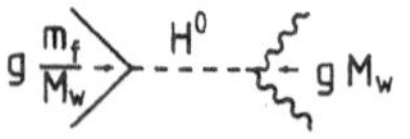

the standard Higgs coupling $\sim g\, M_W$ is necessary and sufficient to produce a contribution to the amplitude $\sim g_W^2 \frac{M_W^2 s}{M_W^4}$ which cancels the remaining offending term. In $f\bar{f} \to W_LW_L$ the trick is done by

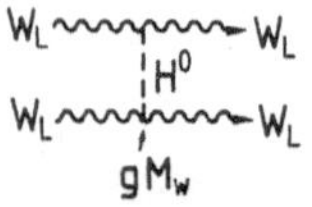

with an $H^0\bar{f}f$ coupling $\sim g \frac{m_f}{M_W}$ producing a term $\sim g \frac{m_f}{M_W} g\, M_W \frac{\sqrt{s}}{M_W^2}$.

Demanding that perturbation theory respects unitarity leads uniquely to a gauge theory spontaneously broken by the Higgs mechanism. Mathematically the Higgs mechanism is needed to generate masses because local gauge invariance (needed to get amplitudes which vanish when $\varepsilon_\mu \to k_\mu$) forbids an ordinary W/Z mass term and ordinary fermion mass terms

$$m\, \bar{\psi}\psi = m(\bar{\psi}_L\psi_R + \bar{\psi}_R\psi_L)$$

are forbidden by SU(2) and U(1) invariance.

In the standard model a doublet Higgs field is needed

$$\Phi = \begin{pmatrix} \phi^+ \\ \phi^0 \end{pmatrix}.$$

With the famous potential

$$V = -\frac{\mu^2}{2}(\Phi^+\Phi) + \frac{\lambda}{4}(\Phi^+\Phi)^2 \tag{7}$$

ϕ^0 develops a vacuum expectation value $<\phi^0> = \sqrt{\frac{\mu^2}{\lambda}}$ at the minimum. When this constant value is put into the kinetic plus gauge interaction part of the Φ lagrangian

$$|(i\partial_\mu + g\vec{T}.\vec{W}_\mu + g\frac{Y}{2}B_\mu)\Phi|^2 \tag{8}$$

$W^\pm$ and Z masses are generated with the mass matrix (4) with $M_W^2 = \frac{1}{2}g^2<\phi^0>^2$ and $\hat{M}_W = M_W$ as required (Higgs fields with $I \neq \frac{1}{2}$ spoil this relation in general). Equivalently, with $<\phi^0> \neq 0$, neutral ϕ's can disappear into the vacuum, thus breaking the SU(2)xU(1) invariance and generating inertia

ϕ^+, ϕ^- and one component of the complex field ϕ^0 provide the longitudinal degrees of freedom for $W^\pm$ and Z (which, like the photon, have only two degrees of freedom in the Lagrangian). A single physical H^0 remains and it is easy to check that there is an H^0WW coupling with strength $g^2<\phi^0> \sim g M_W$, whose necessity we have already derived.

Fermion masses are generated by SU(2)xU(1) invariant interactions

$$\lambda(\bar{\psi}_L\Phi)\psi_R \quad \text{or} \quad \lambda(\bar{\psi}_L\sigma_2\Phi^*)\psi_R. \tag{9}$$

When $\phi^0 \to <\phi^0>$ we get

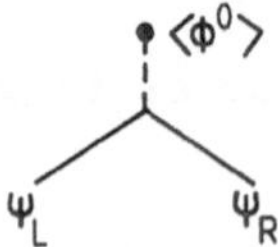

and a mass $m_f = \lambda \langle \phi^0 \rangle$ is generated. The strength of the $H^0\bar{\psi}\psi$ coupling then satisfies $\lambda = m_f/\langle\phi^0\rangle \sim g \frac{m_f}{M_W}$.

Crucial tests of SU(2)xU(1) are

1. To find the W and Z with the predicted masses. The lowest order predictions, which follow from eqn. 3 and $\frac{g^2}{8M_W^2} = \frac{G}{\sqrt{2}}$, obtained by comparing (1) with (5), are

$$M_W = \frac{37.3}{|\sin\theta|}\ , \quad M_Z = \sec\theta\ M_W. \tag{10}$$

Careful analysis of neutral current neutrino data and the parity violating asymmetry in ed → ex measured at SLAC, yields $\sin^2\theta = 0.227\pm.015$ and $0.223\pm.015$ respectively[5], if the electroweak interactions are treated in Born approximation. The first value gives

$$\begin{aligned} M_W &= 78.2^{+2.7}_{-2.5}\ \text{GeV.} \\ M_Z &= 89.0^{+2.2}_{-2.0}\ \text{GeV.} \end{aligned} \tag{11}$$

Electro-weak radiative corrections to the experiments, from diagrams such as

γ etc.

can change the value of $\sin^2\theta$ which is obtained by amounts of order $\frac{\alpha}{\pi}\ \ln Q^2/M_W^2 \sim 0.02$, $\frac{\alpha}{\pi\sin^2\theta} \sim .01$ and α/π . The exact change depends on how θ is defined. Using the $\overline{\text{MS}}$ renormalization scheme, which simplifies the calculation, with a subtraction scale $\mu = M_W$, both the neutrino experiments and the SLAC ed experiment yield a corrected value[10]

$$\sin^2\theta(M_W)_{\overline{MS}} = 0.215\pm.015. \tag{12}$$

Including the second order corrections to eqn. (10), this value of $\sin^2\theta$ then gives the corrected predictions

$$M_W = 83.1^{+3.1}_{-2.8} \text{ GeV.}$$
$$M_z = 93.9^{+2.5}_{-2.2} \text{ GeV.} \qquad (12)$$

The difference between (12) and (11) is clearly measurable, showing that it will be possible to test the renormalizability of SU(2)xU(1) which renders the radiative corrections calculable. Deviations of more than 25 GeV from (12) would be very hard to understand in any sensible gauge theory. Deviations of up to 25 GeV might be explained in gauge theories other than SU(2)xU(1)[11]. Deviations of up to 5 GeV might be explained in SU(2)xU(1) by the existence of new relatively light particles (e.g. the plethora of particles predicted by supersymmetric theories) which would alter the radiative corrections.

2. To find the Higgs boson whose mass, alas, is not predicted[12]. The toponium decay $t\bar{t} \to H^0\gamma$, followed by H^0 decaying to the heaviest particles available (e.g. $b\bar{b}$) because the couplings are proportional to masses, could reveal the H^0 if $M_{H^0} < 0.9\ M_{t\bar{t}}$. $Z^0 \to H^0\bar{e}e$ could lead to H^0 discovery if $M_{H^0} < 45$ GeV. The reaction $\bar{e}e \to ZH^0$ would be a good way to discover a heavy H^0 if superhigh energies were available.

3. To measure the WWZ and WWγ vertices. This will be very difficult but might be done in $\bar{e}e \to WW$ at SUPER LEP.

To summarize, the data are consistent with SU(2)xU(1) but the crucial tests remain to be carried out. Although the global SU(2) plus "γ W_3" mixing observed in the data might be explained otherwise[8] (e.g. by models with composite W's), only gauge theories indisputably lead to a satisfactory tractable description of weak interactions.

3. WHY BELIEVE IN QCD?

There is excellent evidence, summarized below, that hadrons are built of coloured quarks bound by vector gluon exchange. Since coloured hadrons are not observed, the forces must be colour dependent. The only sensible (unitary etc.) theory of vector gluons coupled to colour is QCD[13]. Furthermore QCD seems to account qualitatively for the observed features of hadrons.

The evidence for colour is that

1. It solves the spin statistics problem of the naive quark model, allowing the Δ^{++} to be an s wave $u^\uparrow u^\uparrow u^\uparrow$ state.

2. It is needed to account for $\Gamma(\pi^0 \to \gamma\gamma)$; the prediction given below would be reduced by a factor of 9 without colour.

3. It is needed to account for $\sigma(\bar{e}e \to \text{hadrons})$.

4. It is needed to exorcise the anomaly from SU(2)xU(1). The anomaly is a disease associated with triangle diagrams

which spoils the renormalizability of gauge theories unless the sum of the anomalous contributions of all fermions vanishes. This occurs multiplet by multiplet for some gauge groups but for SU(2)xU(1) it requires a cancellation between the leptonic and the hadronic contribution, which is proportional to the number of colours.

There is excellent evidence for an (approximate) chiral symmetry realized in the Nambu Goldstone mode, which can only be explained by assuming that the strong forces are mediated by vector gluons and that quarks are almost massless (assuming that gluons do not carry weak and electromagnetic charges). The evidence is[14]

	Prediction in symmetry limit	Data
$\frac{M_\pi^2}{M_\rho^2}$	0	0.03
$1 + \frac{2Mg_A}{f_\pi g_{np\pi}}$	0	0.08±.01
$M_\pi^2 a_{\pi N}^{1/2,3/2}$	0.16,-0.079	0.17±.005, -0.088±.004
$\lambda^0_{K_{e3}}$	0.021±.003	0.019±.004
$\Gamma(\pi^0 \to \gamma\gamma)$	7.87eV	7.95±.55eV

where the second line is the Goldberger Treiman relation, $a_{\pi N}^I$ are pion nucleon scattering lengths for different isospin channels and λ^0 is the slope parameter in K_{e3} decay.

A most impressive feature of QCD is that it is asymptotically free, the running coupling constant decreasing, with $\alpha_s = \frac{g_s^2}{4\pi} \sim (\ell nr)^{-1}$, at short distances. Some such behaviour is needed to explain the success of the naive parton model. Furthermore QCD predicted various deviations from the naive parton model, such as scaling violations in lepto-production and three jet events in $\bar{e}e$ annihilation, which it explains semi-quantitatively[15]. The corollary of asymptotic freedom is that α_s increases at large distance, possibly leading to confinement.

The simplest possible colour dependence of the interquark force, which is $\vec{\lambda}\cdot\vec{\lambda}$ where λ^a are the eight Gell-Mann matrices, would explain why colour singlets are lighter (perhaps infinitely lighter) than coloured states[16]. This in turn would explain why only hadrons containing qqq or $q\bar{q}$ (or combinations thereof) are observed, which was a total mystery before the advent of colour. In addition a "chromomagnetic force", such as is generated by one gluon exchange, explains $M_\rho > M_\pi$, $M_\Delta > M_N$ and the old mysteries of why $M_\Sigma > M_\Lambda$ and why the charge radius of the neutron is negative[17].

Asymptotic freedom plays a vital role in grand unification. We shall therefore close this brief exegesis of QCD with a review of the notion of a "running coupling constant" and the behaviour of α_s. Even classical charges depend on the distance at which they are measured. Consider a positive test charge placed in a dielectric:

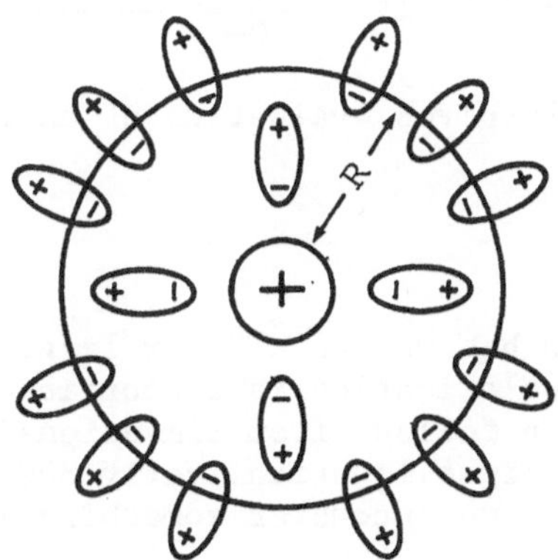

We can define a charge $Q(R) = \int\vec{E}.d\vec{s}$ by integrating over a sphere at radius R. For R << d, the typical intermolecular spacing, it is equal to the free space value Q but for R > d it is screened and tends to Q/ε:

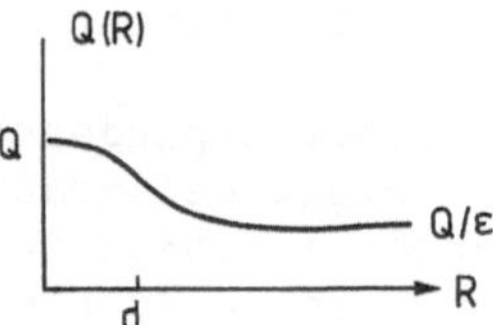

In QED, the effective charge depends on the momentum transfer because of radiative corrections

$$\Longrightarrow \frac{1}{Q^2}(e_0^2 + e_0^4\, f(Q^2) + \ldots) \equiv \frac{e^2(Q^2)}{Q^2} \xrightarrow{Q^2 \to 0} \frac{e^2}{Q^2}$$

Calculation of $f(q^2)$ involves evaluation of an integral

$$\int I(q,k,m_e)\,d^4k$$

in which the integrand behaves as k^{-4} for large k (gauge invariance requires that vacuum polarization is proportional to $-g_{\mu\nu}q^2+q_\mu q_\nu$; the behaviour of I then follows from dimensional analysis). The integral therefore diverges logarithmically and, after removing a finite low energy part, we encounter something like

$$\int_q^{2q}\frac{dk}{k} + \int_{2q}^{4q}\frac{dk}{k} + \ldots + \int_{256q}^{512q}\frac{dk}{k} + \ldots$$

Even if we remove the divergence by cutting off the integral at the mass of the Universe, which seems reasonable, we encounter the problem that the range from one to two times the mass of the sun - at which we clearly do not understand the physics - contributes as much as q to 2q.

This dilemma is resolved by recognizing that attempting to calculate $e(q^2)$ from the bare charge e_0 was too ambitious. It is like attempting to explain the macroscopic properties of water in terms of quarks and gluons! In practice we can study hydrodynamics by introducing a few phenomenological parameters (such as density and viscosity), which are determined by physics on the next (atomic) scale. In turn we can do atomic physics using a few parameters input from nuclear physics and so on. The idea that physics at one scale depends only on parameters derived from the behaviour at neighbouring scales works also in renormalizable field theories. Thus calculating $e(q^2)$ in terms of $e(q_0^2)$ we encounter $I(q,k) - I(q_0,k)$. The contribution from very high energy, where QED may or may not be correct, drops out of the difference and the integral converges, with k effectively running from $O(q)$ to $O(q_0)$. In contrast, in non-renormalizable theories physics at one scale depends essentially on all other scales and it is impossible to understand anything without understanding everything. Luckily it seems that this catastrophic possibility is not realized in nature, at least down to the level we have reached so far, and we can understand one layer of the onion at a time.

Expressing $e(q^2)$ in terms of $e(0) \equiv e$, we have

$$e^2(q^2) = e^2 + e^4 f'(0) q^2 + \ldots$$

where f' (which is of order m_e^{-2} and contributes a very well tested -27 Mc/s to the Lamb shift) is negative i.e. $e(q^2)$ increases for space like q^2:

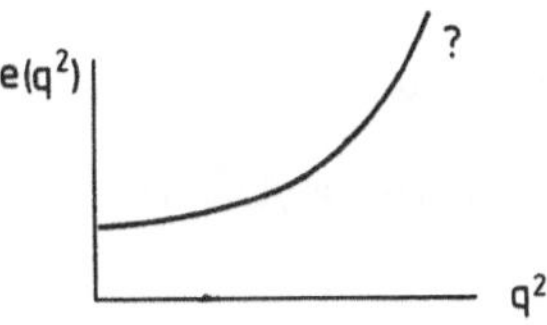

We now consider QCD, taking massless quarks for simplicity. We could define the running coupling constant $g(p^2)$ as the value of the three point function[18]

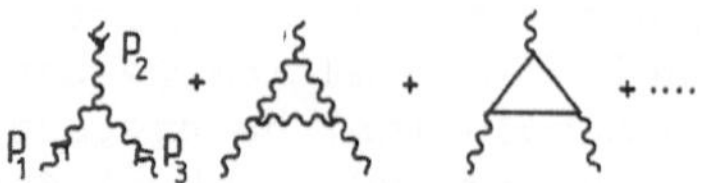

at the symmetric point $p_i^2 = -2p_i p_j = p^2 < 0$, space like (alternatively we could define $g(p^2)$ through the gluon quark vertex). In terms of a bare coupling constant

$$g(p^2) = g_0 + g_0^3 \int d^4k I(k,p)|_{p_i^2=p^2} +$$

where I, which has dimensions M^{-4}, behaves as k^{-4} for large k. Adopting the subtraction philosophy discussed above, we remove the contribution of unknown high energy physics by considering

$$g(p^2) = g(\mu^2) + g^3(\mu^2) \int d^4k (I(k,p)|_{p^2} - I(k,p)|_{\mu^2}) + \ldots$$

thus making the result finite. Differentiating and setting $p^2 = \mu^2 < 0$, we obtain

$$\frac{\partial g(\mu^2)}{\partial \ln\sqrt{-\mu^2}} = \beta(g(\mu^2)), \tag{13}$$

where β can be calculated perturbatively, provided g is small. This "renormalization group equation" expresses mathematically the fact that g only depends on neighbouring scales. Writing

$$\beta = -\beta_0 \frac{g^3}{16\pi^2} + O(g^5)$$

we can integrate (13), obtaining

$$\frac{1}{\alpha_s(Q^2)} - \frac{1}{\alpha_s(Q_0^2)} = \frac{\beta_0}{4\pi} \ln Q^2/Q_0^2 + \ldots \tag{14}$$

an equation which will prove very useful later, or equivalently

$$\alpha_s(Q^2) = \frac{\alpha_s(Q_0^2)}{1+(\beta_0/4\pi)\alpha_s(Q_0^2)\ln Q^2/Q_0^2}$$

$$\equiv \frac{4\pi}{\beta_0 \ln Q^2/\lambda^2}.$$

This equation defines the scale parameter λ of QCD to leading order and exhibits the fact that QCD can be parameterised either in terms of α_s at some reference point (Q_0^2) or in terms of a scale (λ), which crudely speaking represents the energy at which strong interactions become strong.

Calculation of the one loop diagram gives

$$\beta_0 = 11 - \frac{2F}{3}$$

in QCD, which is positive provided the number of flavours (F) is 16 or less, and α_s therefore descreases for $Q^2 \equiv -p^2 \to \infty$. This is asymptotic freedom.

4. CRITIQUE OF SU(3)xSU(2)xU(1) AND ATTEMPTS AT TRUE ELECTRO-WEAK UNIFICATION

The standard SU(3)xSU(2)xU(1) model is not a satisfactory candidate for a complete final theory. It simply does not explain enough. More particularly

1. Charge quantization is not explained; Y and hence Q is quite arbitrary.

2. Electroweak interactions are not truly unified; θ is an arbitrary parameter.

3. There are far too many parameters, mainly connected with the complex Higgs sector which supplies one for every mass and mixing angle.

4. The proliferation of flavours of quarks and leptons is not explained.

Other objections are frequently listed e.g. some find the left-right lopsidedness of the model unacceptable. This can be cured by constructing left-right symmetric models in which parity violation is spontaneous[19]. Point 4) might be answered by introducing "horizontal" symmetries which link different generations but nothing much has come of this idea[1]. Alternatively perhaps 4) hints at substructure for quarks and leptons[20].Point 3) might be answered by doing without Higgs fields, introducing heavy techniquarks with new strong interactions to play their role, but realistic models of this type are very complex and run into grave phenomenological difficulties[2]. Alternatively we might constrain the Higgs bosons by connecting them with fermions in supersymmetric (SUSY) models[21]. SUSY has various attractions, discussed briefly in section 15, but known SUSY models are not particularly economical in parameters. Finally points 1) and 2) might be answered by further unification. This is the route we shall follow, returning to the other points in sections 11 and 15.

The obvious first step of unifying just SU(2)xU(1) into a single group G without unifying quarks and leptons, runs into two immediate difficulties:

1. Since the photon can in this case be chosen to be one of the basic gauge bosons (not a mixture), the electric charge Q is a generator of G and therefore[22] satisfies $\mathrm{Tr}Q = 0$, i.e. the sum of the charges of the fermions in a given representation must be zero. At the least a new quark $\hat{q}$ with $Q = -1/3$ is needed to make $\mathrm{Tr}\,Q = Q_u+Q_d+Q_{\hat{q}} = 0$. This would suggest that G contains SU(3). Likewise a new lepton with $Q = +1$ is needed to accompany the electron. In SU(3) a whole octet would be needed.

2. The Z also couples to a generator (Q_Z) and we require $\mathrm{Tr}(QQ_Z)= 0$. Physically this ensures that if the symmetry were exact the Z and the photon would be othogonal and mixing from

$$\sum_f \; Z \sim\bigcirc^{f}\sim \gamma$$

would vanish. We know that $Q_Z = T_3-Q\sin^2\theta$ so that using

$$\mathrm{Tr}(QT_3) = \mathrm{Tr}((T_3= \tfrac{Y}{2})T_3) = \mathrm{Tr}\,T_3^2$$

the condition $\mathrm{Tr}QQ_Z= 0$ requires

$$\sin^2\theta = \frac{\Sigma T_3^2}{\Sigma Q^2} \tag{15}$$

in the symmetry limit, where the sums run over all members of a multiplet. In SU(3) with u,d and a $Q = -1/3$ quark forming a triplet, eqn. (15) gives the wrong result $\sin^2\theta = 3/4$. This is reduced to 3/8 in $SU(3)_L \mathrm{x} SU(3)_R$, which is probably the minimal model which can accommodate the known left and right handed fermions. However to get $\sin^2\theta$ down to O(.22) clearly requires drastic additions - either new quarks with $Q > 2/3$ or many new quarks with $I_3 = 0$, $Q \neq 0$. The only ways which are known to work are absurd e.g. $(SU(3))^{15}$!

One way out of this problem is partial unification[23] with $SU(3)_c$ i.e.

$$\underbrace{SU(3)_C \mathrm{x} U(1)}_{G_s} \mathrm{x} \underbrace{SU(2)_L}_{G_W}$$

where the new strong and weak groups (G_s and G_W) might be broken at some energy above a TeV. Such schemes are quite interesting in principle but not very appealing in practice. Another possibility is that unification occurs at colossal energy M_u. Recall that

$\tan\theta = g'/g$. The U(1) coupling constant g' increases with energy, like e, whereas SU(2) is asymptotically free, like all non-Abelian gauge theories, and g decreases. θ therefore increases with energy and an asymptotic value of 3/8 might be reconciled with the data. Since the low energy corrections to (15) are of order $\alpha/\pi \ln(E/M_u)$ this would require enormous M_u. As we shall see, a very large scale allows us to "grand unify" all three forces in models in which α_s decreases to merge with α_2 for $E \gtrsim M_u$, and also to unify quarks and leptons by putting them in one multiplet, which allows us to obtain TrQ = 0.

5. GRAND UNIFIED THEORIES

In GUTs there is a single gauge group G which contains SU(3)xSU(2)xU(1). Fundamentally there is therefore no distinction between strong and electroweak interactions which merge at infinite energy, in a way which we shall discuss. Consequently the operational distinction between hadrons and leptons as particles with/without strong interactions has no deep meaning. We can therefore simultaneously unify quarks and leptons, by putting them in the same multiplet.

We would like to put <u>all</u> the quarks and leptons in one multiplet but, alas, this turns out to be difficult. For the time being we pursue the more modest programme, which is known to work, of unifying a single "generation", taking the first generation (e,ν,u,d) as a paradigm. We have become used to linking quarks and leptons thus[24]

$$\nu_e, e(\sim 1/2 \text{ MeV}) \xleftrightarrow{\text{connection?}} u,d(\sim 10 \text{ MeV})$$

$$\nu_\mu, \mu(\sim 100 \text{ MeV}) \longleftrightarrow c,s(\sim 1 \text{ GeV})$$

$$\nu_\tau, \tau(\sim 2 \text{ GeV}) \longleftrightarrow t,b(\sim 15 \text{ GeV})$$

However, it should be remembered that this is an assumption which we must try to check. At the very least it must be corrected by Cabibbo-Kobayashi-Maskawa mixing, which we neglect to begin with.

The usual convention, which we shall follow, is to use left handed fermions and left handed antifermions as the independent degrees of freedom. The first generation then contains the following fifteen objects

$$\begin{array}{lll} \nu_L, e^-_L, e^+_L & u^r_L, u^b_L, u^g_L & d^r_L, d^b_L, d^g_L \\ & \hat{u}^r_L, \hat{u}^b_L, \hat{u}^g_L & \hat{d}^r_L, \hat{d}^b_L, \hat{d}^g_L \end{array} \qquad (16)$$

where ^ denotes charge conjugation ($\hat{u} \equiv u^c$ etc.) and where r,b,g are colour indices ($e^+_L, \hat{u}_L, \hat{d}_L$ are of course the charge conjugates of e^-_R, u_R, d_R and are SU(2) singlets). If there are no more undiscovered first generation fermions, the states (16) must form a representation of G (which may be reducible) and their I_3's and Q's can be substituted into (15) to give

$$\sin^2\theta(E) = \frac{\Sigma I_3^2}{\Sigma Q^2} = \frac{3}{8} + O(\frac{\alpha}{\pi}\ \ln\ E^2/M_X^2)$$

where the $O(\alpha)$ term is produced by symmetry breaking radiative corrections. With

$$\sin^2\theta(1) \simeq 0.22 = \frac{3}{8} + \frac{2c\alpha\ln}{\pi}\ 1/M_X$$

we obtain

$$M_X \sim 10^{15/c}\ \text{GeV}.$$

We shall shortly calculate M_X properly. Anticipating that $M_X \sim 10^{15}$GeV, we find that at the unification scale $\alpha_s(10^{15}) \sim 1/40$, which is not dissimilar from $\alpha_2 \simeq \alpha/\sin^2\theta$, so there is hope that the GUT programme will work. Later we shall see that M_X is the mass of heavy vector bosons which couple quarks to leptons and can mediate proton decay. The lifetime for nucleon decay mediated by single X exchange will be of order $(1/40)^2 \left(\frac{M_p^5}{M_X^4}\right)^{-1} \sim 10^{31}$years and has a chance of being above the experimental limit.

A more accurate calculation can be performed without commitment to a particular group G if we assume that there is a single large scale M_X. We then make the approximation that for $E > M_X$, G is exact (i.e. $M_X = 0$) while for $E < M_X$, SU(3)xSU(2)xU(1) is exact (i.e. $M_X = \infty$). The true curve

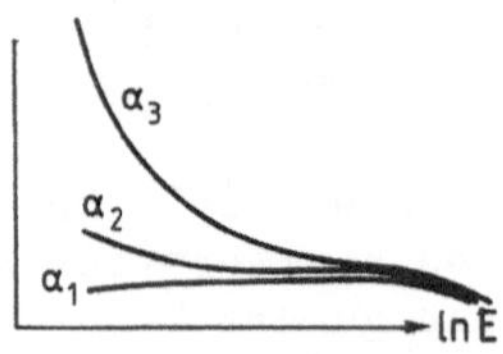

in which the α_i meet asymptotically is replaced by

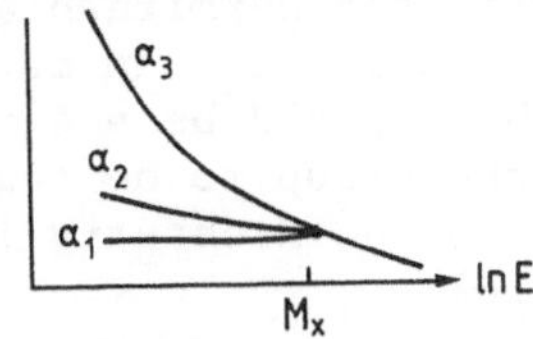

At M_x, we then have $\alpha_2 = \frac{g_W^2}{4\pi} = \alpha_3$ (for SU(3) and SU(2) interactions $g_s \frac{\vec{\lambda}\cdot\vec{A}_\mu}{2}$ and $g_W\ \vec{T}\cdot\vec{W}_\mu$ with common normalization, $Tr(\lambda^a)^2 = Tr(2T^i)^2 = 2$), and $g' = \tan\theta\ g_W = \frac{3}{5} g_W$, so that $\alpha_1 \equiv \frac{5}{3}\frac{g'^2}{4\pi} = \alpha_2$ at M_x. Using (3) and the results of the renormalization group (14) we then obtain

$$
\begin{aligned}
\frac{1}{\alpha_3(E)} &= \frac{1}{\alpha_u} + \frac{1}{6\pi}(4N-33)\,\ell n\frac{M_x}{E} + \ldots \\
\frac{1}{\alpha_2(E)} &= \frac{\sin^2\theta(E)}{\alpha(E)} = \frac{1}{\alpha_u} + \frac{1}{6\pi}(4N-22+\tfrac{1}{2})\,\ell n\frac{M_x}{E} + \ldots \\
\frac{1}{\alpha_1(E)} &= \frac{3\cos^2\theta(E)}{5\alpha(E)} = \frac{1}{\alpha_u} + \frac{1}{6\pi}(4N+\tfrac{3}{10})\,\ell n\frac{M_x}{E} + \ldots
\end{aligned}
\tag{17}
$$

where N is the number of generations. The $4N/6\pi$ contribution is due to fermion loops, $-33(22)/6\pi$ is the contribution of gluon (W) loops and $1/2(3/10)/6\pi$ is due to Higgs loops. The neglected terms in (17), which are of O(1) and depend on G, must (and will) be included to get M_x precisely. Eliminating $\alpha_u^{-1} + 4N/6\pi$ we obtain

$$
\begin{aligned}
\frac{3}{5\alpha(E)} - \frac{8}{5\alpha_3(E)} &= \frac{201}{30\pi}\,\ell n\frac{M_x}{E} \to M_x \sim 2\times10^{15}\text{GeV}. \\
\sin^2\theta(E) &= \frac{3}{8}\left[1 - \frac{109}{18\pi}\,\alpha\ell n\frac{M_x}{E}\right] \to \sin^2\theta(M_W) \simeq 0.21
\end{aligned}
\tag{18}
$$

in encouraging agreement with (12).

6. SU(5)

We must now find a unifying group $G \supset SU(3)\times SU(2)\times U(1)$. Since

SU(3)xSU(2)xU(1) has "rank" (≡ the number of diagonal generators) 2+1+1 = 4, G must have rank ≥ 4. The simplest choice is SU(5), whose generators are 5x5 traceless hermitean matrices, 4 of which can be diagonalized simultaneously[25]. In fact SU(5) is the only group of rank 4 for which there would be a single gauge coupling (technically this requires the group to be semi simple) which contains SU(3)xSU(2)xU(1) with the necessary representaions to accommodate the 15 states (16).

The fundamental representation of SU(5) are 5x5 unitary matrices U with detU = +1 which act on a vector $Q_{i=1..5}$ thus

$$Q_i \xrightarrow{SU(5)} Q_i' = \sum_j U_{ij} Q_i$$

so that

$$\Sigma Q^i Q_i \equiv \Sigma Q_i^* Q_i$$

is an invariant. The matrix U can be parameterized as

$$U = \exp(i\sum_a \omega_a G^a/2)$$

where G^a are 24, 5x5 traceless hermitean generators of the fundamental representation of SU(5). All this is familiar from SU(2) (the rotation group) which acts on wave functions so that

$$\psi \to R(\theta_i)\psi, \quad R = \exp(i\vec{J}\cdot\vec{n}\theta)$$

and SU(3), of colour or flavour, under which a triplet transforms as

$$q \to uq, \quad u = \exp(i\sum_a \omega_a \lambda^a/2),$$

where λ^a are the eight Gell-Mann matrices.

From Q_i and Q^j we can form the bases for higher dimensional representations which can be reduced (or split) into irreducible multiplets which do not mix under SU(5) transformations. For example

$$\begin{array}{llll} \tilde{Q}^i Q_j = & (\tilde{Q}^i Q_j - \frac{\delta_{ij}}{5} \sum_k \tilde{Q}^k Q_k) & + \frac{\delta_{ij}}{5} \sum_k \tilde{Q}^k Q_k & \\ \bar{5} \times 5 = & 24 & + \ 1 & \quad (19) \\ (\bar{3} \times 3 = & 8 + 1) & & \end{array}$$

where the numbersin brackets are the analogous results for SU(3), and

$$\tilde{Q}_i Q_j = \frac{[\tilde{Q}_i Q_j - \tilde{Q}_j Q_i]}{2} + \frac{[\tilde{Q}_i Q_j + \tilde{Q}_j Q_i]}{2}$$

$$5 \times 5 = 10 + 15 \tag{20}$$

$$(3 \times 3 = \bar{3} + 6).$$

To discover how SU(3)xSU(2)xU(1) fits into SU(5) it is convenient to choose the 5x5 generators as follows

$$\frac{1}{2}\begin{pmatrix} \lambda^a & 0 \\ 0 & 0 \end{pmatrix}, \quad \begin{pmatrix} 0 & 0 \\ 0 & T^a \end{pmatrix}, \quad \frac{1}{\sqrt{60}}\begin{pmatrix} 2_{2_2} & 0 \\ 0 & -3_3 \end{pmatrix}, \quad \begin{pmatrix} 0 & \Sigma^a \\ \Sigma^a & 0 \end{pmatrix} \tag{21}$$

where $\frac{\lambda^a}{2}$ $[T^a]$ are the SU(3) [SU(2)] generators, the other diagnol matrix (normalized so that $Tr(2G^a)^2 = 2$) generates U(1) and the 12 Σ^a's are (complex) 2x3 matrices which generate the remaining SU(5) transformations. There are 24 gauge bosons V^a_μ which enter the covariant derivative in the form

$$g \sum_a \frac{G^a}{2} V^a_\mu$$

(in an N dimensional representation, the G^a would be NxN matrices). The eight gluons, the $\vec{W}$ and the U(1) boson B are associated with the the first 12 of the G^a in eqn. 21 and there are six new bosons, which form a colour triplet, isodoublet

$$\begin{pmatrix} X^{r,b,g} \\ Y^{r,b,g} \end{pmatrix}, \tag{22}$$

and their antiparticles, associated with the Σ^a.

We can now assign the 15 fermions in (16) to representations of SU(5). Unfortunately they do not fit the 15 dimensional representation (20). If we think of the G^a (21) acting on a five dimensional vector, we see that the 5 representation contains a colour triplet - SU(2) singlet, (3,1), and a colour singlet - SU(2) doublet, (1,2). The fifteen, which is the symmetric combination of two 5's, therefore contains a colour sextet and is not suitable for the known fermions. We might try to put some of the fermions into a 5 but we have no candidate for the (3,1) as the quarks in (16) transform as $q \sim (3,2)$ while $\hat{q} \sim (\bar{3},1)$. In fact, the correct assignment is to a $\bar{5}$ and a 10. Under SU(3)xSU(2) the $\bar{5}$ splits into $(\bar{3},1) + (1,\bar{2})$ (it is purely a matter of convention that this is called $\bar{5}$ and not 5 but once this convention is fixed the remaining fermions necessarily belong to a 10 rather than a $\overline{10}$). The $\bar{5}$ therefore contains

$$X = \begin{pmatrix} \hat{d}^r \\ \hat{d}^b \\ \hat{d}^g \\ e^- \\ \nu \end{pmatrix}_L \tag{23}$$

which is the only choice with the correct SU(3)xSU(2)xU(1) quantum numbers. Note that the condition TrQ = 0 is automatically satisfied and requires the unit of quark charge to be 1/3 of a unit of lepton charge.

The quantum numbers of the 10 can be found by considering a 5 of hypothetical building blocks, which will be thrown away later,

$$Q = \begin{pmatrix} D^r \\ D^b \\ D^g \\ E^+ \\ N \end{pmatrix}$$

We can then make a 10 from two 5's using (20)

$$\psi_{\alpha\beta} \sim \tilde{Q}_\alpha Q_\beta - \tilde{Q}_\beta Q_\alpha$$

and we read off the quantum numbers e.g.

$$\begin{aligned} \psi_{12} &\sim \tilde{D}^r D^b - \tilde{D}^b D^r \\ \psi_{14} &\sim \tilde{D}^r E^+ - \tilde{E}^+ D^r \\ \psi_{15} &\sim \tilde{D}^r E^+ - \tilde{E}^+ D^r \\ \psi_{45} &\sim \tilde{E}^+ N - \tilde{N}\, E^+ \end{aligned} \tag{24}$$

The 10 is therefore seen to accommodate the remaining states thus

$$\psi = \begin{pmatrix} 0 & \hat{u}^g & -\hat{u}^b & -u^r & -d^r \\ -\hat{u}^g & 0 & \hat{u}^r & -u^b & -d^b \\ \hat{u}^b & -\hat{u}^r & 0 & -u^g & -d^g \\ u^r & u^b & u^g & 0 & -e^+ \\ d^r & d^b & d^g & e^+ & 0 \end{pmatrix}_L \tag{25}$$

The overall phase of e^+ and each flavour of q or $\hat{q}$ is arbitrary but the relative phases for given q or $\hat{q}$ are not; in particular under SU(3) we know that $\varepsilon^{ijk} q_j q_k$, where i,j,k are colour indices, transforms like q^i which explains the phase of $\hat{u}^b$. As an elementary check note that ψ_{45}(eqn. 24) is an SU(2) singlet so the assignment of e_L^+ is correct.

7. BREAKING SU(5)

We now examine simple choices of representations for the Higgs fields and learn how to break SU(5) to SU(3)xSU(2)xU(1) at $M_X \sim 10^{15}$GeV, with a further breakdown to $SU(3)xU(1)_{em}$ at around 100 GeV. The neutral member of a Higgs 5 could get a vacuum expectation value (vev) thus

$$\langle H\rangle = \begin{pmatrix} 0 \\ 0 \\ 0 \\ 0 \\ V \end{pmatrix} \tag{26}$$

which carries I_3 into/out of the vacuum and breaks SU(2). This "$\Delta I = 1/2$" breaking of SU(2) is just what we need. A Higgs 10 contains no neutrals and cannot develop a vev. The 15 has a neutral member which behaves as $\tilde{N}N$ in terms of the hypothetical 5. If it acquired a vev it would break SU(2) with "$\Delta I = 1$" which would spoil the successful relation $M_W = \cos\theta M_Z$. The next possibility is a 24 which turns out to give the required SU(5) breaking at M_X. If we represent the 24 by a 5x5 traceless hermitean matrix Φ, the required vev is

$$\langle\Phi\rangle = c \begin{pmatrix} 2 & & & & \\ & 2 & & 0 & \\ & & 2 & & \\ & 0 & & -3 & \\ & & & & -3 \end{pmatrix} \tag{27}$$

This is a singlet in the SU(3) and SU(2) spaces and also conserves U(1) (formally Φ transforms like Q^iQ_j: $\Phi \to U\Phi U^{-1} = \exp(i\vec{\omega}\cdot\vec{G}/2)$ $\Phi\exp(-i\vec{\omega}\cdot\vec{G}/2)$ so that under an infinitessimal transformation $\delta\langle\Phi\rangle = i\frac{\vec{\omega}}{2}.[\vec{G},\langle\Phi\rangle]$, which is zero for SU(3)xSU(2)xU(1) transformations). $\langle\Phi\rangle$ obviously breaks SU(5) and will give equal masses to X and Y because it conserves colour and isospin.

An appropriate potential for Φ is

$$V(\Phi) = -\frac{\mu^2}{2}\mathrm{Tr}\Phi^2 + \frac{a}{4}(\mathrm{Tr}\Phi^2)^2 + \frac{b}{2}\mathrm{Tr}(\Phi^4) \tag{28}$$

which leads to (27) with $c^2 = \frac{\mu^2}{30a+14b}$ provided $\mu^2>0$ and $a > -\frac{7}{15}b>0$ (to see this diagonalize $\langle\Phi\rangle$ so that $\langle\Phi_{ij}\rangle = \delta_{ij}a_i$ and minimize with respect to a_i imposing $\sum_i a_i = 0$ using a Lagrange multiplier). The Φ kinetic energy term, which contains interactions with the vector fields V_μ^a as it involves the covariant derivative, is

$$L = \frac{1}{2}\mathrm{Tr}|\partial_\mu\Phi + \frac{ig}{2}V_\mu^a[G^a,\Phi]|^2,$$

This generates $M_x = M_y = \frac{25g^2c^2}{2}$ when Φ is replaced by $<\Phi>$.

With a potential

$$V(H) = -\frac{\gamma^2}{2}(H^+H) + \frac{\lambda}{4}(H^+H)^2, \tag{29}$$

H will develop a vev of the required form (26) for $\gamma^2 > 0$. However, we cannot avoid cross-talk between Φ and H and we must add a term

$$V' = \alpha(H^+H)Tv\Phi^2 + \beta H^+\Phi^2 H \tag{30}$$

Even if we set $\alpha = \beta = 0$ initially this term is effectively induced by diagrams such as

At the minimum of the total potential we now find

$$<\Phi> = \hat{c}\begin{pmatrix}2 & & & & \\ & 2 & & & \\ & & 2 & & \\ & & & -3-\varepsilon & \\ & & & & -3+\varepsilon\end{pmatrix}$$

where $\hat{c}^2 = c^2+O(v^2)$, $\varepsilon = O(\frac{v^2}{c^2})$. Since we need $\hat{c} = O(10^{15}\text{GeV}) \simeq c$, $v = O(100\text{ GeV})$ we see that the cross-talk has a negligible effect here and only induces very small SU(2) breaking. Furthermore we find that the colour triplet H^c in the 5 acquires a mass of O(c) - which is desirable since it can mediate nucleon decay. However, not surprisingly, v (eqn. 26) gets an admixture of O(c)

$$v^2 = \frac{1}{\lambda}(\gamma^2 - c^2[15\alpha + 2\beta - 3\varepsilon\beta]) \tag{31}$$

To keep v of order 100 GeV we must adjust α and β so that $15\alpha + 2\beta = O(10^{-26})$! This is technically possible but totally unreasonable (it might perhaps be possible to find a symmetry which made $\alpha = -2\beta/15$ but when higher orders are added the required value is no longer given by such a simple formula).

The very artifical tuning of parameters needed to keep v and hence M_W from getting admixtures of order M_X is the worst aspect of what is known as the hierarchy problem. We shall return to this flaw of GUTs in the final section. Another part of the problem is the origin of the very disparate scales M_W and M_X which we have put in by hand. I find this problem less serious as I can imagine a large ratio being generated automatically by a more complete theory, whereas it is difficult to see how precise numerical relations between α and β could arise.

Proceeding with α, β and other parameters tuned and adjusted to produce the desired c and v, we can now generate fermion masses. First a technical point. With only left handed fields, as in (16), a mass term in the Lagrangian would have the form

$$X_L^T \, CMX_L \tag{32}$$

where X is a vector containing all the fermion fields, M is a matrix in this fermion flavour space and C is the charge conjugation matrix, which acts in Dirac space. Recall that X contains entries such ψ_L and ψ_L^c. Since $(\psi_L^c)^T C\tilde{\psi}_L = \overline{\psi}_R\tilde{\psi}_L$ and $\psi_L^T C\tilde{\psi}_L^c = \tilde{\psi}_R\psi_L$ we see that (32) contains the usual Dirac mass terms as well as Majorana mass terms, generated by $\psi_L^T C\tilde{\psi}_L$ and $(\psi_L^c)^T C\tilde{\psi}_L^c$, to which we return later. An explicit mass term of the form (32) would violate SU(5) and hence spoil renormalizability. We must therefore induce masses from Yukawa interactions of the form

$$g_{ij}^a \, X_L^i \, CX_L^j h^a \tag{33}$$

where some component of the Higgs field h^a develops a vev.

To find candidates for h^a, we examine the SU(5) content of X^iX^j. The $\bar{5}\text{x}\bar{5}$ pieces are of no use since they carry baryon and lepton number. Since $\bar{5}\text{x}10 = 45+5$ we can make an invariant with h^a transforming like $\bar{5}$ i.e. we can put a term

$$\lambda_1 H_\alpha^+ \chi^\beta \psi_{\alpha\beta} \tag{34}$$

in the Lagrangian, where C and Dirac indices are suppressed, χ and ψ are the fermion fields (23 and 25) and H the Higgs 5 already introduced. With $\langle H_5^+ \rangle = v$ this gives

$$\lambda_1 v(\bar{e}e + \sum_i \bar{d}^i d^i)$$

which generates $m_e = m_d \neq 0$. Since $10\text{x}10 = \overline{45}+\bar{5}+50$ we can add the invariant

$$\lambda_2 \varepsilon^{\alpha\beta\gamma\delta\varepsilon} \psi_{\alpha\beta} \psi_{\gamma\delta} H_\varepsilon \tag{35}$$

which generates a mass for the u quark.

There are radiative corrections to the quark masses from diagrams such as

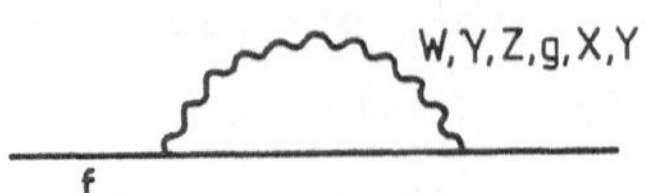

The (divergent) contributions from $|k|>>M_X$ respect SU(5) symmetry and leave $m_d = m_e$ but the low energy parts change this result by a calculable amount, mainly by increasing the mass of the d because it interacts strongly with the gluon field. Summing the leading gluon contributions to all orders the result is

$$\frac{m_d(Q^2)}{m_e(Q^2)} = \left(\frac{\alpha_s(Q^2)}{\alpha_s(M_X^2)}\right)^{4(11-2F/3)^{-1}} (1 + O(\alpha_s(Q^2))) \qquad (36)$$

where F is the number of flavours and $m(Q^2)$ are the values appropriate for experiments at a scale $\sqrt{Q^2}$, assumed to be large enough for QCD perturbation theory to apply. Masses for higher generations and Cabbibo-Kobayashi-Maskawa mixing can be produced by constructing the analogues of (34) and (35) using higher generation fermion fields and appropriate λ's. Relations exactly analogous to (36) are found for each generation and will be compared to the data below.

8. PRE- AND POST-DICTIONS OF SU(5)

SU(5) successfully unifies not only the strong and electro-weak forces but also the two different forms of matter, quarks and leptons. It leads automatically to quantized electric charges, with $Q_q = \frac{1}{N_c} Q_\ell$ (this is very nice but follows from the unpalatable fact that not all the states in (16) are in one representation, which would give TrQ = 0 automatically). Furthermore SU(5) requires the weak interaction to be V-A for quarks (rather than antiquarks) once V-A is put in for leptons. SU(5) leads to baryon violation, which may be desirable for general reasons, as discussed in the next section.

A prediction[26] which is more specific to the minimal version of SU(5) which we have discussed is

$$\sin^2\theta_{\overline{ms}}(M_W) = 0.215\pm.006 \qquad (37)$$

for $\Lambda_{\overline{ms}} = 150^{-100}_{+250}$ MeV, which follows from a careful second order calculation[27]. ($\Lambda_{\overline{ms}}$ is the Λ parameter of QCD, defined to leading order at the end of section 3, in the $\overline{ms}$ renormalization scheme.) The agreement with the data (12) is excellent. Many other GUTs can accommodate the measured value but none make a precise prediction as extra unknown parameters introduce uncertainties.

Another specific and successful prediction is

$$m_b(2m_b) \simeq 5\text{-}5.5 \text{ GeV} \tag{38}$$

which follows from the analogue of eqn. (36) for $\frac{m_b}{m_\tau}$ and preceded the discovery of the b[28]. The corresponding prediction for the second generation is

$$m_s(2m_s) \simeq 0.5 \text{ GeV} \tag{39}$$

which is bad but not terrible given the uncertainties in the values of these current algebra quark masses and the use of perturbation theory at dubiously low energy[2]. For the first generation, we write the prediction in the form

$$\frac{m_d}{m_s} = \frac{m_e}{m_\mu} \tag{40}$$

which is free from radiative corrections to first approximation. This is terrible. The right hand side is 1/200 while the left is determined relatively unambiguously by chiral symmetry perturbation theory to be about $\frac{1}{20}$. One possible response to this disaster is to introduce a 45 of Higgs bosons whose contribution destroys the relation between charged lepton and Q = -1/3 quark masses. Another[29] is to argue that although the prediction for m_d is wrong by a factor of 10, the absolute error is not enormous, since m_d is only a few MeV., and could be patched up by small residual contributions from a more complete theory which subsumes SU(5) at higher energy.

9. BARYON DECAY AND COSMOLOGY

The gauge dogma asserts that all continuous symmetries related to exact conservation laws are local in nature and are therefore associated with long range forces, as in the case of electric charge conservation and electromagnetism. If there is a long range force coupled to baryon number (B), the Eötvös-Dicke experiment requires[1] that the coupling g_B satisfies

$$\alpha_B \equiv \frac{g_B^2}{4\pi} < 6\text{x}10^{-48}\,\alpha$$

The limit for a force coupled to lepton number (L) must be comparable. This strongly suggests that no such long range forces exist, in which case B and L must be violated if the gauge dogma is correct.

Another general argument[30] for B violation follows from the observation that for the universe as a whole

$$\frac{n_B}{n_\gamma} \sim 10^{-9} - 10^{-10}. \tag{41}$$

It is known experimentally that our galaxy is dominantly made of matter. If the Universe contains a large amount of antimatter it must be in clusters of antigalaxies. If we assume that the net baryon number $n_B - n_{\bar{B}} = 0$ and that B is conserved, we must explain how matter and antimatter became separated on such a large scale. No viable separation mechanism is known. Furthermore when baryons and antibaryons fell out of equilibrium as the Universe cooled, annihilation would have given

$$\frac{n_B + n_{\bar{B}}}{n_\gamma} \sim 10^{-20}$$

in gross contradiction to (41). Thus if B is conserved it seems necessary to impose the bizarre initial condition

$$\frac{n_B - n_{\bar{B}}}{n_B + n_{\bar{B}}} \sim 10^{-9} - 10^{-10}, \tag{42}$$

which would explain (41) and the absence of antimatter.

If B is not conserved, a baryon asymmetry can be generated from a state with B = 0 provided

1. CP is violated (otherwise the CP non-symmetric state $B \neq 0$ could not be reached).

2. There is a non-equilibrium phase (in equilibrium B necessarily becomes zero if baryon number is violated, since $m_B = m_{\bar{B}}$ by CPT invariance).

Equilibrium is lost at early times when the density drops to the point where the interaction rate for B non-conserving processes falls below the expansion rate. At this point the decays of X and Y bosons or heavy Higgs particles, all of which we denote generically X, can generate an asymmetry. Suppose there are two decay channels

$$\Gamma(X) \equiv \Gamma(X \to qq) + \Gamma(X \to q\ell)$$
$$\Gamma(\bar{X}) \equiv \Gamma(\bar{X} \to \bar{q}\bar{q}) + \Gamma(\bar{X} \to \bar{q}\bar{\ell})$$

Necessarily $\Gamma(X) = \Gamma(\bar{X})$ by CPT invariance but we can have $\Gamma(X \to qq) \neq \Gamma(\bar{X} \to \bar{q}\bar{q})$, leading to an asymmetry proportional to the difference. The lowest order process which could generate the required CP violation is the interference of

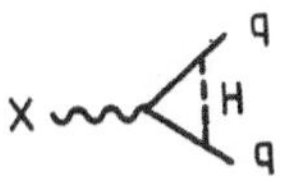

with

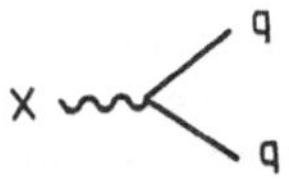

which is capable of producing a value for

$$\Delta = \frac{\Gamma(X \to qq) - \Gamma(\bar{X} \to \bar{q}\bar{q})}{\Gamma(X)}$$

of order 10^{-3}; this would be sufficient to explain n_B/n_γ.

In minimal SU(5), however, the Higgs couplings necessarily conserve CP which is only violated in the Kobayashi-Maskawa matrix. Recall that the charged weak current has the form

$$J_\mu^{cc} = \bar{U}_L \gamma_\mu (KM) D_L$$

where U[D] are column vectors containing the fields corresponding to the mass eigenstates for quarks with charge 2/3 [-1/3] and KM is an NxN unitary matrix for N generations of quarks. After absorbing as many phases as possible in the fields, KM is parameterized by $\frac{N(N-1)}{2}$ rotation angles and $\frac{(N-1)(N-2)}{2}$ non-trivial CP violating phases. For N=2, the only angle is the Cabibbo angle and KM conserves CP. CP violation occurs for $N \geq 3$ but can only be generated by diagrams in which quarks of three different generations enter, since otherwise we are essentially back to the N = 2 case and can define the phase away. CP violation in X decay enters first through the interference of diagrams such as

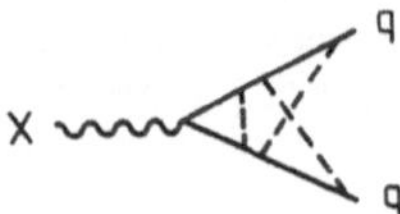

with the Born term. Putting in the relevant Yukawa couplings gives

$$\Delta \sim \frac{1}{10} \left(\frac{\alpha}{\pi}\right)^3 \frac{m_b^4 m_t m_c}{M_W^6} \sim 10^{-14} \frac{m_t}{M_W}$$

which is nothing like large enough.

This problem can be "solved" by adding a second Higgs 5 whose couplings involve CP violating phases relative to those of the first 5. This ad hoc source of CP violation is probably not directly connected with K^0 decays but leads to an electric dipole moment for the neutron within three orders of magnitude of the present limit, a bound which is thought[31] to hold in most models capable of generating the required values of Δ.

10. NUCLEON DECAY IN SU(5)

We now return from the speculative subject of baryon number generation to the expected characteristics of nucleon decay in minimal SU(5). The diagrams

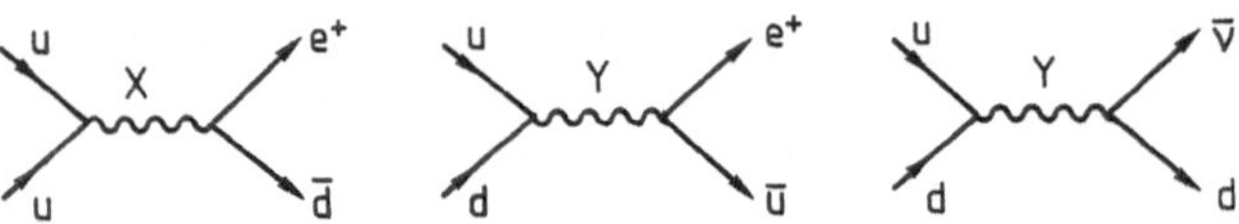

make contributions of order α_u/M_X^2 to the amplitude. There are also contributions from coloured Higgs bosons of order $\alpha_u \frac{m_q}{M_W}^2 \frac{1}{M_H^2}$ which can be ignored unless $M_H \lesssim 10^{11}$ GeV, which seems unlikely as M_H contains radiative contributions of order $\alpha_u M_X$. Two important points about nucleon decay in minimal SU(5) are

1. B-L is conserved, e.g. $p \rightarrow \pi^0 e^+$ is allowed but $p \rightarrow e^-\pi^+\pi^+$ and $n \rightarrow e^-\pi^+$ are absolutely forbidden. This selection rule is expected to be a good approximation in other models without being exact, as we shall see in section 13.

2. Predictions for generation changing decays can be made

when mixing is introduced e.g.

$$\frac{\Gamma(N \to \mu^+ S = 0)}{\Gamma(N \to e^+ S = 0)} \simeq \frac{s^2c^2}{1+(1+c^2)^2}$$

$$\frac{\Gamma(N \to e^+ S = +1)}{\Gamma(N \to \mu^+ S = +1)} \simeq \frac{s\ c}{1+(1+s^2)^2}$$

with $c = \cos\theta_c$, $s = \sin\theta_c$. These predictions are specific to minimal SU(5). The selection rule $\Delta S/\Delta B \leq 0$, which allows $p \to \nu K^+$ but not $n \to e^+K^-$, is exact in minimal SU(5) but is expected to be a good approximation in other models (section 13).

We now consider the calculation of the decay rate in some detail as it is a topical subject. First, a precise second order calculation of M_X is needed which gives[27]

$$M_X \simeq 1.3 \times 10^{15}\ \Lambda_{\overline{ms}} \tag{43a}$$

Uncertainties in low energy $\bar{e}e \to$ hadron data, used in evaluating the evolution of α_{em}, introduce an error of ±.18 in the coefficient 1.3. The error from unknown 3 loop contributions is of order ±.16 (taking an error $\pm\alpha_s(M_W)$ in the two loop equations for $\alpha^{-1}(E)$, which seems reasonable as α_s is expected to enter divided by π). A further error of order ±30% is introduced by allowing the masses of the coloured Higgs bosons to be $10^{\pm 2}M_X$. Thus we find

$$M_X = (1.3^{+.9}_{-.6}) \times 10^{15}\ \Lambda_{\overline{ms}} \tag{43b}$$

Next we must calculate the amplitude for diagrams such as

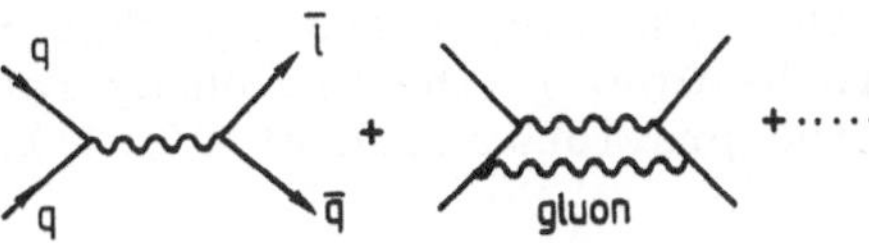

which give

$$\frac{\alpha_u}{M_X^2}\ \psi_q\psi_q\psi_\ell\psi_q\ (1 + A\alpha_s \ell n M_X/E + ..),$$

where E is the typical energy of the external particles. The "large logs" can be summed to all orders to give an effective Lagrangian

$$L_{eff} = \frac{\alpha_u}{M_x^2} \sum_i \left(\frac{\alpha_s(E)}{\alpha_u}\right)^{p_i} O_i, \tag{44}$$

which aborbs all knowledge of physics at a scale M_x, where O_i are four fermion operators and the p_i are calculable[1,2].

It remains to calculate $\langle X|O^i|N\rangle$. Until very recently[32] a large uncertainty was attributed to these calculations because different methods seemed to give very different results. However, the disagreements have now been greatly reduced. We consider the quark calculation of $p \to \pi^0 e^+$. There are two diagrams

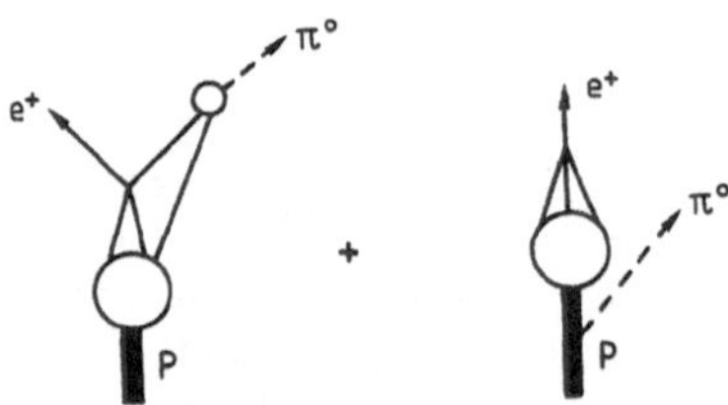

The second diagram was previously ignored on the grounds that the amplitude for finding three quarks at a point is small. However this turns out to be compensated by the large value of the pion-nucleon coupling constant[33,34] and the ratio of the two amplitudes, which interfere constructively, is found[33] to be approximately $|g_A|/1$, the value given by current algebra calculations [35,36]. This increases to O(70%) the branching ratio for this mode, which was previously thought to be O(40%) (the branching ratio for $n \to \pi^- e^+$ also increases from the previous value of O(70%)). The new quark model result is

$$\tau_{p \to \pi^0 e^+} = 3 \left(\frac{M_x}{1.3\text{x}0.15\text{x}10^{15}\text{GeV}}\right)^4 \text{ x } 10^{29}\text{years}$$

$$= 3^{+22}_{-2.75} \left(\frac{\Lambda_{\overline{ms}}}{150 \text{ MeV}}\right)^4 \text{ x } 10^{29} \text{ years.} \tag{45}$$

where (43b) has been used in the second line. A calculation[36] using current algebra and ITEP sum rules gives 0.5 in place of 3 while a bag calculation[34] including both diagrams gives 2 (after removing a guessed form factor suppression of the first diagram, whose presence is not confirmed by quark model calculations[33]).

The error in these calculations is hard to estimate but it seems unlikely that the liftime is substantially longer than the quark model result (45) given that other models give shorter lifetimes.

During the last year a belief that $\Lambda_{\overline{ms}}$ is of order 150 MeV has developed but some recent fits give values of order 300 MeV and 400 MeV is not excluded. The experimental lower limit on the lifetime is currently a few times 10^{30} years with one experiment reporting three candidate events which would correspond to a lifetime $\sim 6 \times 10^{30}$ years[37]. This is clearly compatible with SU(5), taking account of the uncertainties. However, if SU(5) is correct the lifetime could only be pushed beyond the range of experiments ($\sim 10^{32}$ years) by a conspiracy of the uncertainties and there is an excellent chance that nucleon decay will soon be observed.

11. CRITIQUE OF SU(5)

Obvious criticisms of SU(5) are that

1. It involves too many parameters, mainly associated with the Higgs system which is arbitrary, complex and ugly and suffers from the hierarchy problem.

2. The existence of different generations is not explained and it is unsatisfactory that the fermions in a single generation do not belong to a single multiplet. Anomalies from the $\overline{5}$ fermions cancel those from the 10, so both seem to be needed, but this mechanism of anomaly cancellation seems very ad hoc.

3. The minimal model does not generate enough baryons after the big bang and apparently leads to far too many magnetic monopoles[38], although these arguments involve speculations which transcend the boundaries of particle physics.

In addition, the model is just as left-right lopsided as SU(2)xU(1), which some find objectionable. It does not include gravity, which should presumably be subsumed in an ultimate unified model, although it may be possible to neglect gravity in constructing GUTs, leaving complete unification until later, since M_x is small compared to the Planck mass (10^{19} GeV).

In the next section we shall consider attempts to construct GUTs using groups larger than SU(5), which might remove some of these defects. We shall then consider some general properties of nucleon decay and neutrino masses, which are pertinent for these larger GUTs. In the final section we briefly consider what theories might replace or go beyond GUTs.

12. GRANDER GROUPS

There are various approaches to the choice of a grand unified group.

1. Seek a "simple"group G (in which there is necessarily only one coupling constant) such that a) there are no triangle anomalies whatever representations are used for the fermions and b) it is automatically impossible for all the fermions to acquire masses of order M_X. The only groups which satisfy these criteria [39]are the exceptional group E(6) and the orthogonal groups SO(4N+2), with $N \geq 2$ so that the rank is not less than four.

The simplest candidate is SO(10), which obviously can work as it contains SU(5). In SO(10) there are 45 gauge bosons and the fermions of a given generation belong to a 16 dimensional representation which has the decomposition

$$16 \rightarrow (10 + \bar{5} + 1)$$

under SU(5). The 1 accommodates a right handed neutrino, which will be discussed further in section 14. The model is left-right symmetric, parity violation occurring spontaneously because bosons coupled to left and right handed currents develop different masses.

If SO(10) breaks directly to SU(3)xSU(2)xU(1) the low energy predictions would be similar to those of SU(5). However, various other breaking schemes are possible e.g.

$$SO(10) \begin{matrix} \nearrow 4 \times 2 \times 1 \searrow \\ \rightarrow 4 \times 2 \times 2 \rightarrow \\ \searrow 3 \times 2 \times 2 \times 1 \nearrow \end{matrix} 3 \times 2 \times 1 \rightarrow 3 \times 1_{em}$$

M M' 100 GeV

where 4 means SU(4) etc. The freedom to adjust M'/M and the existence of large numbers of Higgs bosons with masses $O(M_x)$ has the consequence that $\sin^2\theta$ can take a large range of values and the proton lifetimes is not precisely determined. B-L is a generator of SO(10) and will be broken at low energy, allowing (in principle) $\bar{N}$-N oscillations, Majorana neutrino masses and $N \rightarrow e^-X$, which will be discussed in the next two sections.

E(6) contains SO(10) so it is clearly a viable group. In E(6) there are 78 gauge bosons and the fermions fill the fundamental 27 dimensional representation with the decomposition

$$27 = (16 + 10 + 1)_{SO(10)}$$

$$\quad 10 \rightarrow (5 + \bar{5})_{SU(5)}$$

An SU(5) invariant mass is allowed for the fermions in $5+\bar{5}$ which will marry to form Dirac particles of mass $O(M_X)$. The SO(10) singlet joins the SU(5) singlet from the 16 to form another particle of mass $O(M_X)$. E(6) can be formulated in a way which leads to

$$m_t \simeq \frac{m_\tau}{m_\mu} m_c \tag{46}$$

which gives $m_t = O(20\text{ GeV})$, taking radiative corrections into account. This interesting relation may have a more general validity.[40] The data exclude the possibility that the mass matrices for the quarks of charge 2/3 and -1/3 and the charged leptons are proportional. The simplest viable relation is therefore that they are linearly dependent and this leads to (46).

2. Seek a semi-simple product group, such as $(SU(4))^4$, and make the theory invariant under interchanging the groups in the product, so that there can only be a single coupling constant. This sort of model has been much discussed by Pati and Salam[41], who were the first to study GUTs and construct a model with an unstable proton. Their model incorporates the interesting idea that leptons are associated with a fourth colour. Above some unification energy (which might be only 10^4 GeV) the strong interactions become $SU(4)_L x SU(4)_R$ and act on quartets such as (u^r, u^b, u^g, ν). In these models quarks and antiquarks do not coexist in the same multiplets. Nucleon decay can therefore only occur if all three quarks decay simultaneously, which allows a low unification scale (in the language of the next section $d \geq 9$ operators are needed). I am no expert on these models but suspect that their virtues do not compensate for their complexities.

3. Find the largest possible, or "maximal", gauge group for a given set of fermions. For a 16 component generation (including ν_R) the maximal group is SU(16). Unfortunately SU(16) with one 16 would have anomalies. Pati, Salam and Strathdee[41] advocate cancelling the anomalies by adding a 16 of "mirror" fermions with V+A couplings. A discrete symmetry must be imposed to prevent the V-A and V+A coupled fermions marrying and acquiring a mass of $O(M_X)$. All fermion masses are then due solely to SU(2)xU(1) breaking and the mirror fermions must be below a few hundred GeV. This is a very interesting possibility, independent of this specific model. In maximal groups there is a unique vector boson V_{ij} coupled to every pair of fermions $(f_i f_j)$ in the 16. Baryon violation is therefore induced by symmetry breaking which produces mixing

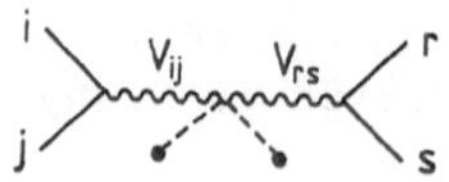

In contrast, in non maximal theories a single V couples to more than one fermion pair and baryon violation is intrinsic.

4. Seek groups which can accommodate all the fermions of the three known generations. Strategy 1) leads[42] to SO(18) while the maximal group is SU(48). Unfortunately little seems to be accomplished except an inflation of degrees of freedom e.g. in SO(18) there are 153 gauge bosons and the fermions belong to a 256 dimensional representation.

13. BARYON NUMBER VIOLATION: GENERALITIES

We saw above that the low energy effects of diagrams which contain a single heavy boson of mass M can be described by a non-renormalizable effective Lagrangian (44) of the form O^6/M^2, where the superscript 6 denotes the mass dimensions of the operator O. Diagrams containing two heavy bosons (M_1,M_2) can produce $O^8/M_1^2M_2^2$ terms etc. The operators O^6, O^8... must be SU(3)xSU(2)xU(1) invariant. Weinberg and Wilczek and Zee have pointed out[43] that the allowed operators of low dimension obey interesting selection rules. Tests of these rules would provide important information about whether there is more than one intermediate mass scale M_i. We consider a few of the properties of some operators O^d:

<u>d = 6.</u> All d = 6 operators which respect SU(3)xSU(2)xU(1) satisfy $\Delta B = \Delta L$ (i.e. we can have $qqq\ell$ but not $qqq\bar{\ell}$) and $\Delta S/\Delta B \leqslant 0$. If there is a single heavy scale, a simple minded estimate is that it must be $O(10^{14}$ GeV) or more to prevent the nucleon decaying too rapidly. With 2 or more scales it might seem that something like their geometric mean would obey the same bound but an example shows that such estimates must be treated cautiously. In supersymmetric (SUSY) GUTs, nucleon decay can be mediated by a spin 1/2 coloured Higgs ($\tilde{H}$), scalar quarks ($\tilde{q}$) and the spinor gluon or "gluino" ($\tilde{g}$) thus:

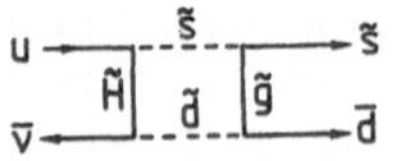

which depends on the heavy scales through $(M_{\tilde{H}})^{-1}x(M_{\tilde{q}}$ or $M_{\tilde{g}})^{-1}$. Since $M_{\tilde{H}}$ is expected to be $O(M_X)$ and $M_{\tilde{q},\tilde{g}}$ < few TeV this appears to be disastrous. However, small Yukawa couplings and numerical factors from the loop integration may save the day, producing a result close to the experimental limit[44]. The main reason for discussing SUSY GUTs here is that they provide a salutary reminder that the favoured SU(5) mode $p \rightarrow \pi^0 e^+$ may not dominate. In SUSY the scalar quarks and gluinos slow down the decrease of α_s postponing the unification point M_X to a higher value[45]. This probably makes the "conventional" X and Y boson mediated decay rate unobservably small. However the new mode fills the breach. The intermediate $\tilde{q}\tilde{q}$ state must be antisymmetric in colour and therefore Bose statistics force the $\tilde{q}$'s to belong to different generations. Thus $p \rightarrow K^+\nu$ is the dominant mode in many SUSY models.

d = 7 operators built of four fermions and a space-time derivative or a boson field can be constructed. It turns out that the allowed operators have $\Delta B = -\Delta L$. In general, therefore, decays $N \rightarrow e^- x$ can occur but they will be suppressed in rate relative to $\Delta B = \Delta L$ processes by a factor $m^2\hat{M}^{-2}$, where $\hat{M}$ is the scale at which $\Delta B = \Delta L$ violation occurs and m is an SU(2)xU(1) violating scale (m_q or M_W) or the energy of one of the particles involved. In SO(10) broken through $SU(2)_L x SU(2)_R x U(1)$, $\hat{M}$ would be the scale at which $SU(2)_R$ breaks (producing parity violation) and U(1) (which couples to B-L) is broken, leaving the standard U(1) intact.

d = 9 operators include qqqqqq which gives rise to neutron-antineutron oscillations, which might be observable if the relevant scale is 10^5 GeV or less. Experiments to look for n-$\bar{n}$ oscillations are underway[46] and the phenomenology is worth considering[47]. The n-$\bar{n}$ mass matrix may be written

$$\begin{pmatrix} V & \delta m \\ \delta m & \bar{V}-iW \end{pmatrix}$$

In free space, CPT requires $V = \bar{V}$, $W = 0$, and the eigenstates are $(n \pm \bar{n})/\sqrt{2}$. The oscillation time $\tau_{osc} = \delta m^{-1}$ and the number of $\bar{n}$'s in an n beam grows like $(\delta mt)^2$. In a nucleus the baryon asymmetry of the environment makes $V \neq \bar{V}$ and $W \neq 0$ (W represents $\bar{n}$ absorption). The oscillation time is very rapid but the amplitude is very small, the decay time being

$$\tau^{-1}_{decay} = \frac{W(\delta m)^2}{(V-\bar{V})^2+W^2}$$

A limit on the nucleon lifetime of 3×10^{30} years combined with plausible values of $V-\bar{V}$ and W gives $\tau_{osc} = \delta m^{-1} > 10^7$ sec for free space. In "free space" experiments it is necessary to shield the Earth's field by a factor 10^3 to reach the $N_{\bar{n}} \sim (\delta mt)^2$ situation. The direct limit on τ_{osc} is now 10^5 secs with the prospect of reaching

3×10^8 secs. It appears to be possible to arrange for observable oscillations in SO(10) but it seems arbitrary to do so. The relevant diagram is

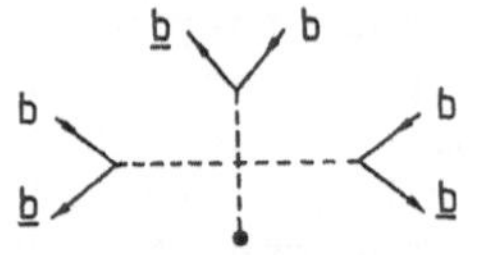

where the dashed line is the propagator of a Higgs boson (H_{qq}) with diquark quantum numbers, belonging to a 126, whose mass must be 10^5 GeV or less to produce observable oscillations. With such a relatively light Higgs boson, it turns out that the low energy data require the left-right unification scale to be at least 10^9 GeV. The mass of H_{qq} is therefore unconnected with any other scale in the model and its introduction is rather artificial and unnatural[48].

d = 12 operators include $qqqqqq\ell\ell$ allowing hydrogen-antihydrogen oscillations[49]! The present direct limit, from the cosmic photon flux, is $\tau_{osc} > 7 \times 10^{10}$ years but the limit on $pp \to e^+e^+$ implies $\tau_{osc} > 6 \times 10^{12}$ years. It is possible to construct models in which these oscillations occur at an observable rate!

14. NEUTRINO MASSES

Massive neutrinos would be welcomed by astrophysicists who could invoke them to explain the "missing mass" problem and other phenomena[50]. As well as Lorentz invariant Dirac mass terms

$$m\bar{\psi}\psi = m(\bar{\psi}_L\psi_R + \bar{\psi}_R\psi_L)$$

we can construct Majorana masses

$$\overline{(\psi^c)}_R\psi_L, \quad \bar{\psi}_R(\psi^c)_L,$$

given that the charge conjugate ψ^c has the same behaviour as ψ under a Lorentz transformation (this is clearly necessary for the existence of Majorana particles for which $\psi^c = \psi$. The Lorentz invariant $\bar{\psi}^c\psi$ may be unfamiliar; it is not encountered in standard text books on QED as such a mass term would clearly violate charge conservation if ψ were the electron field).

If it has zero mass and purely V-A interactions, we can consider the neutrino to be
either a Dirac particle with only a left handed state plus its (right handed) antiparticle
or a Majorana particle with left (L) and right (R) handed states, the latter being what is usually thought of as the antineutrino, which is indentical to its antiparticle i.e.

$$R = \bar{L}, \quad L = \bar{R} .$$

In either case there are two degrees of freedom and these formulations are equivalent. If $m_\nu \neq 0$, a Dirac neutrino would have 4 components and a ν_R field is needed. A Majorana neutrino, however, would have only two components (this clearly requires $\psi = \psi^c$ as the two L and R components mix under Lorentz transformations if $m_\nu \neq 0$).

With four components both Dirac and Majorana masses are allowed, the most general mass term being

$$\overline{((\nu^c)_R \quad \bar{\nu}_R)} \begin{pmatrix} M_L & M_D \\ M_D & M_R \end{pmatrix} \begin{pmatrix} \nu_L \\ (\nu^c)_L \end{pmatrix}$$

In minimal SU(5) there is no ν_R and the Majorana mass term M_L violates B-L, so it can only be generated by adding extra Higgs fields or appealing to small non-renormalizable contributions left over from a more complete theory[29]. In SO(10) there is ν_R. A Dirac mass term M_D, which has $\Delta I_W = 1/2$, may be generated by SU(2) breaking and would be expected to be of $O(m_q)$. A ($\Delta I_W = 0$) Majorana mass M_R is related to the scale $\hat{M}$ at which B-L is broken. The $\Delta I = 1$ mass M_L violates B-L as well as weak isospin ($\Delta I_W = 1$) and would be $O(\frac{m_q^2}{\hat{M}})$. This mass matrix would give ν_R a mass of order $\hat{M}$ and ν_L a mass of order $\frac{m_q^2}{\hat{M}}$. In SO(10), M_R can be generated at tree level if a 126 of Higgs bosons is added. In this case $M_R \sim M_X$ is a reasonable guess, making $m_{\nu_L} \sim 10^{-6}$ eV. If M_R is zero at tree level, a non-zero value is generated by higher order processes such as[51]

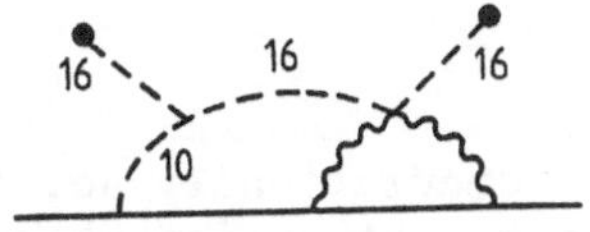

in a minimal model with a Higgs 10 and 16. This gives $M_R \lesssim (\frac{\alpha}{\pi})^2 M_X$ and could generate neutrino masses of order 1 eV.

The possibility of Dirac or Majorana masses is clearly independent of the correctness of GUTs, although they provide a framework for discussing the likely orders of magnitude. Evidently the range of possible masses is enormous. As in the case of nucleon decay, much will be learned if neutrino masses are observed but very little if they are not.

15. THE GRAND PROBLEMS

GUTs are very interesting and have proved fruitful in focussing attention on the possibility of nucleon decay and neutrino masses. However, there are (at least) three grand problems which GUTs do not answer!

1. The problem of flavours. Possibly this is answered by going to bigger groups such as SO(18) and SU(48) although this strategy has not so far been very successful. Alternatively the resolution may lie in substructure for quarks and leptons[20], in which case we have been vainly attempting to construct GUTs from the wrong degrees of freedom unless the scale of compositeness is above M_X.

2. The origin of mass/symmetry breaking. First there is the question of the origin of the vast range of masses, from $m_{\nu_e} < 10^{-7}$ GeV through $m_e \simeq 10^{-3}$ GeV to the Planck mass (10^{19}GeV). Then there is the question of how light particles can coexist with very heavy particles. For fermions, radiative corrections give $\delta m_F = O(g^2)m_F$ so that once a small value is put in, it remains small (this is a residue of chiral symmetry which requires $\delta m_F = 0$ if $m_F = 0$). Scalars, however, tend to get dragged up to the highest scale available. This is the hierarchy problem[52]. Consider the Higgs potential (7). The parameter μ, which we now rechristen μ_0 as it is the bare value, gets radiative corrections given symbolically by

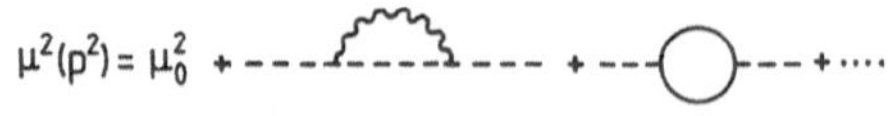

$$\mu^2(p^2) = \mu_0^2 + \text{-- -- --} + \text{-- --}\bigcirc\text{-- --} + \cdots$$

The integrals diverge quadratically so, following the discussion in section 3, we make a subtraction obtaining

$$\mu^2(p^2) = \mu^2(\Lambda^2) + cg^2\int_{p^2}^{\Lambda^2} dk^2 + \ldots$$

where the omitted terms grow at most logarithmically as $\Lambda \to \infty$. If $\mu(\Lambda^2)$ is given at some super unification scale $\Lambda >> 1$ TeV, a miraculous cancellation is needed to get the required value $\mu^2(M_W^2) \sim (300 \text{ GeV})^2$ since the integral contributes $O(g^2\Lambda^2)$. The reasonable requirement that the radiative corrections to μ^2 should not be much bigger than μ^2 itself strongly suggests that there is new physics below 1 TeV - either Λ itself is less than 1 TeV or other new physics somehow cuts off the integral. There are various possibilities:

a) There are no fundamental scalars, in which case new physics is needed to avoid the unitarity crisis discussed in section 2. This is the situation in technicolour models, in which W_L is a heavy $\bar{Q}Q$ state bound by O(1 TeV). In such models any scalar fields are composites whose structure provides a cut off in the diagrams above.

b) Composite structure for W's, Z's quarks and leptons, which would also lead to a cut off.

c) The ϕ's develop strong self interactions at around 1 TeV, which could provide a cut off. Strong interaction phenomena (resonances etc.) would then be observed in $e^+e^- \to W_L^+W_L^-$ at $\sqrt{s} \sim 1$ TeV.

d) Nature is supersymmetric, in which case positive contributions from virtual bosons would be cancelled by negative contributions from their supersymmetric fermion partners (because of the symmetry, $\delta m \sim m$ for bosons as well as fermions). Although the sum of contributions of the form $\int^{\Lambda^2} dk^2$ is finite for $\Lambda^2 \to \infty$, the lower limits of the integrals are effectively controlled by the masses of the particles in the loops and the net contribution grows with the mass splitting in SUSY multiplets. It follows that SUSY can only solve the hierarchy problem if the SUSY partners of the known particles are below 1 TeV. Specific models show that some are very probably at M_W or below, making SUSY easy to test at LEP[21]. The idea of symmetries which relate fermions to bosons is extremely attractive. However, it is not clear that the virtues of SUSY GUTs compensate for the complexities of the models constructed up to now. Furthermore, aesthetic arguments lead on to local SUSY and supergravity rather than the global SUSY which has been the focus of so much attention in the last year.

3. Gravity. The best behaved theories of quantum gravity are various versions of supergravity which provide the possibility of a unification of all forces. Unfortunately the most accommodating scheme available, which is O(8) supergravity, is not large enough to include $SU(2)_L$ and all the known quarks and leptons[53]. A very interesting speculation due to Ellis,

Gaillard and Zumino[2] is that there is an underlying O(8) theory whose quanta are bound at M_{Planck} to form the quarks and leptons and the gauge bosons of SU(5). There are intriguing hints that this could occur but such speculations will be very hard to substantiate. In any case the fact that the vacuum expectation values which break SU(2)xU(1) and chiral symmetry generate a cosmological constant which is at least 50 orders of magnitude too big may be a hint that gravity is intimately connected with the rest of particle physics.

ACKNOWLEDGEMENTS

I wish to thank Tom Ferbel for the pleasant atmosphere and the stimulating company he provided at Lake George. I am grateful to Nathan Isgur for comments on the manuscript.

PROBLEMS

1. Derive eqn.(5) from the preceeding equations. Show that if $\hat{M}_W \neq M_W$ the neutral current term must be multiplied by $M_{\hat{W}}^2/M_Z^2\cos^2\theta$.

2. Consider the Born diagrams for fermion + antifermion → WW, drawn in section 2. Neglecting fermion masses and using the Dirac equation show that when $\varepsilon_\mu^{a,b} \to k_\mu^{a,b}$ the sum of the fermion exchange diagrams is proportional to $[L^a, L^b]_{ij}$.

3. Show that when the SU(2) Higgs doublet develops a vacuum expectation value $\langle\Phi\rangle = \begin{pmatrix} 0 \\ \langle\phi^0\rangle \end{pmatrix}$, W and Z masses are generated with the mass matrix (4) with $\hat{M}_W = M_W$.

4. Show that with $Q_L = \left(\frac{1-\gamma_5}{2}\right)\begin{pmatrix} u \\ d \end{pmatrix}$, $q_R = \left(\frac{1-\gamma_5}{2}\right)$ q the interaction terms

$$(\bar{Q}_L\Phi)d_R + \text{h.c}$$

$$(\bar{Q}_L\sigma_2\Phi^*)u_R + \text{h.c.}$$

are SU(2)xU(1) invariant and generate d and u masses when ϕ^0 develops a vacuum expectation value.

5. Write the Lagrangian

$$L = \bar{\psi}(i\not{\partial} - m + g\phi + g'\phi'\gamma_5 + g_V\not{A})\psi$$

in terms of left and right fields $\psi_{L,R} = \left(\frac{1\mp\gamma_5}{2}\right)\psi$. Show that ψ_L and ψ_R decouple if and only if $g' = g = m = 0$ (allowing "chiral" symmetries which act independently on ψ_L and ψ_R).

Show, equivalently, using the Dirac equation that this condition is necessary and sufficient to ensure that $\partial^\mu A_\mu = 0$ where $A_\mu = \bar{\psi}\gamma_\mu\gamma_5\psi$.

6. Consider the axial current responsible for β decay whose nucleon matrix elements are $\langle p|A_\mu(0)|n\rangle = \bar{U}_p(p')(\gamma_\mu\gamma_5 g_A(q^2)+q_\mu\gamma_5 f_p(q^2))$ $U(n)$ where $q = p-p'$. Show that if $\partial^\mu A_\mu = 0$ <u>either</u> $M_{p,n} = 0$ <u>or</u> there is a pole in f_p at $q^2 = 0$. Interpret this pole as a massless pion and show that $\partial^\mu A_\mu = 0$ leads to the Goldberger-Treiman relation $2M_N g_A + f_\pi g_{pn\pi} = 0$ where f_π is the pion decay constant and $g_{pn\pi}$ the π-p-n coupling constant.

7. Attempt to devise models in which (15) leads to an acceptable value of $\sin^2\theta$ without putting quarks and leptons in the same multiplet.

8. In eqn. (20) a 10 and a 15 dimensional basis for SU(5) were formed from two fives. Combine another 5 with the 10 and 15 and show that

$$5 \times 10 = 10 + 40$$
$$5 \times 15 = 35 + 40.$$

9. Verify that the vacuum expectation value in eqn. (27) gives equal masses to the X and Y bosons, leaving the others massless.

10. Verify that eqns. 34 and 35 give $m_e = m_d \neq 0$, $m_u \neq 0$ respectively when H_5 develops a vacuum expectation value.

11. Show that the SU(5) Lagrangian leads to the B violating interactions exhibited at the beginning of section 10.

REFERENCES

1. P. Langacker, Phys. Rep. 72: 185 (1981).
2. J. Ellis, 1981 Les Houches Lectures, CERN TH 3174 (1981).
3. Pedagogical introductions to the SU(2)xU(1) model of Glashow, Weinberg, Salam and Ward may be found in
   H. Fritzsch and P. Minkowski, Phys. Rep. 73: 67 (1981).
   I.J.R. Aitchison and A.J.G. Hey, "Gauge Theories in Particle Physics" Adam Hilger, Bristol (1982).
4. Limits come from measurements of longitudinal polarization in β decay and μ^+ decay. For reviews see
   J.J. Sakurai, p.457, Neutrino '81 Vol. 2, ed. R.J. Cence, E.Ma and A. Roberts, University of Hawaii (1981).
   F. Sciulli, these proceedings.
5. J.E. Kim et al., Rev. Mod. Phys. 53: 211 (1981).
   For another review see
   P.Q. Hung and J.J. Sakurai, Ann. Rev. Nucl. Sci. 31: 375 (1981).

6. TASSO Collaboration, R. Brandelik et al., DESY 82-032.
7. Perturbative QCD effects go some way towards explaining the data; references may be found in
G. Altarelli et al., Nucl. Phys. B187: 461 (1981).
8. This was first suggested by
J.D. Bjorken, Phys. Rev. D19: 335 (1979).
For a review of non standard models see
J.J. Sakurai, Max Planck, Munich, preprint MPI-PAE/PTh 22/82. To be published in Proc. XII Schladming School and XVII Rencontre de Moriond.
9. For a detailed exposition of the argument below see
C.H. Llewellyn Smith in "Phenomenology of Particles at High Energy" ed. R.L. Crawford and R. Jennings, Academic Press (1974) or "Proc. Fifth Hawaii Topical Conference in Particle Physics (1973)" ed. P.N. Dobson et al., University of Hawaii Press (1974).
10. This is the value obtained by John Wheater and myself. For a review and refereneces see
J.F. Wheater, to be published in Proc. 1982 Paris Conference.
11. For a review and references see
V. Barger, Madison preprint MAD/PH/36 (1982).
12. See Ref. 2 and references therein for a discussion of how to find Higgs bosons.
13. For reviews of QCD and references see e.g.
W. Marciano and H. Pagels, Phys. Rep. 36: 137 (1978).
G. Altarelli, Phys. Rep. 81: 1 (1982).
C.H. Llewellyn Smith, Phil. Trans. Roy. Soc. 304: 5 (1982).
H.D. Politzer, to be published in Proc. 1982 Paris Conference.
14. For reviews of chiral symmetry see e.g.
V. De Alfaro et al., "Currents in Hadron Physics", North Holland (1973).
H. Pagels, Phys. Rep. 5: 219 (1975).
15. For reviews of the experimental status of perturbative QCD see e.g.
D.H. Perkins, Phil. Trans. Roy. Soc. 304: 23 (1982).
F. Eisele, to be published in Proc. 1982 Paris Conference.
16. R.P. Feynman in "Weak and Electromagnetic Interactions at High Energy", ed. R. Balian and C.H. Llewellyn Smith. North Holland (1977).
17. A. De Rujula, H. Georgi and S.L. Glashow, Phys. Rev. D12: 147 (1975).
18. I have suppressed corrections to external lines (which can be normalized to one at the subtraction scale) and ignored technicalities such as the gauge dependence of g.
19. For references to left-right symmetric models see
V. Barger, E. Ma and K. Whisnant, Phys. Rev. Lett. 48: 1589 (1982) and University of Hawaii preprint UH-511-470-82.
For references to left-right symmetric GUTs see
W.E. Caswell, J. Milutinovic and G. Senjanovic, Brookhaven preprint BNL 31046 (1982).

20. For a review see
M.E. Peskin in Proc. 1981 International Symposium on Lepton and Photon Interactions at High Energy, ed. W. Pfeil, University of Bonn (1981).
21. See C.H. Llewellyn Smith, Oxford preprint 44/82, and other contributions to the CERN Supersymmetry Workshop to be published as a Physics Report.
22. This follows from $Tr[T^a,T^b] = 0 = f_{abc}TrT^c$.
23. See e.g.
P.Q. Hung, A.J. Buras and J.D. Bjorken, Phys. Rev. D25: 805(1982).
24. We assume that ν_τ and t exist. It is essentially impossible to make a viable gauge theory without ν_τ. Topless models are not yet all excluded by the data (see e.g. S.S. Grigoryan and S.F. Sultanov, Serpukov preprint IHEP 82-61).
25. The SU(5) model of H. Georgi and S.L. Glashow, Phys. Rev. Lett. 32: 438 (1974) is reviewed in ref. 1.
26. A value of O(.21) was predicted by the calculation of H. Georgi, H.R. Quinn and S. Weinberg, Phys. Rev. Lett. 33: 451 (1974), at a time when the mean experimental value was 0.35.
27. The values given here for $\sin^2\theta$ and below for M_x are taken from C.H.Llewellyn Smith, G.G. Ross and J.F. Wheater, Nucl Phys. B177: 263 (1981) (other authors find essentially the same values).
28. M.S. Chanowitz, J. Ellis and M.K. Gaillard, Nucl. Phys. B128: 506 (1977).
29. J. Ellis and M.K. Gaillard, Phys. Lett. 88B: 315 (1979).
30. For a review of the problem of cosmic baryon number generation in GUTs and references see
J. Ellis, Contribution to Royal Society Discussion meeting, March 1981, Phil. Trans. Roy. Soc. (in press).
31. J. Ellis, et al., Phys. Lett. 99B: 101 (1981).
32. The change post dates the delivery of these lectures whose text therefore differs from the original presentation.
33. N. Isgur and M. Wise, Phys. Lett. (in press).
34. Y. Tomozawa, Phys. Rev. Lett. 46, 463 (1981).
M. Claudson, M.B. Wise and L.J. Hall, Nucl. Phys. B195: 297 (1982).
35. B.S. Berezinsky et al., Phys. Lett. 105B: 33 (1981).
36. J.F. Donoghue and E. Golowich, Amherst preprint UMHEP-166 (1982) and references therein. The number quoted is for the case of a pseudovector pion-nucleon coupling with the form factor removed (N. Isgur, private communication).
37. M.R. Krishnaswamy et al., Phys. Lett. (in press).
For a review of the current situation and future prospects see D.H. Perkins, Oxford Dept. Nucl. Phys. preprint 60/82, to be published in Proc. 1982 Paris Conference.
38. For a review of monopoles and references see N. Cabibbo, these proceedings.
39. For an espousal and explanation of this approach see
D.V. Nanopoulos "Electroweak Interactions and Unified Theories"

ed. J.Tran Thanh Van, Editions Frontieres (Dreux, France), P.427 (1980). A fuller discussion of subsequent remarks about SO(10) and E(6) and references may be found in this paper.

40. S.L. Glashow in "Gauge Theories and Experiments at High Energy" ed. K. Bowler and D.G. Sutherland, Scottish Universities Summer School in Physics (1980).

41. For a review and references see
J.C. Pati, Maryland preprint 82-151, to be published in Proc. International Conf. on Baryon Non-Conservation, (ICOBAN) Bombay, 1982.

42. For reviews see ref. 1 and
A. Zee, University of Washington preprint RLO-1388-873 (1981), to be published in Proc. 4th Kyoto Summer Institute on Grand Unification and Related Topics.

43. S. Weinberg, Phys. Rev. Lett. 43: 1566 (1979) and Phys. Rev. D22: 1694 (1980).
F.A. Wilczek, H.A. Weldon and A. Zee, Nucl. Phys. B173: 269 (1980).
For a simplifed derivation see
H. Lipkin, Phys. Rev. Lett. 45: 311 (1980) and Nucl Phys. B171: 301 (1980).

44. J. Ellis, D.V. Nanopoulos and S. Rudaz, Nucl. Phys. B202: 43 (1982)
S. Dimopoulos, S. Raby and F. Wilczek, Phys. Lett. 112B: 133 (1982).

45. This is compensated somewhat by the slower evolution of α_2 and models can be devised in which M_X is not much increased. See e.g. Y. Igarishi, J. Kubo and S. Sakakibara, Dortmund preprint DO-TH 82/09.

46. For a popular account of the status of these experiments see Physics Today, June 1982, p.19.

47. For reviews see ref 2 and
R.N. Mohapatra, City College New York preprint CCNY-HEP 82/1 to be published in the ICOBAN proceedings, loc. cit.

48. D. Lüst, A. Masiero and M. Roncadelli, Phys. Rev. D25: 3096 (1982).

49. L. Arnellos and W.J. Marciano, Phys. Rev. Lett. 48: 1708 (1982).
R.N. Mohapatra and G. Senjanovic, Phys. Rev. Lett 49: 7 (1982).

50. For reviews of the role of neutrinos in astrophysics and the question of neutrino masses see
F.W. Stecker, Goddard Space Flight Center memo 83909 (1982), to be published in Proc. 1982 Schladming Winter School.
P.F. Frampton and P. Vogel, Phys. Rep. 82: 339 (1982).

51. E. Witten, Phys. Lett. 91B: 81 (1980).
For reviews see refs. 1 and 2.

52. E. Gildener and S. Weinberg, Phys. Rev. D13 3333 (1976) and E. Gildener, Phys. Rev. D14: 1667 (1976).
For a forceful exposition see
M. Veltman, Acta. Physica Polonica B12: 437 (1981).

53. M. Gell-Mann, unpublished talk at the 1977 Washington APS meeting.

GAUGE THEORIES AND MONOPOLES

(A MODEST INTRODUCTION)

N. Cabibbo

Ist. di Fisica and INFN
Sezione di Roma
Piazzale A. Moro 2, I-00185, Rome, Italy

In these lectures I will try to present some of the fundamental geometrical ideas at the basis of gauge theories. I will build up these ideas in a way which will lead to a description of Dirac Monopoles. I will not try to discuss applications, but only those ideas that you are not likely to meet in more "utilitarian" presentations which concentrate on QCD or on the Glashow-Salam-Weinberg model. The reason for choosing this topic for my lecture is the announcement of the possible detection of a Dirac monopole. Rumors received while preparing the manuscript indicate that no new events have been found in an extension of the experiments. Thus the search for monopoles remains one of the most exciting enterprises in particle physics and cosmology.

Superheavy monopoles ($M \simeq 10^{15}$ GeV) may be the only living relics of the first instant of time in the universe. (It has been argued that the net baryon number of the universe is another such relic. The argument is forceful and suggestive, but very indirect.) If relic monopoles are detected, we will have proof that the bold extrapolation, according to which the universe at the beginning was as hot as 10^{15} GeV, is indeed correct.

The existence of monopoles depends on topological features of gauge theories, i.e., on global properties of field configurations which are unique to gauge theories. These are the features I will try to present in my lectures.

Global Symmetry-Local Symmetry

As a starting point, consider the Dirac equation for a set of N fermion fields ψ_i. We assume equal masses, so:

$$\left(i\gamma^\mu \frac{\partial}{\partial x^\mu} - m\right)\psi_\ell(x) = 0 \quad . \tag{1}$$

This equation is invariant under transformations:

$$\psi_\ell(x) \rightarrow \psi'_\ell(x) = u_{\ell m}\psi_m(x) \tag{2}$$

in short:

$$\psi(x) \rightarrow \psi'(x) = u\psi(x) \tag{2'}$$

where u is a unitary N × N matrix. (The need to limit ourselves to unitary transformation is dictated by other physical arguments, such as the wish to keep the Hamiltonian invariant.) What is more important, u has to be a constant, the same for all x. <u>Invariance</u> means that, given the transformation (2), from (1) it follows that ψ' obeys the same equation as ψ:

$$\left(i\gamma^\mu \frac{\partial}{\partial x^\mu} - m\right)\psi'(x) = 0 \ ,$$

as can be easily verified.

Why are we limited to <u>constant</u> u, i.e. to "global" transformations? Why can we not have invariance under "local" or "gauge" transformations"

$$\psi(x) \rightarrow \psi'(x) = u(x)\psi(x) \ . \tag{3}$$

The reason is that Dirac's equation (1) is a <u>differential</u> equation. To reveal the core of the problem, rewrite (1) as a difference equation, i.e., in the form you would use if you wanted to solve it on a computer.

$$i\gamma^o \frac{\psi(t+\Delta t,x,y,z) - \psi(t,x,y,z)}{\Delta t} + \ldots$$

$$\ldots + i\gamma^3 \frac{\psi(t,x,y,z + \Delta z) - \psi(t,x,y,z)}{\Delta z} = m\psi(t,x,y,z) \ .$$

As you see, Dirac's equation gives a relation between the values of ψ at five different points of space-time. It will not be invariant if we apply a <u>different</u> transformation to the ψ at each different point, as suggested by (3).

A gauge transformation (3) tends to destroy the relation between different points of space-time. If we want to build a gauge theory we must counter this tendency. What we need in addition to the "normal fields", such as the ψ_ℓ of Eq. 1, is some new entity which guarantees the "connection" of space-time even in the presence of gauge transformations. I will build this new entity in a global form, and will later verify that its existence is equivalent to the existence of a set of local fields - the gauge fields. The global form of the gauge fields is more fundamental than their local form, and is the formulation of choice in the discussion of such problems as monopoles and lattice gauge theories.

The Connection

The new variable will depend on two points of space time. (As we will see shortly, this concept will have to be enlarged in a rather striking way.) It's value will be an element of the symmetry group G to which the u(x) belong. In the example from which we started, G would be the group of unitary $N \times N$ matrices, $G = U(N) = U(1) \otimes SU(N)$. We will call this variable B(y,x). In mathematical terms it is called a connection. We will specify the transformation properties of B as follows:

$$B(y,x) \rightarrow u(y)B(y,x)u^{-1}(x) \ . \tag{4}$$

What good does B(y,x) do? Consider the quantity

$$B(y,x)\psi(x) \ .$$

Up to a transformation (remember that B is an element of G), this is the value of ψ at the point x. Putting together (3) and (4), we see, however, that this quantity transforms as:

$$B(y,x)\,\psi(x) \rightarrow u(y)B(y,x)u^{-1}(x)\cdot u(x)\psi(x) = u(y)(B(y,x)\ \ (x)) \ ,$$

that is, as the value of ψ at the point y.

The common derivative of ψ

$$\partial_x\psi = \left.\frac{\psi(x+\Delta x) - \psi(x)}{\Delta x}\right|_{\Delta x \rightarrow 0}$$

is a combination of two quantities: $\psi(x)$, $\psi(x+\Delta x)$, which do not transform in the same way; and, as we discussed earlier, this was at the root of the trouble with Dirac's equation. With the help of B(y,x) we can define a new type of derivative, the "covariant" derivative:

$$D_x\psi(x) = \left.\frac{B(x,x+\Delta x)\psi(x+\Delta x) - \psi(x)}{\Delta x}\right|_{\Delta x \rightarrow 0} \tag{5}$$

For any Δx, the two terms on the right transform in the same way, so that

$$D_x\psi(x) \rightarrow u(x)\, D_x\psi(x) \tag{6}$$

The covariant derivative transforms therefore in the same way as $\psi(x)$. Using it we can write a modified Dirac equation:

$$(\gamma^\mu D_\mu)\psi(x) = m\psi(x) \tag{7}$$

which is invariant under the local (gauge) transformation (3),(4), (6).

Note that the modified Dirac equation is physically different from the original one. $B(y,x)$ is a new variable in the theory, a new set of physical degrees of freedom, which are equivalent (as we will see) to a set of vector fields. The requirement of gauge invariance forces the existence of a well defined interaction between these and other fields.

Before proceeding, let us find the relation of $B(y,x)$ to the more usual formalism for the case where G is U(1), namely, the group of phase transformations:

$$U(1) = \{e^{i\alpha}, \quad 0 \lesssim \alpha \lesssim 2\pi\} \tag{8}$$

the relation is simply:

$$B(y,x) = \exp iq \int_x^y A_\mu(\xi)d\xi^\mu \tag{9}$$

Thus, when y is close to x,

$$B(x,x+dx) = 1 - iqA_\mu(x)dx^\mu \tag{10}$$

so that (details left to the reader) the covariant derivative reduces to

$$D_\mu\psi(x) = [\partial_\mu - iqA_\mu(x)]\psi(x) \tag{11}$$

The transformation law:

$$A_\mu(x) \rightarrow A_\mu(x) + \partial_\mu\alpha(x) \tag{12}$$

is equivalent, as it should be, to:

$$B(y,x) \rightarrow e^{i\alpha(y)}\, B(y,x)e^{-i\alpha(x)} \tag{13}$$

Path Dependence and the Gauge Fields

We have defined B as a function of two points. We will now show that B is also a function of a path joining the two points. To see this, consider B for a pair of infinitesimally close points. We can assume that in this situation B is an infinitesimal transformation, i.e. close to the identity. If Q^i are the generators of the group (for SU(2), for example, $Q^i = \sigma^i/2$; for SU(3), $Q^i = \lambda^i/2$; where σ^i and λ^i are the Pauli and Gell-Mann matrices), we can write (with q a constant, that plays the role of a coupling constant):

$$\begin{aligned} B(x+dx,x) &= 1 + iq(\sum_i A^i_\mu(x)Q^i)dx^\mu + O(dx^2) \\ &= 1 + iq\, \underset{\sim}{A}_\mu(x)dx^\mu + O(dx^2) \ . \end{aligned} \tag{14}$$

The consideration of infinitesimal distances shows the equivalence of the B formulation of gauge theory to the usual one in terms of vector fields $A^i_\mu(x)$.

Let us note an important property:

$$B(x,x+dx) = B^{-1}(x+dx, x) \tag{15}$$

This can be shown in a few steps:

$$\begin{aligned} B(x,x+dx) &= B((x+dx) - dx, x+dx) \\ &= 1 + iq\underset{\sim}{A}_\mu(x+dx)(-dx^\mu) + O(dx^2) \\ &= 1 - iq\underset{\sim}{A}_\mu(x)dx^\mu + O'(dx^2) \quad . \end{aligned}$$

The variable B, for points x,y at a finite relative distance, can be built up from the infinitesimal B. To do this, choose a path P joining the two points (see Fig. 1). The B variable can then

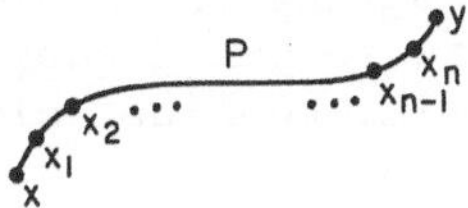

Fig. 1 The variable B depends on the path joining the two points x and y.

be computed by dividing P into n segments (n will tend to infinity), and defining:

$$B(y,x;P) = \lim_{n\to\infty}[B(y,x_n)B(x_n,x_{n-1})\dots \dots B(x_2,x_1)B(x_1,x)] \tag{15}$$

It is left to the reader to verify that if $\bar{P}$ is the path opposite to P (see Fig. 2), then:

$$B^{-1}(y,x;P) = B(x,y;\bar{P}) \ . \tag{16}$$

(A hint: fix the number of segments along P, and use Eq. (15).)

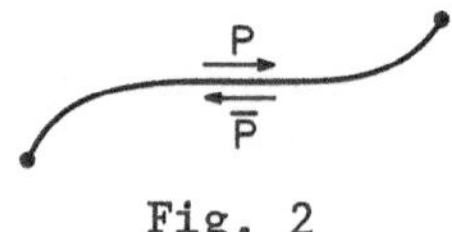

Fig. 2

We conclude that gauge invariance (3) requires the presence of a new dynamical variable B with the following properties:

a) It is defined for an arbitrary path P joining any pair of points.
b) Its value is an element of the symmetry group G.
c) Under a gauge transformation it transforms as given in Eq. (4).

The path dependence of B is the basis for topological properties which are connected, as we will see, with the possible existance of monopoles. Before going into this, we should indicate the way in which we can specify the dynamics for the B variables. The standard way to do this is to specify the action of the B variables (i.e. the lagrangian). We want the dynamics to be gauge invariant, consequently the lagrangian should be gauge invariant. The first task seems therefore to build gauge invariant quantities out of B. The recipe is simple: consider closed paths (see Fig.3)

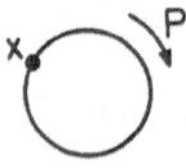

Fig. 3

We note that for a closed path the transformation law is:

$$B(x,x;P) \Rightarrow u(x)B(x,x;P)u^{-1}(x) \quad .$$

The trace of a closed-path B is easily seen to be gauge invariant:

$$\begin{aligned} \mathrm{Tr}\, B(x,x;P) &\Rightarrow \mathrm{Tr}\, u(x)B(x,x;P)u^{-1}(x) \\ &= \mathrm{Tr}\, B(x,x;P)u^{-1}(x)u(x) \\ &\quad \text{(by cyclic property of trace)} \\ &= \mathrm{Tr}\, B(x,x;P) \end{aligned}$$

Tr B is therefore just what we need.

We would actually like to go one step further, and define gauge invariant quantities which are local to the point x. This can be achieved by considering the limit of an infinitesimal path around x, e.g. a small square, defined by two infinitesimal vectors $\delta^{\mu},\varepsilon^{\nu}$ (see Fig. 4).

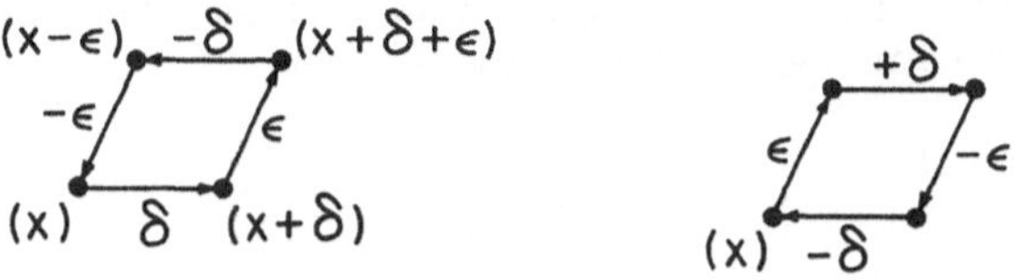

Fig. 4 The infinitesimal path $P(\delta,\varepsilon)$ and its inverse $\bar{P}(\delta,\varepsilon) = P(\varepsilon,\delta)$.

Before proceeding, let me recall how these concepts are used in lattice gauge field theories. These theories are interesting mathematical objects by themselves, and offer an approximation scheme for real-life situations. The idea is to approximate continuum space-time by a lattice of points. A "normal" field is represented by its value at each lattice point. The approximation can be made as good as one wishes by taking a lattice with very small spacing. On the lattice the only acceptable paths are made up of links, i.e., segments connecting adjoining sites. The smallest closed path is a square containing four links: the plaquette.

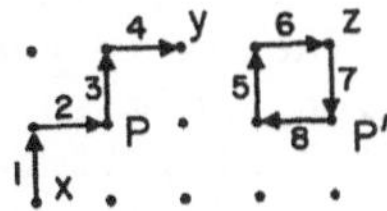

Fig. 5 A path and a plaquette on a two-dimensional lattice.

The general gauge variable B(y,x;P) can be built from elementary variables which correspond to one-link paths. Thus, while "normal" field variables are associated with lattice points, the gauge fields are associated with links. To build the variables associated with composite paths, you multiply link variables. Referring to Fig. 5, we would have

$$B(x,y;P) = B(4)B(3)B(2)B(1)$$

$$B(z,z;P') = B(6)B(5)B(8)B(7)$$

The gauge invariant quantity associated with the closed path P' (a plaquette) is

$$A(P') = \mathrm{Tr}(B(6)B(5)B(8)B(7))$$

Note that the cyclical property of traces implies that A(P') is a function of the plaquette, but not of the starting point of the path, z.

To build a dynamical system out of these variables we should specify the action S. For a continuum system, the action is the integral of the Lagrangian density:

$$S = \int d^4x\, L(x)$$

Since in a lattice gauge theory the smallest element of the lattice (equivalent to the space-time element d^4x) on which we can define a gauge invariant quantity is the plaquette, K. Wilson proposed the following form for the action:

$$S = \sum_{\substack{\text{plaquettes}\\ P}} K(n - \mathrm{Re}A(P))$$

Note:

1) A(P), the trace of a unitary matrix, is in general complex, we therefore take its real part.
2) The maximum value for ReA(P) is equal to n, dimension of the unitary matrices. This value is reached when the link variables coincide with the unit matrix. The action is minimum when all link variables are the unit matrix.
3) K plays the role of a coupling constant (actually the inverse of a coupling constant). Since the action tends to remain "as small as possible", a large value of K will force the link variables close to unity, while a small value will allow large fluctuation.

(A very interesting aspect of lattice theories is that they involve only discrete variables; they lend themselves therefore to being "solved" by "computer simulation". This has given rise to a flourishing theoretical industry that I will not elaborate on here.)

To go to the continuum limit, imagine a given physical situation approximated by a denser and denser lattice, so that the plaquettes become - in the limit - infinitesimal. Since the action is a sum of more and more plaquettes, the contribution of each plaquette must become smaller and smaller. This means that the link variable for an infinitesimal plaquette, such as the one in Fig. 4, must deviate only infinitesimally from the unit matrix. One can show that the form of this limit for the square plaquette in Fig. 4 is:

$$B(x,P(\vec{\delta},\vec{\varepsilon})) = \exp iq\underset{\sim}{F}_{\mu\nu}(x)\delta^{\mu}\varepsilon^{\nu} \tag{17}$$

where $\underset{\sim}{F}_{\mu\nu}$ is an antisymmetric tensor <u>and</u> a combination of the group generators Q_i. The number q defines the normalization of the $\underset{\sim}{F}_{\mu\nu}$. In the case of the U(1) gauge invariance of electromagnetism, q is identified with the unit of electric charge.

<u>Problem</u>: Using Eq. (10), show the (usual) relation in the case of electromagnetism:

$$F_{\mu\nu}(x) = \partial_{\mu}A_{\nu} - \partial_{\nu}A_{\mu}$$

In the case of non-abelian groups, the components $\underset{\sim}{A}_{\mu}(x)$ are matrices and do not commute among themselves. Using Eq. (14) prove the Yang-Mills relation:

$$\underset{\sim}{F}_{\mu\nu} = \partial_{\mu}\underset{\sim}{A}_{\nu} - \partial_{\nu}\underset{\sim}{A}_{\mu} + iq[\underset{\sim}{A}_{\mu},\underset{\sim}{A}_{\nu}] \tag{18}$$

The B formulation is thus seen to be equivalent to the usual formulation in terms of vector fields. We have seen that it is the simplest one for lattice theories. The most interesting feature of the B variable is that for any pair of points it depends on the path. This is the basis of interesting topological properties which lead to the possibility of monopoles.

Topology

Most properties of gauge theory depend on local properties of groups, i.e. on the commutation relations of the group generators:

$$[Q_i, Q_j] = if_{ijk}Q_k \tag{19}$$

These commutation relations determine the properties of group elements which are close to the identity, i.e., they can be used to compute the product of two such elements. The commutation relations also enter the definition of quantities such as $\underset{\sim}{F}_{\mu\nu}(x)$ (see Eq. (18)).

There are issues in field theory which depend on properties of the group in the large, i.e., on topological properties. The existence of monopoles is one such issue. The set of elements of a group G will be called the group space. The first step in defining a topology is that of defining the concept of closeness. In the case of a group this is done in two steps:

1) Define a set of group elements close to the identity.
2) Define closeness in general.

For the first step we start from the usual expression for small transformations:

$$g = 1 + i \sum \varepsilon_i Q_i + O(\varepsilon^2)$$

If we restrict the ε_i to be "small", i.e., with a suitably small ε:

$$\sum \varepsilon_i^2 < \varepsilon \quad ,$$

we have defined a small "ball" of elements close to the identity. The dimensionality of the ball is equal to the number of parameters ε_i, i.e., to the number of group generators (1 for SU(1), 3 for SU(2), 8 for SU(3), etc.).

This definition of closeness is now extended to any element of the group as follows:

Closeness: g_1 and g_2 are "close" if $g_1 g_2^{-1}$ is close to the identity.

What is important for the monopole issue is the connectivity of the group manifold. This concept is best explained by examples.

A) SU(1) - the group gauge transformations in QED.

$$SU(1) = \{e^{i\alpha};\ 0 \leqslant \alpha \leqslant \pi \quad \text{with } e^{2\pi i} = 1\} \quad .$$

The group space can be represented by a circle: it is connected, in the sense that we can join any pair of points, but not simply connected, in the sense that there are paths in group space which cannot be shrunk to zero continuously, such as P_2 of Fig. 6.

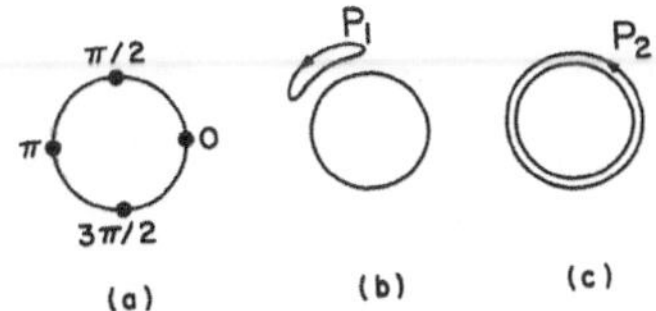

Fig. 6 Group space for U(1).

In (b) and (c) we have two examples of paths on group space (think of them as additions to the circle). P_1 can be shrunk to zero, P_2 cannot.

B) SU(2). This is the group of isospin or of spin 1/2 rotation. It can be defined as the set of 2×2 unitary matrices of the form:

$$u(\vec{\alpha}) = \exp i \frac{\vec{\alpha}\cdot\vec{\sigma}}{2}$$

This is interpreted as a rotation of an angle $|\vec{\alpha}|$ around the axis $\vec{\alpha}$. Remember that a 2π rotation around any axis changes the sign of a spinor, so that we can further specify

$$|\vec{\alpha}| \lesssim 2\pi,$$

if $|\vec{\alpha}| = 2\pi, \; u(\vec{\alpha}) = -1$

Thus the group space is a sphere in $\vec{\alpha}$ (3-dimensional) space, and all points on the surface of the sphere are to be considered as the same point.

It is easily seen that in this case the group space is both <u>connected</u> (any pair of points can be connected) and <u>simply</u> connected (any closed path can be shrunk to zero).

<u>Topology and Monopoles</u>

The problem of monopoles is connected to the following topological problem illustrated in Fig. 7. One may establish a correspondence between surfaces in normal space and closed paths in group space. This is illustrated in the top part of Fig. 7. You select two points on S as "poles", and a path P_1 joining them (keeping with our metaphore, a meridian). To P_1 there corresponds

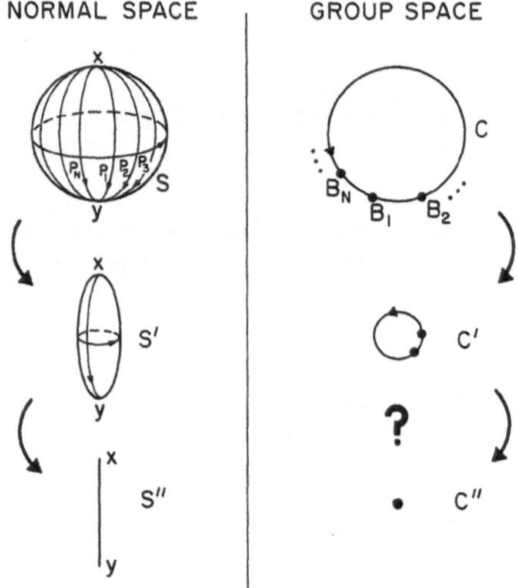

Fig. 7 When P is moved around a surface S the corresponding B(y,x;P) goes around a closed loop C. If we shrink S to a smaller surface S', then to a single line S", does C shrink <u>continuously</u> to a single point C"?

a point in the group space, $B_1 = B(y,x;P_1)$. If we move P_1 in such a way as to sweep S and come back to the starting position, the corresponding point moves around a closed path C in group space. If we now shrink S to a single line S", C will shrink to a single point, C". Will this happen continuously? If the answer is yes, all is well and normal. If not, the surface S contains a monopole.

A gauge theory can have monopoles if the group space is not simply connected, so that there are closed loops in group space which cannot be shrunk continuously. An example of such a group is the U(1) of QED (example A given above, see Fig. 6). Another example is the U(1) × SU(3), the combined exact symmetry of particle physics, to which we will return later. The corresponding monopoles are called Dirac Monopoles. We will now outline the argument which shows that these particles are indeed isolated magnetic charges, and that their magnetic charge g is related to the minimum unit of electric charge, q, by Dirac's relation:

$$g\, q = \frac{1}{2} \qquad (\text{for } \hbar = 1) \tag{20}$$

Dirac Monopoles

Under a gauge (in this section we use "gauge" in the normal QED sense) transformation, the fields (or wave functions) of particles with charge q_i transform as:

$$\psi_i(x) \to e^{iq_i\alpha}\,\psi_i(x) \quad .$$

If charge is quantized

$$q_i = n\, q \qquad (n \text{ integer})$$

the gauge group is defined as

$$\{e^{iq\alpha}; \quad 0 \lesssim q\alpha \lesssim 2\pi\}$$

This is not simply connected as discussed above, and can support monopoles.

If charge is not quantized, e.g., if there are two charges q_1 and q_2 whose ratio is irrational

$$q_1/q_2 = \text{irrational number}$$

there is no pair of values of α which lead to the same transformation on both ψ_1 and ψ_2 . In this case, the group space is not a circle, but the full α line, and is simply connected - no monopoles.

Let us assume that charge is quantized. Looking back at Fig. 7, we first want to compute the change in B when P moves from a meridian P_1 to an adjacent one, P_2. This can be done by a series of plaquette-like moves (Fig. 8). The change for one such move is (see enlargement in Fig. 9) obtained by multiplying the original B(P) by the B corresponding to the closed path of sides $\vec{\delta},\vec{\varepsilon},-\vec{\delta},-\vec{\varepsilon}$. Using Eq. (17):

$$B(P') = \exp i[q \underset{\sim}{F}_{\mu\nu}(x)\delta^{\mu}\varepsilon^{\nu}]\cdot B(P)$$

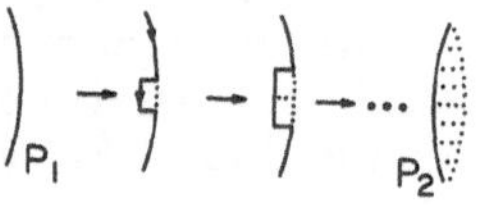

Figure 8

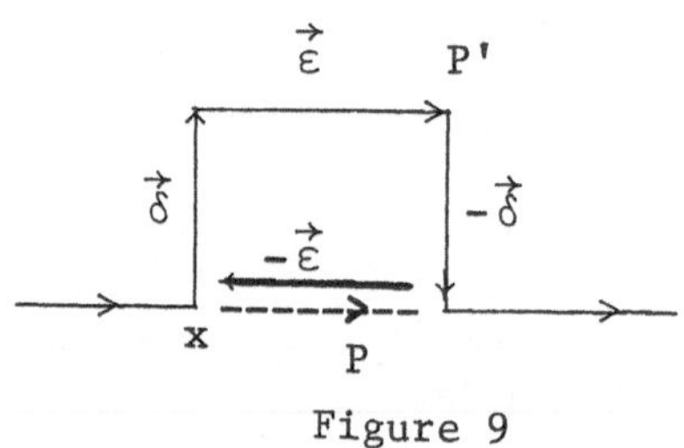

Figure 9

If we choose a spacelike surface, the added phase is simply the magnetic flux through the plaquette:

$$B(P') = \exp[i\ q\ \vec{B}\cdot d\vec{S}]\cdot B(P)$$

If we move P from the original position P_1 around S (back to Fig. 7) back to P_1, we pick up a phase factor

$$B(P_1) = \exp[iq \int_S \vec{B}\cdot d\vec{S}]\cdot B(P_1)$$

For a complete sweep of S we must come back to the original value $B(P_1)$, so that

$$q \int_S \vec{B}\cdot d\vec{S} = 2\pi n$$

The integral represents $4\pi \times$ the magnetic charge within S, so this must be quantized according to Dirac's condition in Eq. (20). The relation of n to our topological discussion is simple: n is the number of times B(P) goes around the circle (Fig. 7) in the anticlockwise direction when P goes around S in one sweep.

Case of SU(3) × U(1)

This is the exact symmetry of particle physics. Since quarks have fractional charges, the minimum charge unit is e/3, so we would expect a quantization condition:

$$\frac{e}{3} \times \int \vec{B}\cdot d\vec{S} = 2\pi n$$

or

$$e\,g = \frac{3n}{2}$$

i.e., the minimum monopole magnetic charge should be three times as large as that computed from the electron charge.

This conclusion is incorrect for a subtle reason. There are three elements of SU(3) which are also elements of U(1): they are the 3×3 matrices

$$\mathbb{1}\,, Z = e^{\frac{2\pi i}{3}} \times \mathbb{1}\,, \quad Z^+ = e^{-\frac{2\pi i}{3}} \times \mathbb{1}$$

Remember that elements of SU(3) are defined to be unitary matrices with unit determinant; $\mathbb{1}$, Z and Z^+ obey these conditions. Since an element of SU(3) × U(1) is the product of an element of SU(3) and an element of U(1), the identity in SU(3) × U(1) can be represented in three different ways:

$$\mathbb{1}\big|_{SU(3)} \times 1\big|_{U(1)} = \mathbb{1}$$

$$Z\big|_{SU(3)} \times e^{-\frac{2\pi i}{3}}\Big|_{U(1)} = \mathbb{1}$$

$$Z^+\big|_{SU(3)} \times e^{\frac{2\pi i}{3}}\Big|_{U(1)} = \mathbb{1}$$

These three elements of SU(3) × U(1) are <u>the same element</u>, so a path in group space joining one to another is a <u>closed path</u>. It can be checked that it cannot be shrunk to a point, so it corresponds to a possible monopole. Along such a path the QED phase changes by 1/3 of 2π, so the quantization condition becomes:

$$\frac{e}{3} \times \int \vec{B}\cdot d\vec{S} = \frac{2\pi}{3}\, n$$

i.e., the original Dirac condition!

A monopole of SU(3) × U(1) has gluonic as well as electromagnetic fields. If color is - as we believe - confined, the gluonic fields would extend only to characteristic hadronic distances, so that such a monopole would look similar to a "normal" Dirac monopole.

The 't Hooft-Polyakov Monopole

Since Dirac's work we know that, if charge is quantized, monopoles can exist, with a magnetic charge bound by Dirac's condition (Eq. (20)). They can exist, but they are extraneous to theories such as QED. In straight QED, or in straight SU(3) × U(1) monopoles do not exist.

The turning point in monopole theory was the discovery, by 't Hooft and Polyakov, that a large class of theories implied the existence of monopoles. This class includes all theories where a gauge symmetry corresponding to a semi-simple group is spontaneously broken through the Higgs-Kibble mechanism to an exact group containing a U(1) factor. The class contains most popular grand unified theories, including those based on SU(5) and O(10). If GUT is correct, we must have monopoles.

A 't Hooft-Polyakov monopole is not an elementary particle, but a composite of Higgs fields and superheavy vector boson fields. Its mass is expected to be larger than the mass of superheavy vector bosons, thus in the neighborhood of 10^{16} GeV. In spite of this large mass the lightest monopole must be absolutely stable - a simple consequence of Maxwell's equations. The structure of such a monopole is rather complex: at large distances (larger than a few fermis) it is a "normal" Dirac monopole. Going inside we would find a gluonic core, extending over typical hadronic distances. Further in, we would have Z and W fields. At the very center ($\sim 10^{-27}$ cm) a hard core of superheavy Higgs and vector fields, where most of the mass is concentrated.

If GUTs and big-bang cosmology are both correct, such monopoles were produced during the first instant of the universe, and some of them must be around now.

LECTURES ON NEUTRINO PHYSICS

Frank Sciulli

Department of Physics
Columbia University
New York, NY 10027

NEUTRINO SOURCES

The studies of the intrinsic nature of neutrinos as elementary particles, their interactions, and their important use as probes of hadron structure have been extremely important over the past two decades. These areas of experimentation have been possible because of the availability of a diversity of sources, providing neutrinos over a wide range of energies with wide flexibility in experimental configurations.

Table 1 shows a few examples of these neutrino sources. At one extreme are the solar neutrinos, isotropic in their spatial dependence and in the MeV energy range. Measurement of the solar neutrino flux is an important check on the standard model for solar energy production, given that neutrinos have all the properties generally attributed to them (which we will discuss). This detection of neutrinos from the sun is a tour-de-force involving a complex of chemical, physical, and nuclear technique (R. Davis et al 1978). The agreement of the observed rate with the standard model is still not good (J. Bahcall 1981); new experiments that measure different parts of the neutrino energy spectrum may, however, shed some light on the problem.

The more detailed study of neutrinos in laboratory environments, under controlled conditions, has involved reactors and high energy accelerators. At lower energies

Table 1. Some Sources of Neutrinos for Experimentation

Source	Energy Range	Source Process	Distance from Source	Detector Mass	Event Rates	Detection Process	Beams
Sun	.2–8 MeV	β decay $p+p\rightarrow D+e^{+}+\nu_e$	1.5×10^{11}m	600 tons	1/2 cnt/day	$\nu_e+{}^{37}Cl \rightarrow e^{+}+{}^{37}Ar$	isotropic
Reactor Bombs	.2–10 MeV	β decay	10–40m >500m	10 "	50/day 10/explosion	$\bar{\nu}_e+p\rightarrow e^{+}+n$	"
LAMPF (800 MeV)	150 MeV	$\mu^{+}\rightarrow e^{+}+\nu_e+\bar{\nu}_\mu$ $\pi^{+}\rightarrow\mu^{+}+\nu_\mu$	200m	20 "	70/day	$\nu_\mu+n\rightarrow\mu^{-}+p$ $\bar{\nu}_e+p\rightarrow e^{+}+n$	wide band focus
BNL PS (30 GeV)	1 GeV	$\{^{\pi}_{K}\}\rightarrow\mu+\nu_\mu$	100m	30 "	~10/min	$\nu_\mu+N\rightarrow\mu+N'+\pi$	"
SPS FNAL (400→1000GeV)	100→300 GeV	$\{^{\pi}_{K}\}\rightarrow\mu+\nu$	1000m	600 "	~10/min	$\nu_\mu+N\rightarrow\mu+X$	narrow band focus

this means using $\bar{\nu}_e$ from nuclear β decay; at higher energies, the use of ν_μ and $\bar{\nu}_\mu$ from the decays of π and K mesons. The interaction rates become higher as the energy increases. This is due to several factors: (1) the interaction cross section increases; (2) detector sizes typically increase and detector granularities decrease as the final states energy to be detected increases; and (3) the solid angle of the neutrino beam becomes smaller due to the Lorentz transformation from decays in flight. This last factor has been amplified even further by the addition of devices to focus the meson beam.

These increases in interaction rates have permitted construction of beams in which event rate is sacrificed to the benefit of increased information about the neutrinos. The narrow band beams indicated in the table do just this. Although the neutrino flux is lower than other beams, the energies of the neutrinos are known to about 10% from the optics and the decay kinematics of the meson decay (see, e.g. Fisk & Sciulli 1982).

NEUTRINO TYPES

We have known for two decades that neutrinos come in different types or flavors (Danby 1962). Our presumption is that each charged lepton is associated with its own uncharged, spin 1/2 neutrino of relatively low mass. The assignment of lepton number is conventionally made on the basis of electron, muon, and tau number, as shown in Table 2. The degree to which these lepton numbers are conserved, and the manner in which they are conserved are legitimate and important areas of experimentation.

Table 2. Conventional Lepton Number Assignments

Charged Lepton	L_e	L_μ	L_τ	Neutrino
e^-	1	0	0	ν_e
μ^-	0	1	0	ν_μ
τ^-	0	0	1	ν_τ

Absolute conservation of individual lepton number would require, for example, muon neutrinos to produce only μ^- in collisions with nucleons ($\nu_\mu + N \to \mu^- + X$). Limits on direct violation of lepton number are illustrated in Table 3. At levels of sensitivity less than 1% in the most experimentally accessible case, the lepton assignment of Table 2 works, and those lepton numbers are conserved.

Table 3. Limits on Forbidden Transitions Relative to Allowed

Reaction	Upper Limit	Reference
$\nu_\mu + N \to e^- + X$	3×10^{-3}	Cnops 1978
$\nu_\mu + N \to \tau^- + X$	2.5×10^{-2}	Cnops 1978
$\nu_e + N \to \tau^- + X$	0.35	Fritze 1980

Other more exotic lepton conservation schemes are possible. The usual (additive) approach is the hypothesis that

$$\Sigma L_i = \text{constant},\ i = e, \mu, \tau \ .$$

The most famous competitor is the multiplicative hypothesis:

$$\Sigma(L_e + L_\mu + L_\tau) = \text{constant}$$

and

$$\Pi(-1)^{L_i} = \text{constant},\ i = e,\mu,\tau \ .$$

Reactions like $\nu_\mu + N \to \mu^- + X$ are allowed by both; reactions like $\nu_\mu + N \to e^- + X$ are forbidden by both. Reactions allowed only by the multiplicative scheme are:

$$\bar{\nu}_\mu + e^- \to \mu^- + \nu_e \qquad \mu^+ \to e^+ + \bar{\nu}_e + \nu_\mu$$

Experimental results now limit these processes to about 10% of those permitted by both schemes (Jonker 1980, Willis 1980). This clearly demonstrates that this multiplicative rule is not the correct one. We are left with the additive hypothesis as the simplest lepton conservation rule consistent with experiment. More exotic lepton assignments are, of course, still possible. (See review by Primakopf and Rosen 1981).

There are some wide open experimental questions in the context of neutrino types that we have not discussed, but that should be mentioned.

1. Does the ν_τ exist? Presently, we infer its existence in precisely the same way that the ν_e was originally inferred. Seeing ν_τ interactions will settle this question (see section on beam dump expts.).

2. How many neutrino types are there? There are cosmological arguments, based on the expansion rate of the universe during synthesis of He^4, which give limits on the number of low mass neutrino types from the presently observed abundance of helium. This is thought now to be ≤ 4, but it is subject to some assumptions. (See, for example Turner 1981). The number of low mass neutrinos may also be measured in the laboratory by careful comparison of the Z^0 mass and width (see, for example, lectures by C. H. Llewellyn-Smith).

3. Are the neutrinos that we already know massless? This used to be the traditional assumption, but it should be checked. As we shall see, the experimental limits are not even close to the limits for the other "zero-mass" particle that we know and love. The present limit on the photon mass is 6×10^{-16} eV/c^2.

NEUTRINO MASS

Who wants massive neutrinos, anyway? A sampling of particle theorists will elicit answers from: "Neutrino mass is ugly!" to "Why not?" to "Neutrino mass is beautiful!". You will probably hear more on the positive side in your lectures on Grand Unified Theories.

The principle motivators for finite neutrino mass has come from the cosmologists. They are confronted with a "Missing Mass Problem". Put simply, there is strong prejudice and even evidence from nearby astronomical objects that the mass of a structure is proportional to its light output. Studies of the motions of very large scale structures, particularly groups of galaxies, indicate that they have much higher mass than can be identified with the measured light emission. That is, a factor of 10 to 100 more mass is indicated from the kinematics of the structure motion than is obtained from the light emission. (See, for example, Faber & Gallagher 1979).

There are many possible explanations for this anamoly. They include massive black holes, low mass stars and planets, and gas or dust. There are arguments against some of these, but they cannot be ruled out. An alternative explanation, which has many adherents, is that the extra mass is contributed by primordial neutrinos of low mass, like 10 eV. This is an exciting idea! The number density would be of the order of that of photons in the background radiation. The kinetic energy of each neutrino would be very small, thousandths of an electron volt. The detection of such primordial neutrinos would be a great triumph! But the experimental difficulties of directly detecting the presence of even large numbers of very low energy neutrinos are enormous. The problem requires a clever, inventive approach.

Particle experimentalists can, however, look hard at their present limits on neutrino mass. They can also look at alternative signatures of finite neutrino mass. The most accurate measurement of the mass comes as a consequence of the kinematics of the weak decay processes producing the neutrino. The energy of the muon from pion decay at rest, in $\pi \to \mu^+ + \nu_\mu$, for example, is given by:

$$E_\mu = \frac{m_\pi^2 + m_\mu^2 - m_\nu^2}{2m_\pi} \quad .$$

The precision to which the muon neutrino mass can be measured depends directly on how well the muon energy can be measured. The scale of the accuracy depends on the energy release in the decay process. Table 4 shows the present mass limits on the neutrino types, the reactions used for the measurement, and the characteristic momentum in the final state. Clearly, the processes have been chosen to set limits which give the smallest energy release.

The three-body decays require a different signature, since there are a continuum of charged electrons. Here, the end point of the energy spectrum is the sensitive quantity. In this case, the effects of resolution are of paramount importance, since they may produce a systematic shift in the end point. Of great interest is a recent experiment, done in the Soviet Union, which indicates that the electron neutrino has finite mass, of order $14 < m_{\nu_e} < 46$ eV (Lubimov 1980). More experiments are clearly needed, and are being done, on this important question.

Neutrino	Mass Limit	Reaction	p_ν	Reference
ν_e	55 eV	$H^3 \rightarrow He^3 + e^- + \nu_e$	$\leq$18keV/c	Berquist 1972
ν_μ	.55 keV	$\pi \rightarrow \mu + \nu_\mu$	37MeV/c	Daum 1973
ν_τ	250 MeV	$\tau^- \rightarrow \nu_\tau + e^- + \bar{\nu}_e$	<750MeV/c	Bacino 1979

Table 4. Limits on Neutrino Mass

Besides the question of mass for the known neutrinos, we might ask if there are other neutrinos around, perhaps more massive. If such particles were coupled to the known ones, they might be observable in neutrino mass experiments as additional peaks in two-body decays, or as "kinks" in the spectra from three-body decays (Schrock 1980). For example, the electron neutrino state would be:

$$|\nu_e\rangle = A|\nu_1\rangle + B|\nu_2\rangle$$

where A and B are the amplitudes for the light and heavy neutrino eigenstates, respectively. The available data on specific decay channels defines limits on the branching ratio, $(B/A)^2$, as a function of the heavy neutrino mass as shown in Fig. 1.

Notice that this analysis yields mass limits in the mass range of MeV or higher. Rarely produced very heavy neutrinos are still possible, but the limits are reasonably stringent. For much lower masses, mixing of neutrino states will produce coherent effects. This is the subject of the next section.

OSCILLATIONS

The questions of the existence of finite neutrino mass and of the possible violation of individual lepton number come together in the phenomenon of neutrino oscillations (Pontecorvo 1957). A plane wave of muon neutrinos might consist, when created, of a superposition of two mass eigenstates, $|\nu_1\rangle$ and $|\nu_2\rangle$, as follows:

$$|\nu_\mu\rangle = \cos\theta|\nu_1\rangle + \sin\theta|\nu_2\rangle \;.$$

The orthogonal state would be:

$$|\nu_X\rangle = -\sin\theta|\nu_1\rangle + \cos\theta|\nu_2\rangle \;.$$

The assumptions made are that there are only two states mixing in the ν_μ state and that laboratory energies are large compared to rest masses.

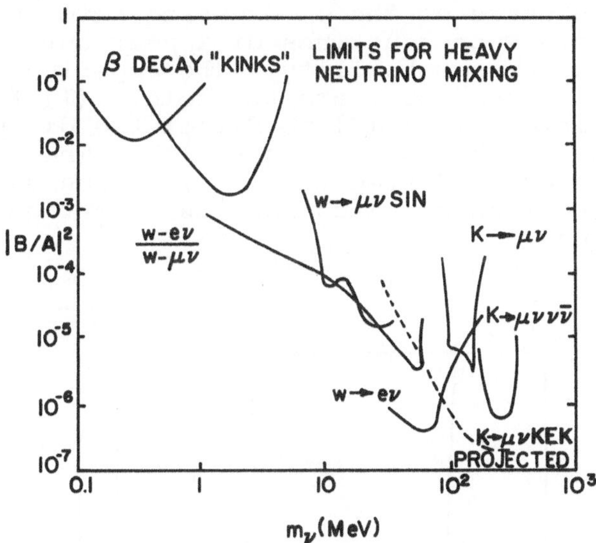

Fig. 1. Mass Limits on Heavy Neutrinos Mixing (Boehm 1982)

In the lab, the eigenstates will evolve with amplitudes $e^{i\phi}$, where ϕ = Et - pℓ at distance ℓ. Since $\ell \approx$ ct and p $\approx$ E-m^2/2E at high energy, then for each eigenstate, the phase $\phi \approx m^2\ell/2E$, and the originally pure ν_μ state becomes at a distance ℓ from its origin:

$$|\Psi(\ell)\rangle = \cos\theta e^{i\phi_1}|\nu_1\rangle + \sin\theta e^{i\phi_2}|\nu_2\rangle \;.$$

The amplitude for finding the orthogonal state at distance, ℓ, is such that the probability for finding the state $|\nu_X\rangle$ is given by:

$$P_{\nu_x}(\ell) = \sin^2 2\theta \sin^2\left(\frac{\Delta^2 \ell}{4E}\right) \quad .$$

Here $\Delta^2 = m_1{}^2 - m_2{}^2$ is the difference in the squares of the eigenstate masses. After taking care of units with appropriate use of Planck's constant and c, the distance dependent phase becomes $1.27\ \Delta^2\ell/E$, with mass in units of eV, distance in m(km), and energy in MeV (GeV).

In the process of the ν_μ evolving into the new state, ordinary muon neutrinos disappear out of the beam with the same probability. The experimental possibilities, therefore, are two-fold: (1) to detect the new particle ν_X, or (2) to detect the disappearance of neutrinos of the type we began with. These two possibilities are referred to as (1) exclusive and (2) inclusive experiments. Each has its advantages and its limitations.

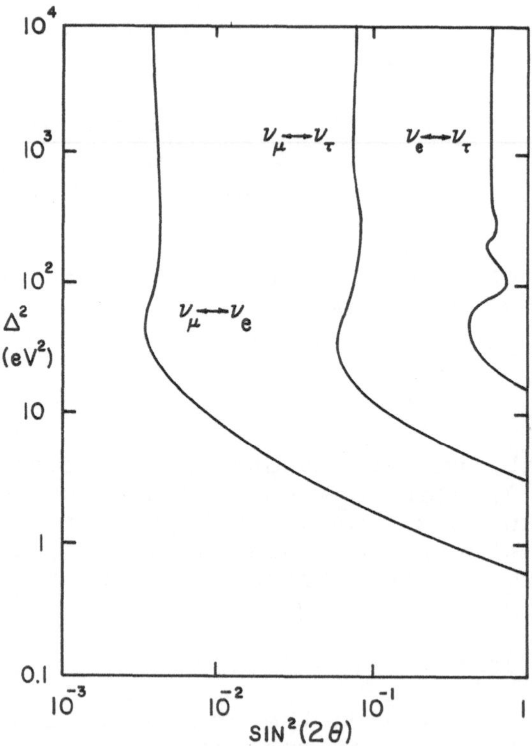

Fig. 2. Limits on Transitions Among Known Neutrinos.

Exclusive experiments can only look for neutrinos that we know about. On the other hand, such experiments can be extremely sensitive. For example, detection of electron neutrinos (by observing final state electrons from their interactions) in a beam of initially pure muon neutrinos would be direct evidence for an oscillation. Very pure ν_μ beams are used at high energy accelerators, so the experimental sensitivity to this particular transition has been quite good.

Figure 2 shows the present limits on transitions among known neutrinos that come from the Fermilab bubble chamber (Baltay 1981). Similar limits come from CERN bubble chamber exposures (see review by Baltay 1981). The figure shows substantially better limits with ν_μ in the initial state, and an especially good limit for $\nu_\mu \rightarrow \nu_e$. Finding interactions of ν_τ is not so easy, because the final state τ has many decay modes, and all of these have at least one unobserved neutrino. In addition, present energies of high flux neutrino beams are such that a substantial threshold factor exists for the charged current reaction making the τ. Even so, a clear signature of ν_τ interactions provides a more precise limit because the separation of such events is much easier. The Fermilab emulsion experiment, with the ability to see the track of the short lived τ, gives a limit for $\sin^2 2\theta$ that is roughly a factor of five smaller than shown in Fig. 2 (Stanton 1981).

Inclusive experiments are sensitive to neutrino evolution to any state. It should be noted that Pontecorvo's original suggestion of this phenomenon discussed it in the context of $\nu \rightarrow \bar{\nu}$ transitions. On the other hand, the experiments are somewhat more difficult. Detection as a function of ℓ/E requires observation of either distance or energy dependence. But the detection capability (i.e. the cross section) and the neutrino flux both depend on energy strongly. The only unassailable demonstration of inclusive oscillatory behavior must, therefore, show dependence of interaction rate with distance at fixed energy. This usually requires the use of more than one detector.

The two types of experiments also differ in the range of sensitivity in Δ^2 covered. Maximum sensitivity occurs for $\ell/E = 2.54\ \Delta^2/\pi$. For ℓ/E much smaller, the wavelength is very long, and both types of experiments are limited to a minimum value of Δ^2 set by geometry and neutrino source energy. For ℓ/E much larger, the wavelength is very short: inclusive experiments are limited by resolution on energy and distance, while exclusive experiments remain sensitive to the average over the oscillations. This means that a wide range of energies and distances are desirable to cover as wide a range as possible of mass differences. Reactors, low and high energy accelerators play important roles in these measurements.

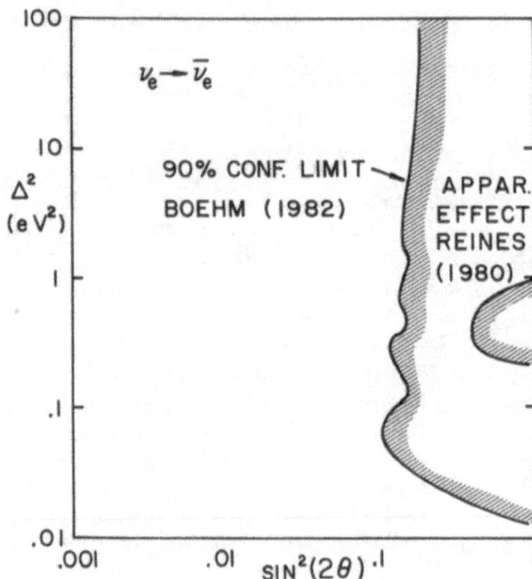

Fig. 3. Inclusive Measurements Involving ν_e

Inclusive experiments involving $\bar{\nu}_e$ have been the source of much discussion. A possible effect involving oscillations was reported with large values of $\sin^2 2\theta$ and $\Delta^2 \approx 1$ eV^2 (Reines 1980). Subsequent experiments have not corroborated this effect, shown from the 90% confidence limits shown in Fig. 3 (Boehm 1982). It may be possible to make apparently conflicting experiments agree by a judicious choice of parameters (Silverman & Soni 1981), but such analyses are subject to systematic differences among experiments. In any case, better experiments will ultimately decide the question. At present, we must conclude, there is no incontrovertible evidence that inclusive oscillations exist for the ν_e.

Searches for inclusive oscillations of muon neutrinos are also being done. Because these neutrinos are made at higher energies, present limits are primarily at larger values of the mass parameter. Figure 4 shows as a solid curve (Shaevitz 1981) the limits obtained from total cross section measurements, assuming that the cross section depends linearly on neutrino energy and that the neutrino flux is as measured. Roughly 20% limits on the mixing parameter for $20 < \Delta^2 < 500$ eV^2 are found.

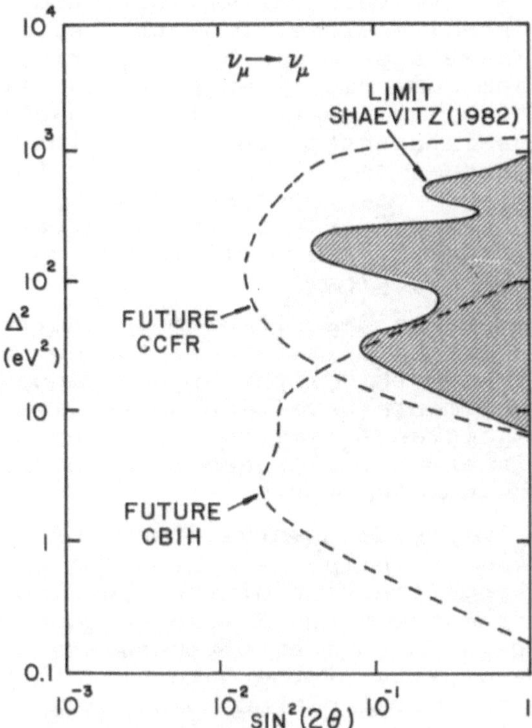

Fig. 4. Limit on Inclusive ν_μ Oscillations (Solid Curve)

New experiments are being planned to extend all of these limits beyond what you see here (Chen 1982). Some improvement in the mixing limits and in the mass parameter are probable in the short term using presently planned experiments and existing accelerator facilities. For example, Fig. 4 shows as dashed curves the sensitivity expected from a run just completed at Fermilab by the CCFR group using two neutrino detectors. Also shown is the sensitivity at lower mass expected from an approved experiment at BNL also using two detectors which should begin running during the next year. In general, improvements in sensitivity will probably be limited to about an order of magnitude over that which presently exists, unless new facilities come into operation to provide much higher fluxes, with correspondingly higher rates at large distances from the sources.

BEAM DUMP EXPERIMENTS

Experiments have been carried out, and continue, which look for novel sources of neutrinos, or for other kinds of particles which have neutrino-like properties. Beam dump experiments are sensitive to any particle which can penetrate large quantities of matter, yet have interaction and production cross sections large enough to be observable in the large neutrino detectors. One important result that we expect from such experiments is the observation of interactions of the ν_τ. We will discuss this further at the end of this section.

These experiments deposit high intensities of machine energy protons on a target, follow this with a filter, and follow this with the neutrino detector. High energy experiments have been done primarily at CERN, where the distance from the neutrino source to the detectors is about 600 m, and the protons are dumped at 0° to the detector.

Such experiments do observe plenty of interactions, and we will review briefly their progress and interpretation. A large fraction of the neutrinos that are observed come from the π and K mesons produced in the hadronic showers. The number of these are determined by changing the density of the dump; i.e., by expanding the spacing of the metal plates comprising the dump. Since the long lived mesons have decay lengths long compared to their interaction lengths, the contributions from such decays will scale inversely with density. The prompt rate is obtained by extrapolating the rate as a function of inverse density to zero (infinite density).

Prompt neutrino rates have been observed in several experiments (Alibran 1978, Bosetti 1978, Hansl 1978). This comes as no great surprise. Prompt muons had been observed from hadron collisions for some time. An appreciable fraction of these come from Drell-Yan production of muon pairs. But about a third of them come from another source (Olsen 1982). These are muons unaccompanied by another muon, and so presumably are accompanied by a ν_μ. Figure 5 shows a comparison of ν_μ and μ from beam dump experiments

(Wojcicki 1980). At the 30% level, the numbers are equal.

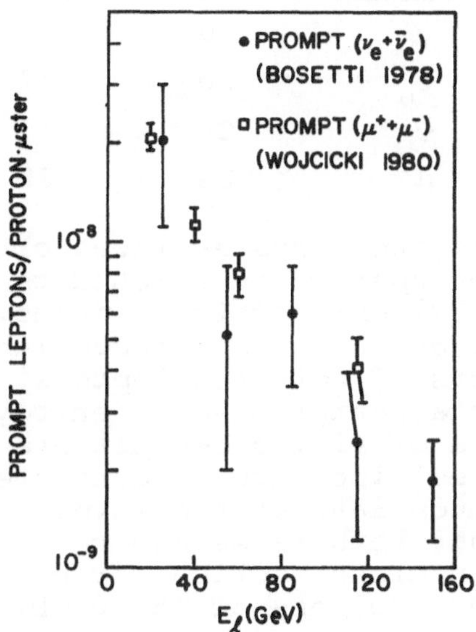

Fig. 5. Comparison of Muons and Neutrinos from Dumps

We know of a source for these prompt leptons; indeed, we expect them. Charmed particles, principally neutral and charged D mesons will leptonically decay. Although these are weak processes, the large available phase space gives lifetimes of order 10^{-13} sec, or decay lengths smaller than 1 mm. After establishing that prompt production of neutrinos exists, do the observed prompt leptons have characteristics that are consistent with this hypothesis, and what is the production cross section for the associated charm production required to explain it? The latter question is somewhat difficult to extract unambiguously, since experiments tend to be sensitive in restricted kinematic ranges. Obtaining a total production cross section requires a model assumption to extrapolate over the entire phase space. It also requires knowledge of the D decay properties, which come from measurements made in e^+e^- experiments, and a few assumptions in applying them.

Table 5. Comparison of D Production from μ and ν Beam Dump Experiments

Reference	Particle	Target	σ/nucleon in μb
BEBC	ν_μ	Cu	11 - 22
CDHS	ν_μ	Cu	7 - 14
CHARM	ν_μ	Cu	14.5 - 29
CSFR	μ	Fe	22 $\pm$ 9

Table 5 shows the cross sections obtained from several beam dump experiments which detect neutrinos and one (Olsen 1982) which detects muons. There is general agreement on the charm production necessary to explain the numbers of detected leptons. These data were all taken at a center-of-mass energy of 28 GeV. These and the data of Fig. 5 demonstrate that the observed prompt neutrino rate is consistent with the observed prompt muon rate at the level of factor two precision, and that both rates can be explained with a total production cross section for charm produced by protons hitting nucleons of 15-30 μb.

There are, however, some interesting anomalies from the beam dump experiments that are not yet explained. The two features are the number of events without muons, presumably neutral current interactions, and the ratio of events with electrons in the final state to those with muons in the final state.

Figure 6 illustrates the first of these. The CHARM (Niebergall 1980) experiment, which has the capability to measure shower energies that are very low, sees an excess of events at very low energy. The dots in the figure indicate the expectations for $D\bar{D}$ production, normalized to shower energies above 20 GeV. The excess below this energy is not explainable in simple models. The CDHS data, shown in the same figure, only exists above 5 GeV; it is consistent both with standard assumptions and with the CHARM data (see Wojcicki 1980).

The other unexplained feature from neutrino beam dump experiments is the ratio of electron to muon neutrinos. Table 6 shows the numbers from three experiments, using several techniques for extracting them. There is a systematic difference from unity in the direction of relatively fewer ν_e than ν_μ. A recent experiment at Fermilab (Reeder 1982) also sees this effect on the average, but most of the effect in their preliminary result appears to come from lower energies. In the lower energy region, there is particular sensitivity to systematic problems, including non-prompt sources not measured by dump expansion and from systematic uncertainties in trigger thresholds. This last experiment differs from those in the table in that it is only 53 m from the dump.

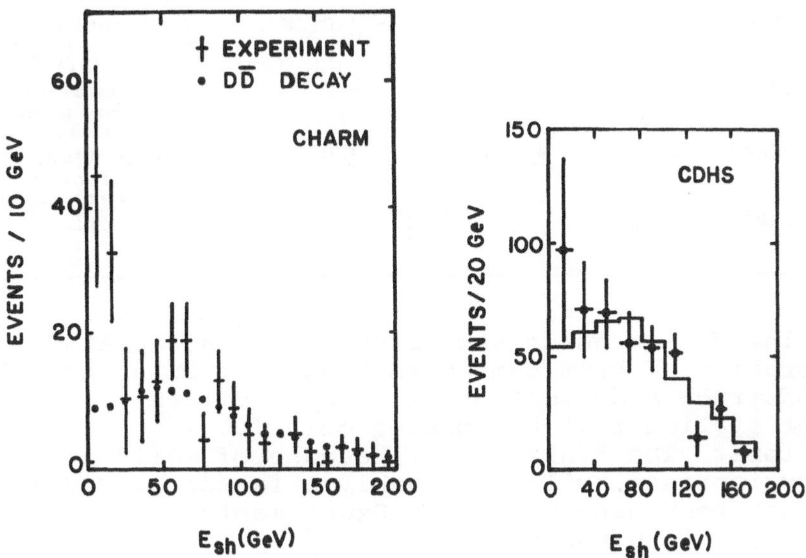

Fig. 6. Shower Energy from Muonless Events (CHARM and CDHS)

This ratio is an important test of our understanding. If the only new threshold in this energy regime involves charm and bottom production, then μ-e universality requires that the number of ν_μ and ν_e be nearly equal. The phase space difference between 3-body decays with muons or electrons is very tiny. Also, the 2-body leptonic decays are strongly suppressed by the vector nature of the weak interaction and the high velocity of the final state leptons. Differences of more than a percent or so in numbers of electron and muon neutrinos would be very surprising. We need to see unambiguous measurements of this ratio in the future.

Table 6. Ratio of ν_e to ν_μ Observed in Various Experiments

ν_e/ν_μ Ratios from CERN Beam Dump Experiments

Group	ν_e/ν_μ Ratio	Statistical Error	Systematic Error	Method
CDHS	0.77	$\pm$ 0.18	$\pm$ 0.24	Extrapolation
CDHS	0.58	$\pm$ 0.07	$\pm$ 0.19	Subtraction
CHARM	0.48	$\pm$ 0.12	$\pm$ 0.10	Subtraction (CCν_e from prod. model)
CHARM	0.49	$\pm$ 0.21		Extrapolation (CCν_e from prod. model)
CHARM	0.44	$\pm$ 0.11	$\pm$ 0.03	Subtraction (CCν_e directly identified)
BEBC	0.59	$^{+0.35}_{-0.21}$		Subtraction

The future holds some promise for systematic exploration of beam dump phenomena. Fermilab has plans to construct a dedicated beam dump facility. The target station will be located only 100-200 m from the detectors. The sensitivity, because of solid angle and energy increase, will be roughly a factor of 200 above what has been done to date. Experiments at this facility will pursue:

1. The discovery of ν_τ interactions. The dump should provide a reasonable flux of tau-like neutrinos from $p+p \to F+X$, with subsequent decay of the charm-strange meson: $F \to \tau + \nu_\tau$ (Sciulli 1978). Holographic techniques in bubble chambers should allow direct observation of τ tracks between production and decay that come from the interaction of the neutrino (Baltay 1980, Pless 1980). This will provide necessary and important direct observations of the ν_τ.

2. Observation of ν_μ and ν_e that come from charm production in the dump. It is hoped that some of the apparent anomalies discussed above will be better understood.

3. Search for new particles and neutrinos from the decays of new particles. This may include looking

for odd topologies, like those that would come from decays of metastable neutral particles (see e.g., Mo 1980). Alternatively, one may look in regions where we expect very few neutrinos from conventional sources, like at high transverse momenta, for penetrating particles that have their origin in the decays of high mass objects. The Higgs boson is the obvious example of a conjectured high mass particle, which is very difficult to find, and would have somewhat unorthodox decay modes (Ellis 1976).

NEUTRINO INTERACTIONS

Neutrinos are known to interact with other fermions only through the weak interactions; these are divided into the charged current and neutral current types. The former (CC) are responsible for interactions which change charge (i.e. $\nu_\mu \to \mu$); the latter (NC) responsible for non-charge changing reactions ($\nu \to \nu$).

The charged current interaction is, to the best of our knowledge, a pure V-A interaction (Feynman & Gell-Mann 1958, Marshak & Sudarshan 1958, Sakurai 1958). How good is the best of our knowledge? (See, for review, Sakurai 1981). Our prejudices press us to seek other vector type interactions, since spin-1 carries most of the other forces familiar to us. Our knowledge of limits on a V+A coupling is good, but not as impressive as popularly thought. The small branching ratio for the $\pi \to e + \nu_e$ decay, for example, is predicted for any combination of V and A. This small rate is a consequence of the preferred helicity for high velocity fermions to have left (right)-handed helicity and for antifermions the opposite when born in V-A (V+A) interactions. The only data which tell us that the particular combination V-A is preferred come from two sources: the helicity of electrons in β decay experiments; and the polarization of muons and the decay asymmetry of the electrons from their decay, $\mu^+ \to e^+ + \bar{\nu}_e + \nu_\mu$.

Experiments on nuclear β decay show that electrons prefer to have negative helicity, and that the magnitude of this helicity is close to the electron velocity as predicted (Koks & vanKlinken 1976). Until recently, the most accurate measurement of the electron asymmetry in μ decay indicated deviation from the expected V-A form by a few percent (Akhmanov 1967). But, more recent experiments on polarized muons indicate complete agreement with predictions (Egger 1980). In summary, the experiments are consistent with the pure

V-A form, but also allow V+A couplings at the level of 1% in decay rates. A simple model with universal weak dimensionless coupling, but with a higher mass V+A propogator, is permitted so long as the new vector boson has a mass more than three times larger than the mass of the familiar boson.

Recently, additional checks on the universal V-A form of the charged current interaction have been borne out. The hypothesis that the decay $\tau^- \rightarrow \nu_\tau + e^- + \bar{\nu}_e$ might have opposite helicity at the τ vertex is, for example, not true (Bacino 1979, Kirkby 1979). Direct observation and measurement of the polarization of muons from $\nu_\mu + e^- \rightarrow \mu^- + \nu_e$ is in good agreement with expectations (Armenise 1979, Büsser 1981).

Until about ten years ago, the neutrinos had only been observed to interact through the charged current weak interactions. We now know that they also interact through a neutral current mechanism; today the prediction of this mechanism, SU(2) x U(1), stands as the theoretical framework that best describes them (Weinberg 1967, Salam 1968). There was no opportunity to see the neutral current in decay processes, because any decay which could occur through neutral currents would be overwhelmed by electromagnetic decays. They were seen initially and studied earliest in production processes initiated by neutrinos: $\nu_\mu + N \rightarrow \nu_\mu + X$, $\nu_\mu + e^- \rightarrow \nu_\mu + e^-$, $\nu_\mu + p \rightarrow \nu_\mu + p$.

The SU(2) x U(1) currents which describe these interactions have vector and axial vector parts that depend on the properties of the particles:

$$g_V = I_3 - 2Q\sin^2\theta_W , \quad g_A = I_3$$

where I_3 and Q are the component of weak isospin and the electric charge, respectively, of the interacting particle. For the neutrino, or any neutral elementary fermion, $g_V = g_A$, and the current structure for it is pure V-A, as in the charged current interactions. The Weinberg angle, θ_W, enters explicitly into the structure of the currents for charged particles.

The differential cross sections for the elastic scattering of neutrinos and antineutrinos from a fermion target at rest in the laboratory is:

$$\frac{d\sigma}{dy} = \frac{G^2 ME}{2\pi} \rho^2[(g_V+g_A)^2+(g_V-g_A)^2(1-y)^2]$$

and

$$\frac{d\bar{\sigma}}{dy} = \frac{G^2ME}{2\pi} \rho^2[(g_V-g_A)^2+(g_V+g_A)^2(1-y)^2] \quad .$$

Here 1-y = (1+cosθ)/2, where θ is the center-of-mass scattering angle. The parameters, g_V and g_A refer to the target particle. In this scheme, scattering from different targets produces very different structure in the angular distributions. For example, scattering from electrons gives (for $\sin^2\theta_w \approx 1/4$) nearly equal amounts of V-A and V+A, while scattering from nucleons (quarks) is dominantly V-A. The parameter, ρ, is an overall normalization which is predicted to equal one identically in the Weinberg-Salam-Ward model, where it is proportional to the ratio of the vector boson masses of charged and neutral currents. The validity of this hypothesis allows us to predict the boson masses with some confidence (see lectures by Llewellyn-Smith).

Table 7 shows several different determinations of the Weinberg angle from different kinds of experiments (see reviews by Baltay 1979, Hung & Sakurai 1981).

Table 7. Determination of $\sin^2\theta_W$ From Various Reactions

Reaction	$\sin^2\theta_w$
$\nu_\mu(\bar{\nu}_\mu)+N\rightarrow\nu_\mu(\bar{\nu}_\mu)+X$	0.230 ± 0.023
$\nu_\mu+p\rightarrow\nu_\mu+p$	0.26 ± 0.06
$\nu_\mu+N\rightarrow\nu_\mu+N+\pi^0$	0.22 ± 0.09
$\bar{\nu}_\mu+N\rightarrow\bar{\nu}_\mu+N+\pi^0$	$0.15-0.52$
$\nu_\mu+e^-\rightarrow\nu_\mu+e^-$	$0.22^{+0.08}_{-0.05}$
$\bar{\nu}_\mu+e^-\rightarrow\bar{\nu}_\mu+e^-$	$0.23^{+0.09}_{-0.23}$
$\bar{\nu}_e+e^-\rightarrow\bar{\nu}_e+e^-$	0.29 ± 0.05
$e^++e^-\rightarrow e^++e^-,\mu^++\mu^-$	0.25 ± 0.15
$e^-+d\rightarrow e^-+X$	0.224 ± 0.020

The last two entries are not from neutrino experiments. It should be noted that the most accurate determinations are from the first and last entries; i.e. deep inelastic neutrino scattering and the deep inelastic electron scattering. In the latter case, the interference of the neutral weak current with the electromagnetic current is observed. Even though the other reactions do not lead to precise measurements of the relevant parameter, it is non-trivial that all experimental results agree.

A two parameter fit to all available data (Kim 1981) yields:

$$\rho = 1.002 \pm 0.015 \pm 0.011$$

$$\sin^2\theta_w = 0.234 \pm 0.013 \pm 0.009$$

in good agreement with the hypothesis $\rho = 1$. Fixing this parameter yields a best value

$$\sin^2\theta_w = 0.233 \pm 0.009 \pm 0.005 ,$$

with $\chi^2 = 33.1$ for 45 degrees of freedom. It should be noted that the couplings assumed are for fermions in left-handed doublets and right-handed singlets. For right-handed doublets, the χ^2 would be significantly worse.

There is, therefore, strong experimental foundation for the currents used in calculating both the charged current interactions and the neutral current interactions. The square of the effective coupling in the former is known to be pure V-A to about 1%, while the predicted deviation from V-A for the SU(2) x U(1) neutral currents has been definitely observed and has been measured to about 5% of its value.

NEUTRINOS AS PROBES OF NUCLEONS

The well-understood charged current interaction of neutrinos with elementary fermions has been a useful tool in determining the structure of nucleons. The technique has been especially important in the context of the parton model for nucleon constituents (Feynman 1969, 1974; Bjorken & Paschos 1969, 1971). The data have corroborated this picture, and have demonstrated that the pointlike objects within the nucleon have the properties expected of quarks. Since this situation has been reviewed in some detail as of January 1982 (see review by Fisk & Sciulli 1982), we will address here only the broad outlines and an update with some recent data.

Figure 7 shows the relevant variables. In the high statistics experiments, E_μ, θ_μ, and ν are directly measured. In some cases, the neutrino energy, E, is also redundantly measured. The measured quantities are used to calculate the scaling variables: x and y; these are the most physically intuitive parameters and the most convenient to utilize.

$$x = Q^2/2M\nu \qquad y = \nu/E \quad .$$

Here, M is the mass of the nucleon target. In the parton picture, y is directly related to the center-of-mass scattering angle between the incident neutrino and the target parton. At very high Q^2 in the same picture, the variable x approximates the fraction of momentum, ξ, carried by the interacting parton inside the target nucleon. This hold in a Lorentz frame where the nucleon is traveling at high momentum.

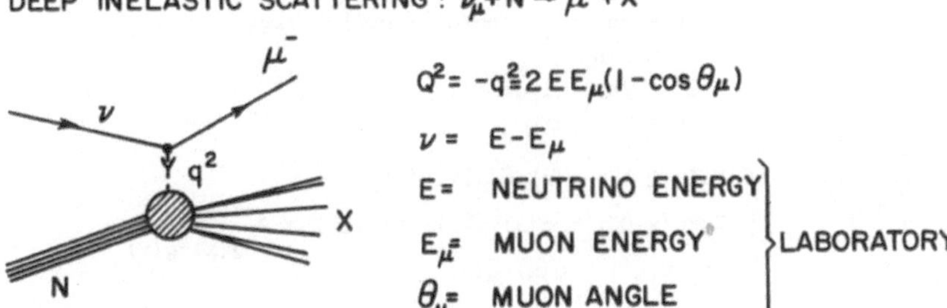

Fig. 7. Variables Describing Neutrino-Nucleon Collisions

Table 8 gives the differential cross sections as a function of the scaling variables in terms of the structure functions, F_2, xF_3, and R. These relations are completely general in the context of the V-A theory at energies well below those where propagator effects begin. In the quark model, the structure functions are interpretable in terms of distribution functions of quark fractional momenta inside the high momentum nucleon. In the simplest quark model, these distributions should be largely independent of Q^2 (the scaling hypothesis) and the function R should be nearly 0. This latter prediction, called the Callan-Gross relation, is a test of the dominance of spin 1/2 partons inside the nucleon.

Included also in Table 8 is the expression for the differential cross section for charged leptons (e or μ) scattering electromagnetically from nucleons. The relation between the two structure functions, (F_2^{em} and F_2^{wk}) should be simply related by the mean square charges of the scattering constituents. The structure function xF_3 is unique to neutrino scattering. It is proportional to the difference between the quark and antiquark distribution functions. If we conceive of the substructure of nucleons as consisting of valence and sea contributions; the former being the quarks which give the nucleon its quantum numbers, and the latter being an ocean of quark and antiquark pairs, then xF_3 isolates the valence contribution. This presumes that the x-dependence of the quarks and antiquarks in the sea is the same.

Table 8. Formulae of Deep Inelastic Scattering

$x \equiv Q^2/2M\nu$ $\quad$ E = beam (neutrino) energy

$y \equiv \nu/E$ $\quad$ E_μ = outgoing lepton (muon) energy

$$\nu = E - E_\mu$$
$$Q^2 = 2EE_\mu\,(1-\cos\theta_\mu)$$
$$s = M^2 + 2ME$$

scaling 2-quark model:
$$\begin{cases} F_2(x,Q^2) = q(x)+\bar{q}(x)+2k(x) \\ 2xF_1(x,Q^2) = q(x)+\bar{q}(x);\quad R(x,Q^2) = 2\dfrac{k(x)}{q(x)+\bar{q}(x)} \\ xF_3(x,Q^2) = q(x)-\bar{q}(x) \end{cases}$$

Neutrino:
$$q(x) = u(x)+d(x)+s(x)+c(x)$$
$$\bar{q}(x) = \bar{u}(x)+\bar{d}(x)+\bar{s}(x)+\bar{c}(x)$$

Muon-electron:
$$q^{em}(x) = 4/9\ u(x)+1/9\ d(x)+1/9\ s(x) + 4/9\ c(x)$$
$$\bar{q}^{em}(x) = 4/9\ \bar{u}(x)+1/9\ \bar{d}(x)+1/9\ \bar{s}(x)+4/9\ \bar{c}(x)$$

Neutrino:
$$\frac{d^2(\sigma^\nu+\sigma^{\bar{\nu}})}{dxdy} = \frac{G^2 s}{\pi}\left[(1-y-\frac{Mxy}{2E})F_2^{wk}(x,Q^2)+\frac{1}{2}y^2(2xF_1^{wk}(x,Q^2))\right]$$
$$\frac{d^2(\sigma^\nu-\sigma^{\bar{\nu}})}{dxdy} = \frac{G^2 s}{\pi}\left[y(1-\frac{1}{2}y)xF_3^{wk}(x,Q^2)\right]$$

Muon-electron:
$$\frac{d^2\sigma^{em}}{dxdy} = \frac{4\pi\alpha^2}{Q^4}s\left[(1-y-\frac{Mxy}{2E})F_2^{em}(x,Q^2)+\frac{1}{2}y^2(2xF_1^{em}(x,Q^2))\right]$$
$$\frac{2xF_1}{F_2} = \left(1+\frac{4M^2x^2}{Q^2}\right)/(1+R) \qquad 0 \le R \le \infty$$

These features of structure functions permit certain predictions in the context of the quark model for nucleon structure:

1. Approximate scaling: $F_i(x,Q^2) \approx F_i(x)$. It follows from this hypothesis that the total cross section for neutrino-nucleon collisions is almost independent of energy. The magnitude of the cross section slope ($\alpha = \sigma/E$) is proportional to the fraction of the nucleon momentum carried by interacting quarks.

2. Callan-Gross relation: $R^{wk} \approx R^{em} \approx 0$ (Callan & Gross 1969, Bjorken & Paschos 1969).

3. Quark charge requirement: $F_2^{em}/F_2^{wk} \approx 5/18$. More precisely, because there is an expected asymmetry between charm and strange sea quarks, we expect for negligible charm content,

$$\frac{F_2^{em}}{F_2^{wk}} = \frac{5}{18}\left(1 - \frac{3}{5}\,\frac{s+\bar{s}}{q+\bar{q}}\right)$$

at each value of x.

4. Gross-Llewellyn Smith sum rule (Gross & Llewellyn Smith 1969):

$$\int_0^1 \frac{xF_3(x)}{x}\,dx = 3 \quad .$$

These predictions are not exact. The scaling hypothesis is, of course, only approximate in principle. (We will return to the questions of scale violations in a later section.) Experimentally, they are valid at the 20% level or better.

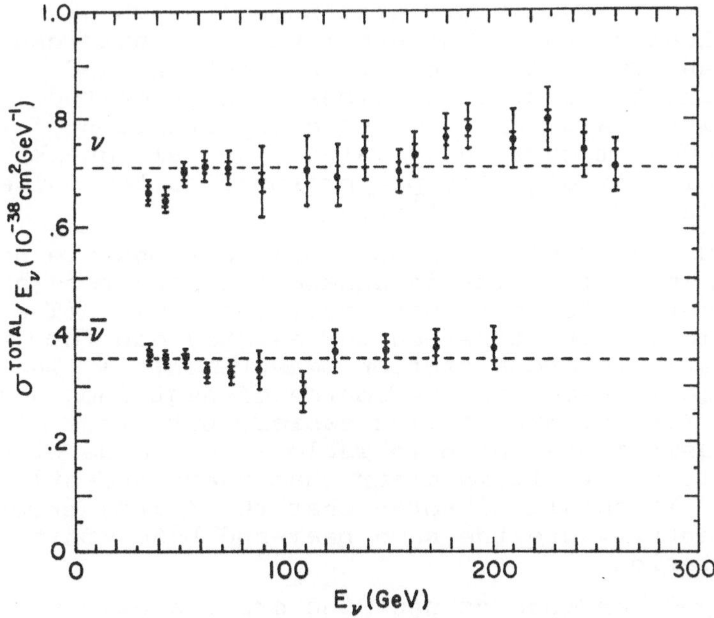

Fig. 8. Neutrino and Antineutrino Cross Section Slope vs. Energy

Figure 8 shows the energy dependence of the cross sections from the CCFRR data. These data were taken in the dichromatic beam at Fermilab, and represent the total charged current cross sections for neutrinos and antineutrinos incident on isoscalar nucleon targets. Although there is some point-to-point variation, the hypothesis of cross section slope constant with energy is acceptable. (The assumption of this constancy leads to the limits on inclusive oscillations shown in Fig. 4.) Other experiments also are consistent with a linear rise of cross section with energy.

Table 9. Measurements of Cross Section Slopes and Resulting Integrated Structure Function Values

Experiment	Energy	$\alpha(\nu)$	$\alpha(\overline{\nu})$	$\int xF_3dx$	$\int F_2dx$
deGroot 1979	$30-200\,Gev$	$.62 \pm .05$	$.30 \pm .02$	$.30 \pm .05$	$.44 \pm .03$
Jonker 1981	$25-260\,Gev$	$.60 \pm .03$	$.30 \pm .02$	$.28 \pm .03$	$.43 \pm .02$
Rapidis 1981	$30-230\,Gev$	$.70 \pm .03$	$.35 \pm .02$	$.33 \pm .04$	$.50 \pm .02$
Fritze 1981	$10-200\,Gev$	$.66 \pm .03$	$.30 \pm .02$	$.34 \pm .03$	$.46 \pm .02$

There exists some controversy regarding the magnitude of the cross section slope. There are at present measurements that differ by as much as 20%. Table 9 illustrates this. Also shown are the integrated values of the structure functions that result from the sum and difference of the neutrino and antineutrino total cross sections. The magnitudes of these integrals, which measure the fraction of the nucleon momentum carried by valence and all quarks respectively, is directly proportional to the values of the measured cross sections.

Measurements of neutrino cross sections are nontrivial. The difficulty is enhanced by the necessarily indirect method for obtaining neutrino flux. (The beam particles cannot be counted as they can for charged particle cross section measurements.) The use of dichromatic beams as the source of neutrinos has made the measurements of flux easier, but since all of the measurements shown in Table 9 were made with dichromatic beams, it is clear that there are still problems. It should be noted that the first, second, and last entries use the same beam and beam monitoring instrumentation.

The y-dependence of neutrino cross sections generally agrees in all experiments with the dominance of spin 1/2 scattering centers. Figure 9 shows typical data, which illustrates that the y-dependence is dominantly flat for neutrinos and dominantly $(1-y)^2$ for antineutrinos. More precisely, the prediction $R = 0$ requires that these data be described by

$$\frac{d\sigma}{dy} = A + B\,(1-y)^2 \qquad \frac{d\bar{\sigma}}{dy} = A(1-y)^2 + B \quad .$$

The ratio B/A measures the relative antiquark to quark content. The Callen-Gross relation forces y distributions (at fixed x and Q^2) to have this very specific dependence on y. In practice, quantitative measurements of R are very difficult, since they depend on the detailed shapes of y distributions at fixed x and Q^2. By the time the data are divided so finely, the statistical accuracy is not very good. There does seem to be general agreement, however, that the value of R lies below about 0.25 (see review by Fisk & Sciulli 1982).

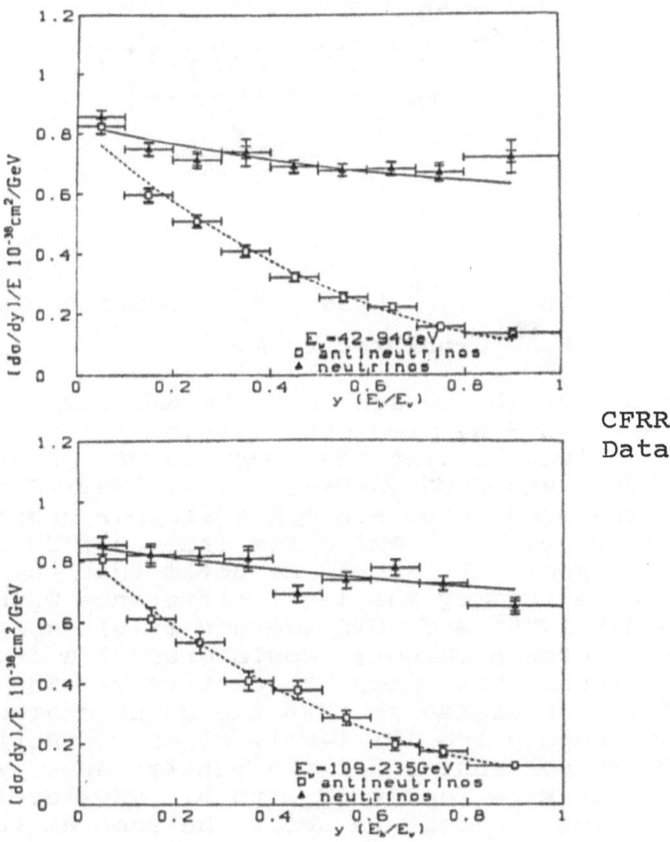

Fig. 9. Y Dependence of Neutrino and Antineutrino Data

The quark charge test requires comparison of normalized structure functions from both neutrino and charged lepton scattering experiments. There exist high statistics data from the latter (Aubert 1981, 1981a), but there are normalization inconsistencies with previous experiments (Bodek 1979, Gordon 1979) that amount to about 10%. There are also normalization differences from neutrino experiments, as seen from the total cross section measurements in Table 9.

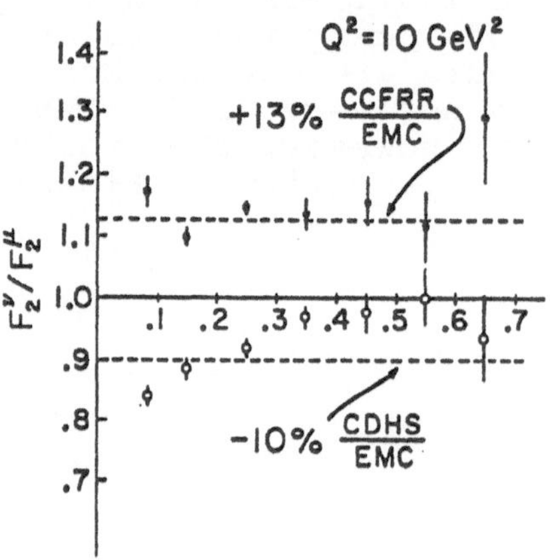

Fig. 10. Ratio of F_2^{wk} Measured in Neutrino Experiments to F_2^{wk} Predicted by F_2^{em}

From Fig. 10, we see that the EMC muon scattering data (Aubert 1981a) predicts neutrino structure function values between the measured values of CCFRR (Bodek 1982) and CDHS (deGroot 1979, 1980; Eisele 1981). The prediction assumes a strange quark sea which is 50% of the $\bar{u}$ and $\bar{d}$ sea (the half SU(3) symmetric case). It should be noted that the EMC data renormalized by the level difference with the SLAC (Bodek 1979) and CHIO (Gordon 1979) experiments, all with hydrogen targets, would predict a level 8% and 10% higher than shown, respectively. The conclusion of this exercise is that the quark charge test works at roughly the 10% level, clearly discriminating against integer charge for the constituents, which would differ by a factor of 2 to 3. The level of agreement for the test is about the same as the agreement among the data from both kinds of experiment.

The Gross-Llewellyn Smith sum rule is a direct test of the quark model prediction for the number of valence quarks in the nucleon. There have been several published values (Bosetti 1977, deGroot 1979, Fritze 1981) for the result of the integration; all are consistent within errors of quark model expectations. There is one operational difficulty having to do with the very small x region. Since the integrand involves division of the experimental values of xF_3 by x, the result depends critically on the manner in which the experimental values rise from 0 near $x = 0$. There are predictions from Regge theory for the behavior near $x = 0$, which argue that $xF_3 \approx x^{\alpha}$ in this region, with α obtainable from the lowest lying trajectory intercept, J; i.e. $\alpha = 1-J$ (Reya 1981). Thus, if the ρ trajectory is dominant, we expect $\alpha \approx 0.5$.

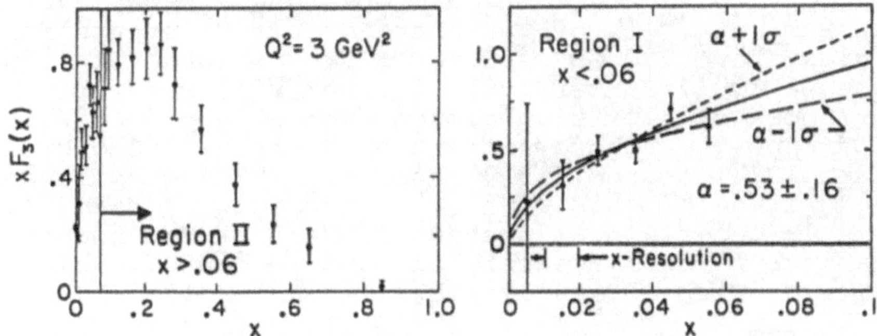

Fig. 11. xF_3 at $Q^2 = 3$ GeV^2 for Entire x-Range and at Small x

Figure 11 shows measured values of the xF_3 structure function (Bodek 1982); this represents the fractional momentum distribution of valence quarks in the model. Shown also is the behavior in the very small x region. A fit to the data for $x < 0.06$ is consistent with a power law rise, giving $\alpha = 0.53 \pm 0.16$. The result for the sum rule evaluation is insensitive to whether the raw data or an extrapolation of this fit is used at very small x. The result is:

$$\int_0^1 \frac{xF_3}{x}\,dx = 2.95 \pm 0.33 \pm 0.15 \quad .$$

The error has approximately equal contributions from systematic and statistical uncertainties. Because the integral gets major contribution from the small x region, the typical Q^2 is low (≈ 4 GeV^2), but because the laboratory energies are high, the typical invariant mass of the final state hadron system is large ($W^2 \approx 200$ GeV^2).

Of important interest in addition to the single muon charged current data are multimuon data, which measure the strange sea. Because the charged current process changes charge and flavor, charmed particles will be made primarily from strange quarks. The muonic decays of these charmed particles will then give events which have two muons of opposite sign. The relative numbers of opposite sign dimuon events provides a measure of the strange composition of the quark-antiquark sea in the nucleon (see, e.g. Knobloch 1980, Jonker 1981, Fisk 1981). Charmed particles produced by neutrinos have also been directly observed. Such experiments (see, e.g. Ushida 1980) have permitted early measurements of charmed particle lifetimes (for review, see Fisk 1981, Fisk & Sciulli 1982).

Opposite sign dimuons are observed at the level of approximately 1% of the charged current events for energies above 100 GeV. Several groups (Shaevitz 1980, Nishikawa 1981, Jonker 1981, Trinko 1981) see rare occurrences of same sign dimuon events, approaching 10% of opposite sign events at high energy. One group (Holder 1977, deGroot 1979, Knobloch 1981) questions the magnitude of this signal. The origin of same sign events is uncertain at present. There are no known mechanisms which can easily account for this level (see, e.g. Fisk & Sciulli 1982).

Events with single muons of the wrong sign have not yet been observed. In the past, limits on the occurrence of such events has been used to set limits on the production and decay of opposite sign heavy leptons (Barish 1974, Cnops 1978). A recent limit (Holder 1978) sets a limit of 1.6×10^{-4} for visible energy above 100 GeV in a dichromatic neutrino beam set at 200 GeV. This provides a limit on charm changing neutral current interactions, as well as associated charm production by neutral currents at a level of 2.6%.

SCALING VIOLATIONS IN INELASTIC NEUTRINO-NUCLEON SCATTERING

This topic has received a considerable amount of attention over the years. When approximate scaling was found to describe inelastic structure functions (Miller 1972, Bodek 1973, Eichten 1973) at $Q^2 \approx 5\ \text{GeV}^2$, it was thought possible that at high Q^2, scaling might become nearly exact.

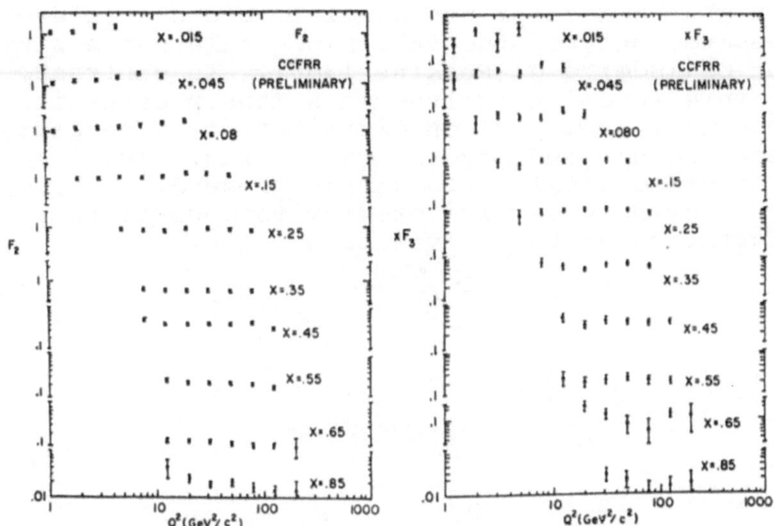

Fig. 12. Structure Functions at Fixed x from Recent CFRR Measurements (Bodek 1982)

At low and moderate energies, the structure functions do have explicit dependence on Q^2, as illustrated in Fig. 12. Other measurements from both muon and electron scattering (Bollini 1981, Aubert 1981, Eisele 1981) in the 100 GeV range demonstrate this as well. Mechanisms which can produce scaling violations include the intrinsic transverse momentum carried by quarks in the nucleon and lightly bound states of pairs of quarks. These effects should disappear as the momentum transfers become very large in comparison to the constituent momenta, and they should fall as a power of Q^2.

The theory of quantum chromodynamics (QCD) has many important qualitative successes in particle physics as an appropriate context for description of the strong interactions between quarks (see lectures by C. Llewellyn Smith). One prediction of perturbative QCD is that scaling violations will occur at very large momentum transfers, falling nearly logarithmically. Reviews of the comparison of prediction with experiment are readily available (Söding & Wolf 1981, Perkins 1980). Here, we will comment generally on the comparison in inelastic scattering, and update somewhat to the present situation.

Neutrino scattering provides some unique ways of approaching this problem. In particular, the structure function xF_3 is a color non-singlet, and so its evolution in Q^2 depends only on the values xF_3 takes at a single value of Q^2 (see, e.g. Perkins 1980). In contrast, the F_2 structure function evolves in a manner depending on itself as well as the gluon distribution. The latter, of course, is not independently measured. The theory cannot at present calculate any of these distributions from first principles; its predictive power lies in the description of their evolution.

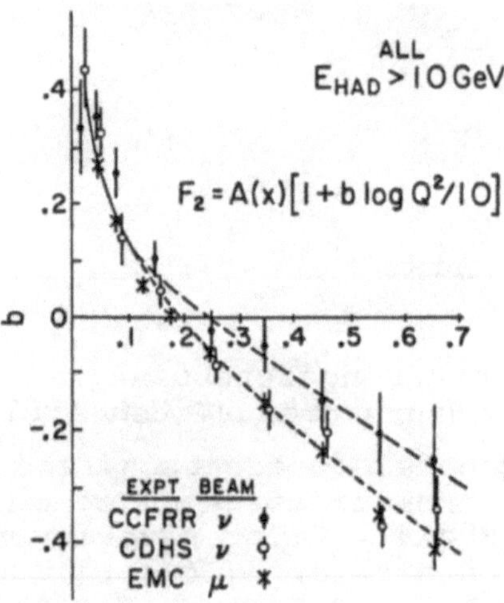

Fig. 13. Logarithmic Derivatives vs. x for Various Data Sets

The fact of scale breaking in this energy regime is not disputed. Figure 13 illustrates this. All experiments see a decreasing structure with Q^2 at large x, and an increasing one at small x. The differences among the experiments are within the quoted accuracies of the experiments. The major question is, "Are the observed scaling violations due to effects calculable from perturbative QCD?"

The evolution of the structure functions is described by the quark gluon coupling constant which, in first order perturbative QCD, is given by:

$$\alpha_s(Q^2) = \frac{12\pi}{(33-2N_f)\log(Q^2/\Lambda^2)}$$

where N_f is the number of quark flavors and Λ is a parameter to be fit by the data. There are hopes that this latter parameter will be predicted from other strong interaction parameters with, for example, lattice gauge theory calculations. The value of Λ has been extracted from both inelastic neutrino and inelastic muon scattering data. The earliest evaluations were near 700 MeV, while more recent data indicates values of order 150 MeV.

Table 10. Values of Λ (GeV) Obtained from Fits to Inelastic Data

Ref-Year	Group		Moments	Buras-Gaemers	Altarelli-Parisi
Bosetti et al 1978	BEBC	νN	$.74 \pm .05$ GeV ($Q^2 > .15$ GeV^2)		
de Groot et al 1979	CDHS	νN (Fe)		$.47 \pm .1 \pm .1$ ($Q^2 > 3$ GeV^2)	
Aubert et al 1981A	EMC	μN (Fe)			$.122^{+.022+.114}_{-.020-.070}$ ($Q^2 > 3.2$ GeV^2)
Aubert et al 1981	EMC	μp (H_2)			$.110^{+.058+.114}_{-.020-.070}$ ($Q^2 > 2.5$ GeV^2)
Bollini et al 1981	BCDMS	μN (C)	$.080^{+.130+.100}_{-.080-.070}$ ($Q^2 > 25$ GeV^2)	$.136^{+.050+.090}_{-.040-.080}$ $.3 < x < .7$	$.085^{+.060+.090}_{-.040-.080}$
Eisele et al 1981	CDHS	νN (Fe)			$.19 \pm .08$ ($Q^2 > 2$ GeV^2) ($W^2 > 11$ GeV^2)
Fritze et al 1981	BEBC	νN	$.245^{+.130}_{-.145}$ ($Q^2 > 1.5$ GeV^2)		$.210 \pm .095$ ($Q^2 > 2$ GeV^2)

Table 10 illustrates the historical evolution of the parameter fits. It is clear that the value of the parameter is directly linked to the degree of scale breaking, especially at large x. The values of Λ have changed primarily by including more high statistics data at high Q^2. Several hundred MeV for Λ seem quite reasonable: i.e., the pion mass sets a distance scale for the strong interactions near such values. It would be surprising to some if the actual value of Λ were substantially smaller than the pion mass, giving a distance parameter, Λ^{-1}, substantially larger than one fermi.

If we have reached the regime where perturbative QCD is the dominant mechanism for scale breaking, there is no problem. But, do we know this to be the case? Is it possible that, with even higher Q^2, the value of Λ will get even smaller? We cannot answer this question until we measure the structure functions in that regime. This should happen when the Tevatron becomes operational, when the available Q^2 range will increase by a factor of 3. The advent of new electron proton colliders will provide an even larger step.

CONCLUSIONS

In this brief overview, we have touched a few highlights on the neutrino properties, its interactions, and the information it helps provide on the structure of nucleons. Experiments are still being done to delineate these more clearly: neutrino mass, coupling to other neutrinos, searches for exotic neutrino types, more precise measurements of the neutral current parameters, and important tests of QCD, the theory describing the binding of quarks.

REFERENCES

Akhmanov, V.V. et al, Soviet Journal of Nuclear Physics 6, 230 (1968).
Aliban, P. et al, Phys. Lett. 74B, 134 (1978).
Armenise, N. et al, Phys. Lett. 84B, 137 (1979).
Aubert, J.J. et al, Phys. Lett. 105B, 315 (1981).
Aubert, J.J. et al, Phys. Lett. 105B, 322 (1981a).
Bacino, W. et al, Phys. Rev. Lett. 42, 749 (1979).
Bahcall, J., Neutrino 81, Maui, Hawaii, 1, 1.
Baltay, C., Proc. 19th Int. Conf. HEP, Tokyo, p. 882 (1979).
Baltay, C. et al, Fermilab Proposal P 646 (1980).
Baltay, C., Neutrino 81, Maui, Hawaii, 2, 295.
Barish, B. et al, Phys. Rev. Lett. 32, 1387 (1974).
Berquist, K.E., Nucl. Phys. B39, 317 (1972).
Bjorken, J.D., Paschos, E.A., Phys. Rev. 185, 1975 (1969).
Boehm, F., Novel Results in Particle Physics, Vanderbilt U (1982).
Bodek, A. et al, Phys. Rev. Lett. 30, 1087 (1973).
Bodek, A. et al, Phys. Rev. D20, 1471 (1979).
Bodek, A. et al, Neutrino '82 (to be publ.).
Bollini, D. et al, Phys. Lett. 104B, 403 (1981).
Bosetti, P.C. et al, Phys. Lett. 74B, 143 (1978).
Busser, F. et al, Neutrino 81, p. 328.
Callen, C.G., Gross, D.J., Phys. Rev. Lett. 22, 156 (1969).
Chen, H., 3rd Workshop on Grand Unification, U of N. Carolina, 1982 (to be published).
Cnops, A.M., Phys. Rev. Lett. 40, 144 (1978).
Danby, G. et al, Phys. Rev. Lett. 9, 36 (1962).
Daum, M. et al, Phys. Lett. 74B, 126 (1978).
Davis, R. et al, Neutrino '78, Purdue, p. 53.
deGroot, J.G.H. et al, Z. Physik, C1, 143 (1979).
Egger, J., Nucl. Phys. A335, 91 (1980).

Eichten, T et al, Phys. Lett. 77B, 274 (1973).
Eisele, F. et al, Neutrino '81, 1, p. 297.
Ellis, J. et al, Nucl. Phys. B106, 292 (1976).
Faber, S.M., Gallagher, J.S., Ann. Rev. Astron. Astrophysics 17, 135 (1979).
Feynman, R.P., Gell-Mann, M., Phys. Rev. 109, 193 (1958).
Feynman, R.P., Phys. Rev. Lett. 23, 1415 (1969).
Feynman, R.P., Neutrino '74, p. 299.
Fisk, H.E., Proc. Int. Symp. on Leptons & Photons at High Energies, Bonn (1981), p. 703.
Fisk, H.E., Sciulli, F., Ann. Rev. Nucl. Sci. 32, 499 (1982).
Fritze, P. et al, Phys. Lett. 96B, 427 (1980).
Fritze, P. et al, Neutrino '81 2, 195.
Gordon, B.A. et al, Phys. Rev. D20, 2645 (1979).
Gross, D.J., Llewellyn Smith, C.H., Nucl. Phys. B14, 337 (1969).
Hansl, T. et al, Phys. Lett. 74B, 139 (1978).
Holder, M. et al, Phys. Rev. Lett. 30, 433 (1977).
Holder, M. et al, Phys. Lett. 74B, 277 (1978).
Hung, P.Q., Sakurai, J.J., Ann. Rev. Nucl. Sci. 31, 375 (1981).
Jonker, M. et al, Phys. Lett. 93B, 203 (1980).
Jonker, M. et al, Phys. Lett. 99B, 265 (1981); erratum 100B, 520.
Kim, J.E. et al, Rev. Mod. Phys. 53, 211 (1981).
Kirkby, J., Proc. Int. Symp. on Lepton & Photons at High Energies, Batavia (1979), p. 107.
Knobloch, J. et al, 20th Int. Conf. HEP, Madison (1980), p. 769.
Knobloch, J. et al, Neutrino '81, 1, 421.
Koks, F.W., vanKlinken, J., Nucl. Phys. A272, 64 (1976).
Lubimov, V. et al, Phys. Lett. 94B, 266 (1980).
Marshak, R.E., Sudarshan, E.C.G., Phys. Rev. 109, 1860 (1958).
Miller, G. et al, Phys. Rev. D5, 6528 (1972).
Mo, L. et al, Fermilab proposal P635 (1980).
Niebergall, F. et al, 20th Int. Conf. HEP, Madison (1980), p. 242.
Nishikawa, K. et al, Phys. Rev. Lett. 46, 1553 (1981).
Olsen, S.L. et al, Moriond Workshop on New Flavors, XVIIIth Rencontre de Moriond (to be published).
Perkins, D., Techniques and Concepts of High Energy Physics, NATO Summer School, p. 279.
Pless, I. et al, Fermilab Proposal P636 (1981).
Pontecorvo, B., Sov. Phys. JETP 6, 429 (1957).
Primakopf, H., Rosen, S.P., Ann. Rev. Nucl. Sci. 31, 145 (1981).

Rapidis, P. et al, Proc. Smr. Inst. Part. Phys., SLAC (1981), p. 641; see also Bodek 1982.
Reeder, D. et al, Fermilab Reports (1982).
Reines, F. et al, Phys. Rev. Lett. 45, 1307 (1980).
Sakurai, J.J., Nuovo Cimento 7, 649 (1958).
Sakurai, J.J., Neutrino '81, 2, p. 457.
Sciulli, F., Neutrino '78, Purdue (1978), p. 863.
Shaevitz, M. et al, SLAC Summer Study (1980), p. 475.
Shaevitz, M. et al, Neutrino '81, p. 311.
Shrock, R., Phys. Lett. 96B, 159 (1980).
Silverman, D., Soni, A., Phys. Rev. Lett. 46, 467 (1981).
Soding, P., Wolf, G., Ann. Rev. Nucl. Sci. 31, 231 (1981).
Stanton, N., Neutrino '81, 1, p. 491.
Trinko, T. et al, Phys. Rev. D23, 1889 (1981).
Turner, M., Neutrino '81, 1, p. 95.
Ushida, N. et al, Phys. Rev. Lett. 45, 1049, 1053 (1980).
Willis, S.E., Phys. Rev. Lett. 44, 522 (1980).

This research supported in part by the National Science Foundation.

TOPICS IN ELECTRON-POSITRON INTERACTIONS

Paul Söding
Deutsches Elektronen-Synchrotron DESY
Hamburg, Germany

BASIC CONCEPTS

Introduction

In the collision of an electron and a positron in a high energy storage ring a large energy

$$Q = W = \sqrt{s} = 2\ E_{beam} \tag{1}$$

is dumped into a tiny region of space-time. If the electron and positron annihilate each other almost all of this energy becomes concentrated in a single field quantum (Fig. 1). The distance scale relevant to a process of four-momentum transfer Q is 1/Q. For a storage ring with $E_{beam} \approx 20$ GeV like PETRA, $1/Q \approx 5 \cdot 10^{-16}$ cm which is <1 % of the radius of a proton. The energy density in the field is therefore about 10^8 times the energy density in a nucleon. This strong field polarizes the vacuum. Virtual particle pairs fluctuating in the vacuum can be broken up and the particles become liberated.

As a consequence three important facts emerge.

1) All flavored particles existing in nature are expected to be pair-produced provided their mass is not larger than W/2. This holds for charged leptons and quarks as well as for neutrinos, and for the conjectured weak bosons as well as for photons.

2) The pair production process acts as an effective 'filter' for fundamental (i.e. pointlike) particles. In our present energy region where electromagnetic forces dominate over the weak ones, the characteristic e^-e^+ production rate for light pointlike particle pairs is given by the lowest order QED cross section.

$$\sigma_{tot} = \frac{4\pi}{3W^2}\alpha^2 e^2 \approx 0.05 \text{ nb} \cdot e^2 \quad \text{at } W \approx 40 \text{ GeV},$$

where α is the electromagnetic fine structure constant and e the charge of the particle. An integrated luminosity of 400 nb^{-1} per day in PEP or PETRA will let us observe about 20 $\mu^-\mu^+$ and 80 $q\bar{q}$ pairs per interaction region. By contrast the production of composite objects of size >> 1/Q is strongly damped by form factors. Put in a different way, concentrating energy into a space-time region much smaller than the size of a composite object will scarcely lead to pair-creation of this object. Indeed "prompt" pair production of mesons or hadrons is not observed in high energy e^-e^+ storage ring experiments. (For production via intermediate $Q\bar{Q}$ bound states different rules apply.)

3) Particles without flavor, i.e. the gluons, are not directly pair-produced in e^-e^+ interactions. Electron positron colliders nevertheless offer unique possibilities to study gluons. This is due to the large energy-momentum transfer Q involved in the interactions. It leads to large accelerations of the particles. When quarks are created the acceleration of their color charge causes color field quanta to be radiated. These can manifest themselves as gluon jets accompanying the quark jets (Fig. 2a). When quarks annihilate as in the decay of a heavy quarkonium $Q\bar{Q}$

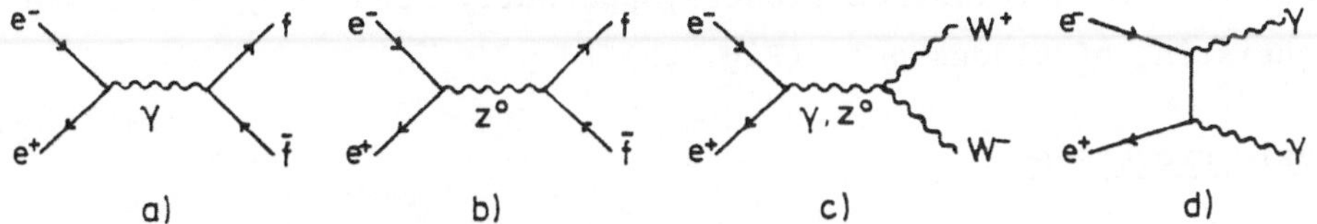

Fig. 1 Fundamental fermion and boson pair production processes in e^-e^+ annihilation

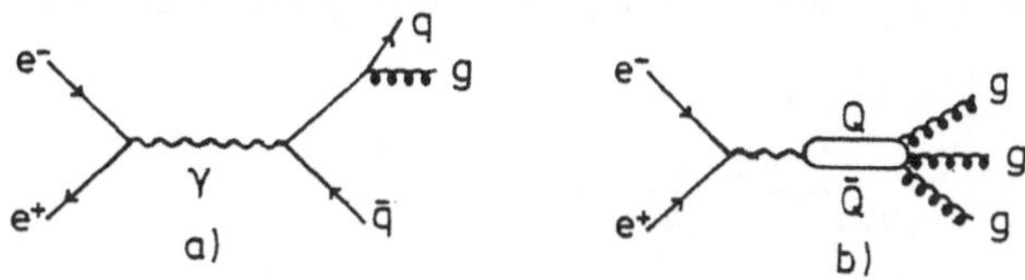

Fig. 2 Gluon radiation from accelerated color charge in a) continuum $q\bar{q}$ production
b) quarkonium decay

e^- ⟹ ⟹ e^+ $\qquad$ e^- ⟸ ⟸ e^+ $\quad$ Z ⟶

$\lambda = \lambda_{e^-} - \lambda_{e^+} = +1$ $\qquad$ $\lambda = \lambda_{e^-} - \lambda_{e^+} = -1$

Fig. 3 Addition of angular momenta in one-photon e^-e^+ annihilation

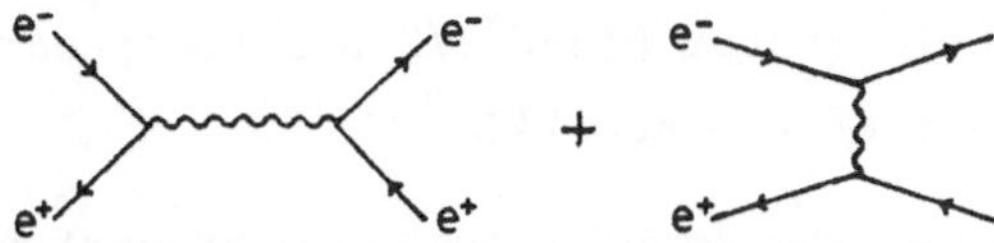

Fig. 4 Lowest order diagrams for Bhabha scattering

bound state formed in the e^-e^+ interaction, the gluon field can survive the annihilation process, apparently resulting in a state consisting of gluons only (Fig. 2b).

Basic Processes

We will discuss the reaction

$$e^-e^+ \rightarrow \mu^-\mu^+$$

as the prototype of the basic one-photon annihilation process. At a total cm energy $\sqrt{s} = W = \sqrt{Q^2}$ the one photon diagram (Fig. 1a) gives

$$\sigma_{tot} = \frac{4\pi}{3} \frac{\alpha^2}{s} \beta \frac{3 - \beta^2}{2} \tag{2}$$

and

$$\frac{d\sigma}{d\Omega} = \frac{\alpha^2}{4s} \beta \left[(1 + \cos^2\Theta) + (1-\beta^2) \sin^2\Theta\right] \tag{3}$$

Here β is the cm velocity of the outgoing muons. The electron mass m_e has been neglected ($\beta = 1$ for the incoming e^-e^+). Equations (2) and (3) hold for electromagnetic pair production of any pointlike spin-$\frac{1}{2}$ particles except electrons (see below). The 1/s dependence simply arises for dimensional reasons. For $q\bar{q}$ pair production the naive quark parton model gives an additional factor

$$R = 3 \sum_{i=1}^{n_f} e_i^2 \tag{4}$$

from color and flavor multiplicity and from electric charge. This holds for high energies ($\beta \gtrsim 1$) of the produced quarks; if β is small the threshold factor $\beta(3-\beta^2)/2$ has to appear in the corresponding term of the sum (4).

We next discuss the angular momentum properties of one photon exchange. As one easily verifies the vector current interaction $\bar{\psi} \gamma^\mu \psi$ conserves electron helicity up to terms of $o(m_e/Q)$.

The same is true for an axial vector current interaction. Since an incoming e^+ is equivalent to an outgoing e^- with oppositely directed momentum the interacting e^- and e^+ have opposite helicities. The total angular momentum component in the beam direction is therefore $\lambda = \pm 1$ (Fig. 3). Thus, the intermediate virtual photon is dominantly in a state

$$J^{PC} = 1^{--}, \ \lambda = \pm 1, \tag{5}$$

the helicity $\lambda = 0$ state being suppressed by factors of $o(m_e/\sqrt{s})$. Applying the same argument to the final state we can write down the angular distribution for production of spin $\frac{1}{2}$ fermion antifermion pairs (of negligible mass) using rotation matrices:

$$\frac{d\sigma}{d\Omega} \sim |d^1_{11}(\theta)|^2 + |d^1_{1-1}(\Theta)|^2 = \frac{1}{2}(1 + \cos^2\Theta) \tag{6}$$

For pair production of light scalar particles we get correspondingly

$$\frac{d\sigma}{d\Omega} \sim |d^1_{01}(\theta)|^2 + |d^1_{0-1}(\Theta)|^2 = \sin^2\Theta \tag{7}$$

The total cross section integrated over angles for charge 1 scalar pair production is calculated from the one-photon exchange diagram to be

$$\sigma_{tot} = \frac{\pi}{3}\,\frac{\alpha^2}{s}\,\beta^3 \tag{8}$$

which asymptotically (for $\beta \to 1$) equals 1/4 of the asymptotic spin $\frac{1}{2}$ fermion pair production cross section (2).

For Bhabha scattering e e $\to$ e e there is a t-channel exchange diagram in addition to the one photon annihilation (s channel) diagram (Fig. 4). The differential cross section to $o(\alpha^2)$ is

$$\frac{d\sigma}{d\Omega} = \frac{\alpha^2}{4s} \left(\frac{3 + \cos^2\Theta}{1 - \cos\Theta} \right)^2 , \tag{9}$$

diverging in the forward direction. This reaction is important for monitoring the luminosity of a storage ring.

The dominant annihilation reaction is the annihilation into two photons (Fig. 5). In contrast to the previously discussed reactions to which in addition to the electromagnetic current also the weak neutral current contributes in every order, two photon annihilation in lowest order is a purely electromagnetic process. Quantum electrodynamics (QED) gives for this process

$$\frac{d\sigma}{d\Omega} = \frac{\alpha^2}{s} \frac{1 + \cos^2\Theta}{\sin^2\Theta} , \qquad \sigma_{tot} = \frac{2\pi\alpha^2}{s} \ln \left(\frac{s}{m_e^2} \right) . \tag{10}$$

A comparison of the different basic cross sections is shown in Fig. 6, for an energy $W = \sqrt{s} \approx 32$ GeV. Note that all these cross sections scale as $s^{-1} \equiv W^{-2}$. Typical sizes of total cross sections at W = 32 GeV are

$$\begin{aligned}
&\sigma_{tot}(e\bar{e} \to e\bar{e}) && \sim 1 \text{ nb} \quad (\text{"large angle region"}) \\
&\sigma_{tot}(e\bar{e} \to \mu\bar{\mu}) && \sim 0.1 \text{ nb} \\
&\sigma_{tot}(e\bar{e} \to \tau\bar{\tau}) && \sim 0.1 \text{ nb} \\
&\sigma_{tot}(e\bar{e} \to \text{hadrons}) && \sim 0.3 \text{ nb}
\end{aligned} \tag{11}$$

Probing QED and Lepton Structure

If leptons and/or quarks were composite their interactions with the electromagnetic field would be damped by form factors. The reactions

$$e^-e^+ \to e^-e^+,\ \mu^-\mu^+,\ \tau^-\tau^+,\ \gamma\gamma \tag{12}$$

can therefore be used to obtain limits on possible sizes of these leptons and on the validity of

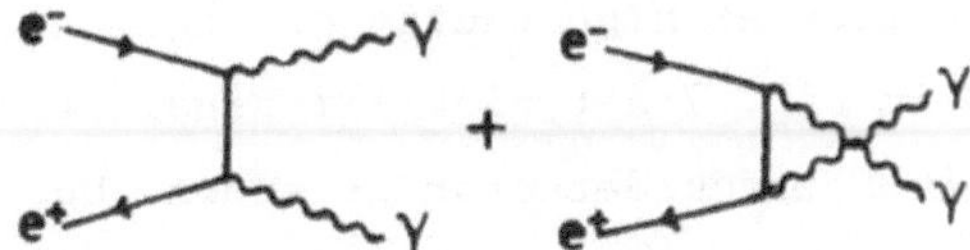

Fig. 5 Lowest order diagrams for the reaction $e^-e^+ \rightarrow \gamma\gamma$

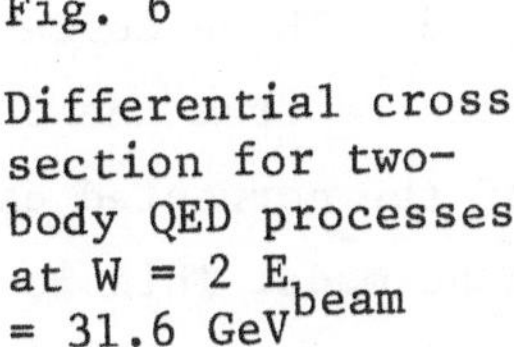

Fig. 6

Differential cross section for two-body QED processes at $W = 2\ E_{beam} = 31.6$ GeV

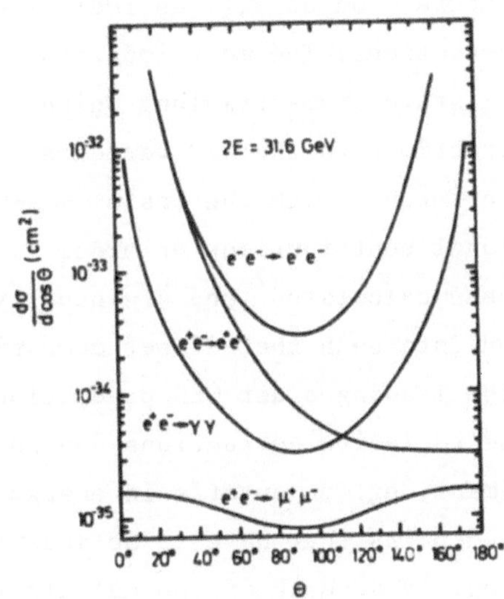

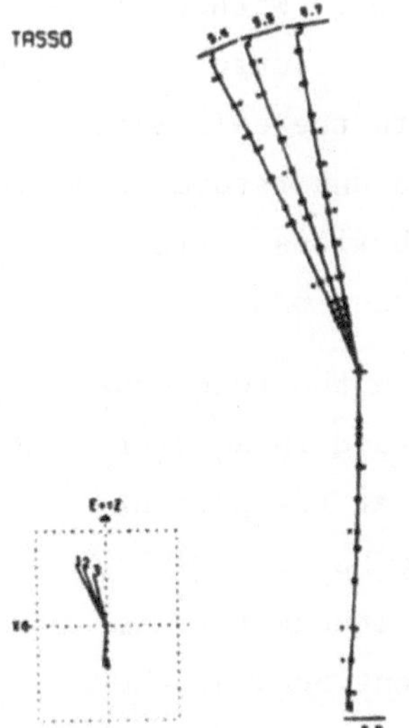

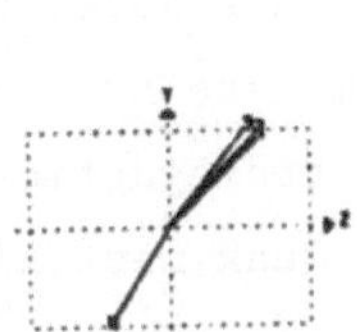

Fig. 7

An event of the reaction $e^-e^+ \rightarrow \tau^-\tau^+$ observed at PETRA at W = 32 GeV. The main view is along the beam axis; two side views are shown as insets.

quantum electrodynamics at high energy or, equivalently, at small distances[1]. In Fig. 7 a typical $\tau\bar{\tau}$ final state as observed at high energies in a track detector is shown. Due to the small mass of the τ this reaction is easily identified once the cm energy W exceeds about 10 GeV.

Discussing QED tests one has to consider the effects of the weak interaction as well as radiative and other higher order QED corrections. The most important of these latter effects is initial state bremsstrahlung which enters in the order α^3, as well as corrections to the QED vertex and propagator functions which by interference with the basic one-photon exchange diagram give additional contributions of order α^3 (Fig. 8). These corrections have been calculated[2] and are usually subtracted from the cross section data such that direct comparison of the corrected data with the leading order QED predictions can be made. This is done because radiative corrections depend on details of the individual experiment, eg. on cutoffs in energy and acollinearity of the produced pairs that have to be made in order to <u>define</u> the reaction. As a check of the calculation of the radiative corrections to Bhabha scattering Fig.9 shows a distribution of the acollinearity angle for Bhabha scattering $e\bar{e} \to e\bar{e}$. Without radiative corrections the acollinearity would, of course, identically vanish. The excellent agreement with the calculation demonstrates that the radiative corrections are understood. Fig. 10 shows measured Bhabha scattering angular distributions[3] after radiative correction, compared with leading order QED.

Next, the effects of the weak interaction on the reactions (12) have to be considered. This will be discussed in a later section. We only note two points here. First, at the present energies the purely weak terms are too small to be observable. Second, the leading weak-electromagnetic interference term is the one involving the <u>axial</u> weak neutral current, because the

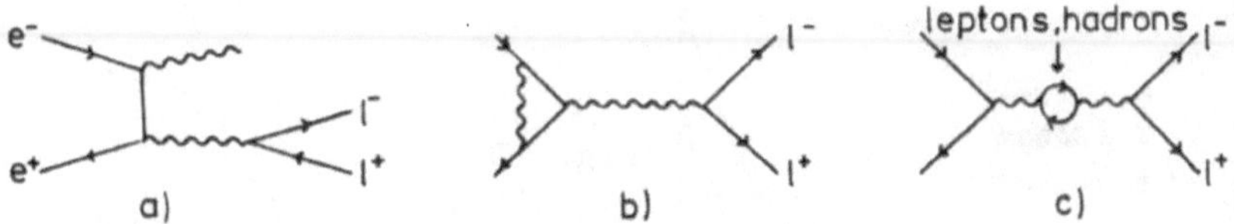

Fig. 8 Higher order QED corrections to lepton pair production by e^-e^+ annihilation
a) initial state bremsstrahlung
b), c) vertex and propagator corrections

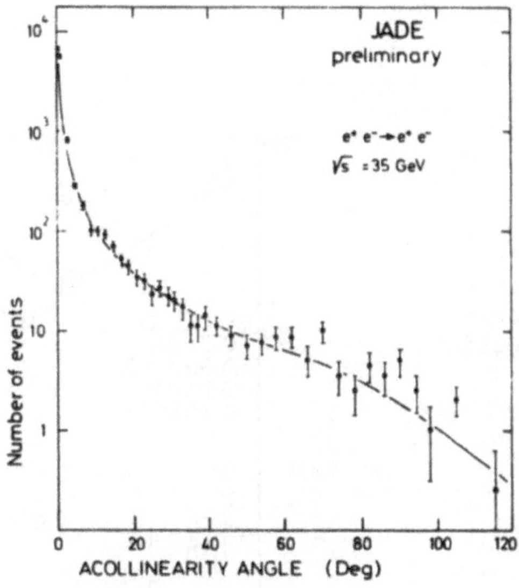

Fig. 9 Acollinearity angle distribution for Bhaba scattering. The curve shows the QED prediction (folded with the angular resolution of the shower detector). From Ref. 1.

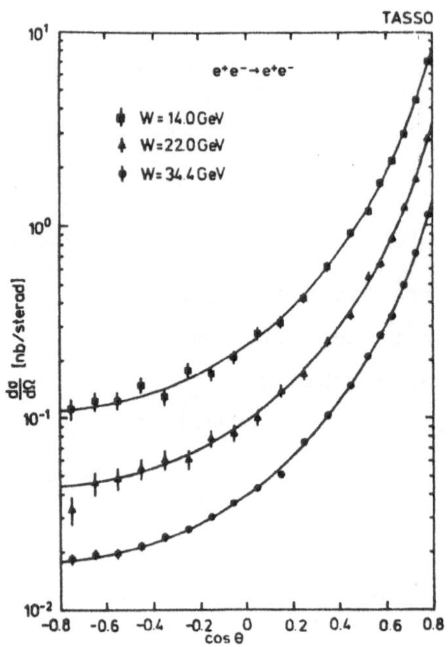

Fig. 10

Differential cross section for Bhabha scattering at cm energies W of 14, 22, and 34.4 GeV. The curves are the predictions from QED.

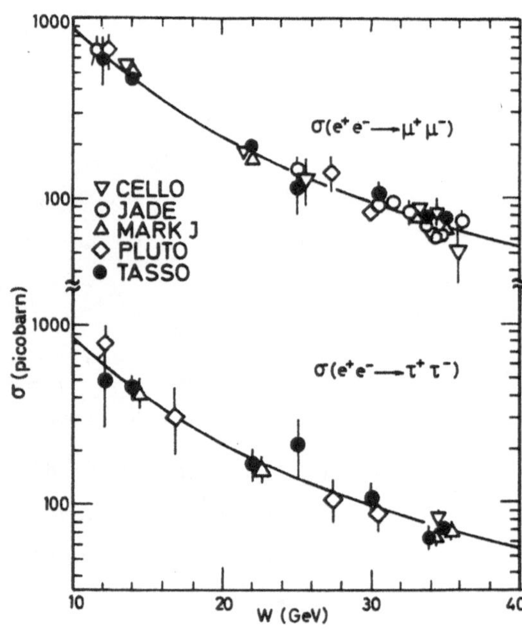

Fig. 11

Total cross sections for muon and tau pair production as a function of cm energy W. The curves show the QED prediction.

square of g_V^2 for leptons is apparently quite small. The axial current interference contribution can be made to vanish by integration over production angles, leaving only the very small weak vector term to contribute by interference with the electromagnetic amplitude to the total cross section. At present energies and accuracies this contribution has not been observed; in fact the W dependences of the total cross sections for the reactions (12) agree very well with the prediction of QED. This is shown in Fig. 11, in which radiatively corrected $\mu\bar{\mu}$ and $\tau\bar{\tau}$ production cross sections are compared with the prediction (2) from leading-order QED.[4]

The degree of agreement between the data and QED allows us to set limits on the validity of QED and on possible finite lepton size. A hypothetical lepton structure can be parametrized by a form factor. A useful analogy is provided by the pion form factor, shown in the spacelike (s<0) and timelike (s>0) regions in Fig. 12. The form factor is measured in the two regions by measuring πe scattering (electroproduction on virtual pions) and e^-e^+ annihilation into pion pairs, respectively. The form factor is thought to arise as a consequence of the composite nature of the pion as a $q\bar{q}$ bound state. It is dominated by the ρ^o, a state composed of the constituents of the pion with quantum numbers such that it couples directly to the electromagnetic current (Fig. 13). The pion form factor is therefore approximately given by the ρ^o pole

$$F_\pi(s) \approx \frac{m_\rho^{\ 2}}{m_\rho^{\ 2} - s - im_\rho \Gamma_\rho} \qquad \text{for } s > 4\, m_\pi^{\ 2}$$

$$F_\pi(s) \approx \frac{m_\rho^{\ 2}}{m_\rho^{\ 2} - s} \qquad \text{(purely real) for } s \leq 0 \qquad (13)$$

normalized to 1 at s = 0. The electromagnetic radius of the pion is determined by the ρ^o propagator to be

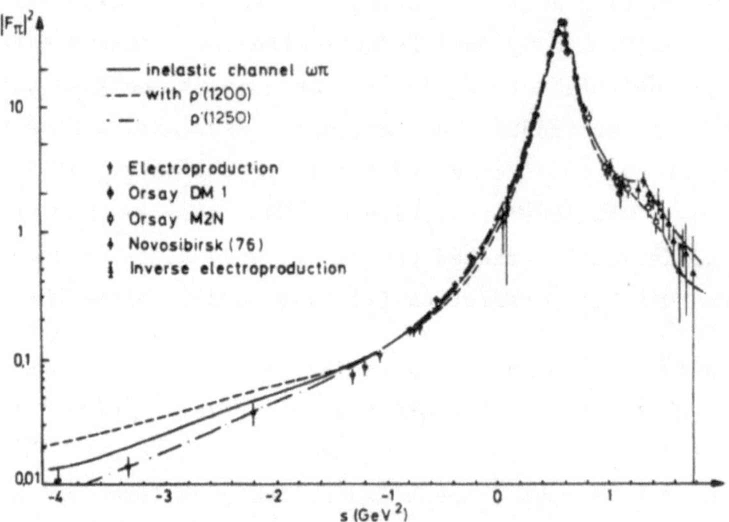

Fig. 12 The square of the pion form factor in the spacelike and timelike regions (from Ref. 5)

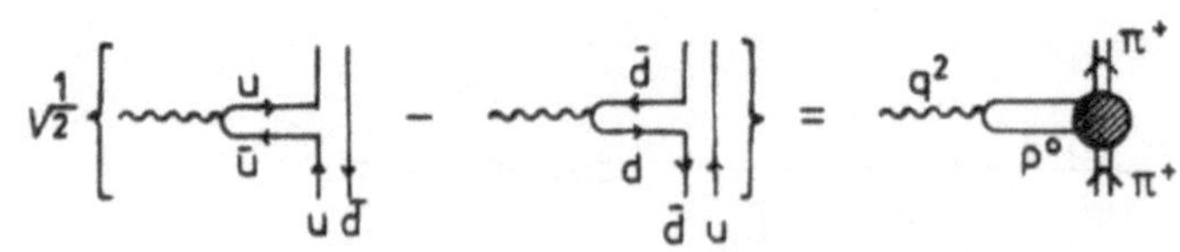

Fig. 13 Pion structure and ρ^o dominated form factor

$$\langle r_\pi^2 \rangle^{1/2} \approx \frac{\sqrt{6}}{m_\rho} \tag{14}$$

Lepton structure may be parametrized in a similar manner, introducing a propagator factor

$$F_+(Q^2) = \frac{\Lambda_+^2}{\Lambda_+{}^2 - Q^2} = 1 - \frac{Q^2}{Q^2 - \Lambda_+{}^2} \approx 1 + \frac{Q^2}{\Lambda_+{}^2} \tag{15}$$

at the lepton photon vertex. The pole at some large mass Λ_+ would correspond to a tightly bound state of the hypothetical constituents of the lepton. Such a term may also be thought of as a modification of the photon propagator by a factor $F_+(Q^2)$, effectively replacing the QED propagator $1/Q^2$ by

$$\frac{1}{Q^2}\left(1 - \frac{Q^2}{Q^2 - \Lambda_+{}^2}\right) = \frac{1}{Q^2} - \frac{1}{Q^2 - \Lambda_+{}^2} \tag{16}$$

This propagator modification leads, however, to a difficulty that becomes obvious in Bhabha scattering: Introducing (16) into the diagrams of Fig. 4 the modification is equivalent to a term describing the exchange of a heavy vector particle with mass Λ_+, coupled to leptons with an imaginary coupling constant ie. The corresponding interaction would be non-hermitian. This illustrates that it is not trivial to modify QED without destroying its fundamental properties. A way out is to change the sign in (15) and (16), defining

$$F_-(Q^2) = 1 + \frac{Q^2}{Q^2 - \Lambda_-{}^2} \approx 1 - \frac{Q^2}{\Lambda_-{}^2} \tag{17}$$

This modification effectively amounts to adding an exchange of a heavy vector particle with mass Λ_- and coupling constant e. At low energies ($Q^2 \ll \Lambda_-{}^2$) it is also equivalent to exchanging a vector particle of mass $(g/e)\Lambda_-$ with effective lepton coupling constant g. The form factor $F_-(Q^2)$ has therefore a qualitatively similar effect as Z^o exchange.

It appears from this discussion that the form factors $F_\pm(Q^2)$ are not much more than convenient parametrizations of upward or downward deviations from the Q^2 dependences predicted by QED. The lower limits on the cutoff parameters $\Lambda_\pm$ deduced from the observed $Q^2 = W^2$ dependences of the reactions (12) give a phenomenological measure of the length scale $\sim 1/\Lambda_\pm$ to which the experiments are sensitive under the assumption that the structure concerned couples with a strength $\approx$ e.

Data like those in Figs. 10 and 11 yield cutoff parameters $\Lambda_\pm$ in the range of 100 to 300 GeV. Interpreted in the above sense we conclude that leptons interact like structureless Fermi-Dirac objects when probed at a resolution of $1/\Lambda_\pm \lesssim (0.7 \text{ to } 2) \times 10^{-16}$ cm. This represents the best spatial resolution achieved in an experiment.

For two-photon production (Fig. 5) the limiting values on the cutoff parameters so far obtained are somewhat smaller, in the 50 GeV range[6]. The propagator modification would in this case correspond to the exchange of a heavy electron. The existence of such an object with unit charge can therefore be excluded up to a mass of about 50 GeV.

Hadron production at High Energies

In the large majority of the cases the hadrons produced in high energy e^-e^+ annihilation emerge as two back-to-back jets. The dominant two-jet structure was first found in experiments at SPEAR[7] at a cm energy of W=7.4 GeV. At the higher cm energies of PEP and PETRA the jet structure became strikingly evident (Fig. 14). Can we be assured that the underlying process is quark pair production (Fig. 1a)? The evidence rests on four basic observations:

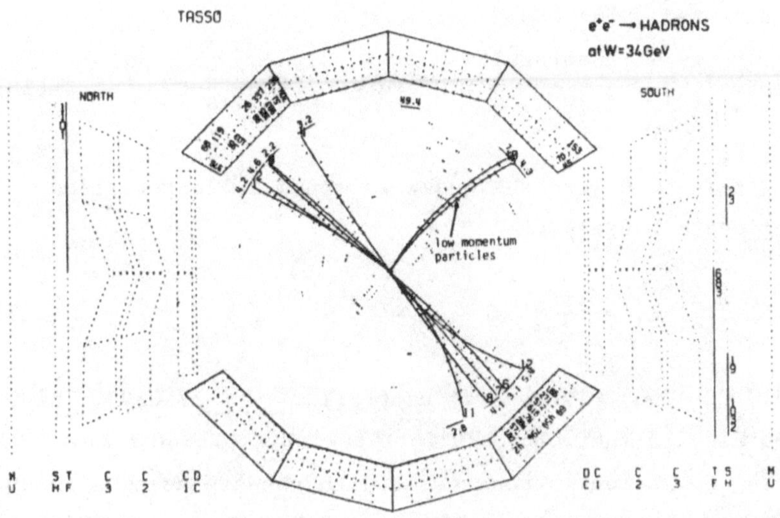

Fig. 14 A typical two-jet event observed at PETRA. The view is along the beam direction. The two tracks going towards the upper right corner have very low momenta. Neutral energy is shown by hits in the shower calorimeter surrounding the tracking region.

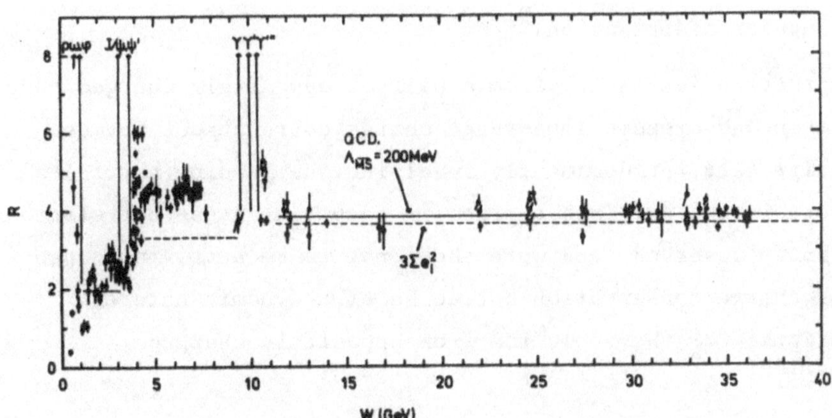

Fig. 15 Ratio R of the total cross section for $e^-e^+ \to$ hadrons to the lowest order electromagnetic $e^-e^+ \to \mu^-\mu^+$ cross section, as a function of cm energy W.

i) The measured value of

$$R = \frac{\sigma_{tot}(e^-e^+ \to \text{hadrons})}{\sigma_{\mu\mu}^{(0)}} \quad \text{where} \quad \sigma_{\mu\mu}^{(0)} = \frac{4\pi}{3}\frac{\alpha^2}{s} \tag{18}$$

agrees to within 10 % with the quark parton model expectation (neglecting quark masses) of

$$R = 3\sum_{i=1}^{5} e_i^2 = \frac{11}{3} \tag{19}$$

in the whole PEP and PETRA energy range ($12 \leq W \leq 37$ GeV). This is shown in Fig. 15 (Ref. 8); the dotted line represents eq. (19). The remaining deviation is consistent with the expectation for QCD radiative corrections as will be discussed below.

ii) Apart from ~10 % three-jet states the events show a two-jet shape (Fig. 14) that gets more pronounced the higher the cm energy W.

iii) The angular distribution of the jet axis with respect to the $e^\pm$ beam direction is in agreement with the expectation $(1 + \cos^2\Theta)$ from the quark parton model for spin-$\frac{1}{2}$ partons (Fig. 16); on closer investigation there are indications for $o(\alpha_s)$ QCD corrections to this angular distribution.[9]

iv) If the two jets arise from a pair of oppositely charged parent partons then one expects long-range charge correlations between the opposite jets, predominantly involving the leading particles (which contain the original quarks) in each jet. These correlations were recently observed, and were shown not to be a trivial consequence of charge conservation but to be of a dynamic nature, characteristic for the evolution from oppositely charged primaries.[10]

How does one identify a hadron production event in the detector? The most important criteria are based on particle multiplicity and total energy of the final state. One further checks charge balance, vertex position, and the possibility of confusion with

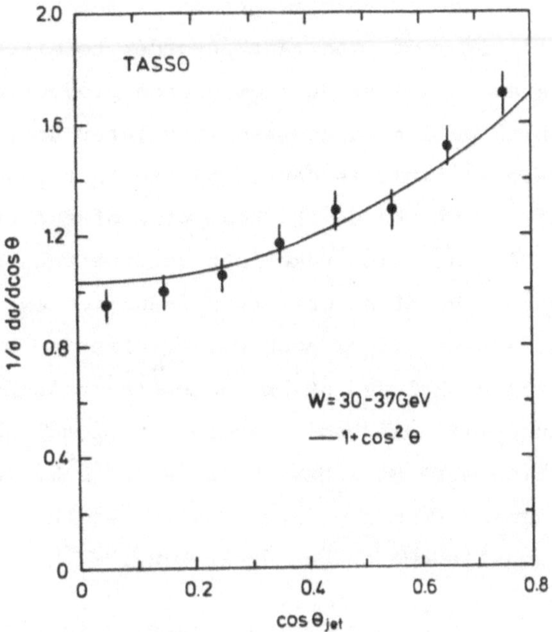

Fig. 16 Angular distribution of the jet axis for two-jet events $e^-e^+ \rightarrow$ hadrons at W = 30-37 GeV.

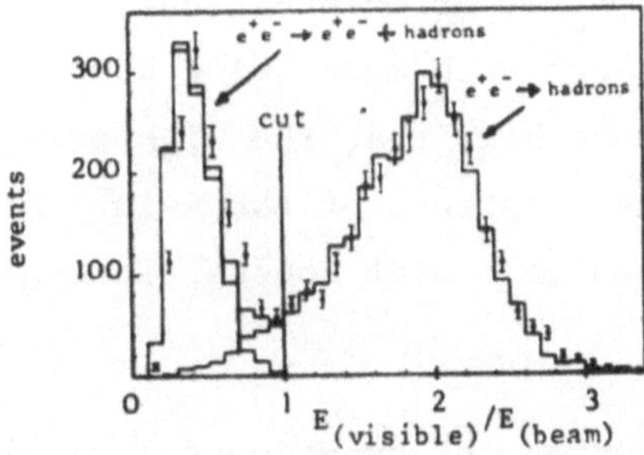

Fig. 17 Use of a cut in the total observed energy of a hadronic final state, in order to distinguish one-photon annihilation from two-photon interactions. Points are JADE data, histograms show expectation. From ref. 12.

$\tau^-\tau^+$ production or with Bhabha scattering followed by electromagnetic showering of one or both electrons in the detector. Energy balance (Fig. 17) is most important in order to distinguish hadron production by e^-e^+ annihilation from hadron production by two photon interactions which will be discussed in a later section. Uncertainty in applying these criteria is caused by the fact that particles travelling close to or inside the beam pipe of the storage ring will be missed. One must also take into account the radiative corrections which cause a boost as well as a reduction in total energy of the hadronic system. These problems are treated by Monte Carlo simulation of the events and of the detector including the subsequent data analysis. One can determine the total hadron production cross section with an accuracy of about 5 %, including the uncertainties coming from the determination of the storage ring luminosity by small-angle Bhabha scattering.[8]

Two further remarks on the experimental results for R should be made. The first is that the existence of a top quark with charge 2/3 in the relevant mass range would have caused an increase of R by $3\ e_t^2 = 4/3$ (times the threshold factor given in equ. (2)). From the measurements[8] of R, coupled with searches for narrow $t\bar{t}$ bound states at PETRA by scanning a range of cm energies in small steps of W, the existence of a t quark with mass below about 19 GeV can be excluded. The second remark is that the high energy data on R give direct evidence for the existence of three internal degrees of freedom of the quarks which are, of course, interpreted to be the three colors.

So far we have compared the experimental results on R (equ. 18) with pure QED, combined with the quark parton model. Real and virtual gluon emission by the quarks (Fig. 18) is expected to add a logarithmically varying quantum-chromodynamic correction[11] to R, which at $W \approx 30$ GeV should amount to about 6 %:

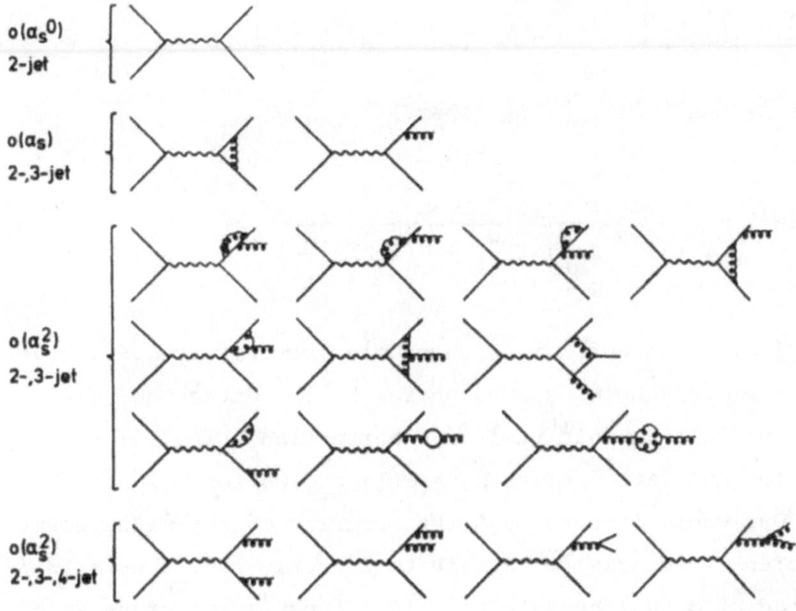

Fig. 18 QCD diagrams relevant for $e^-e^+ \rightarrow$ hadrons

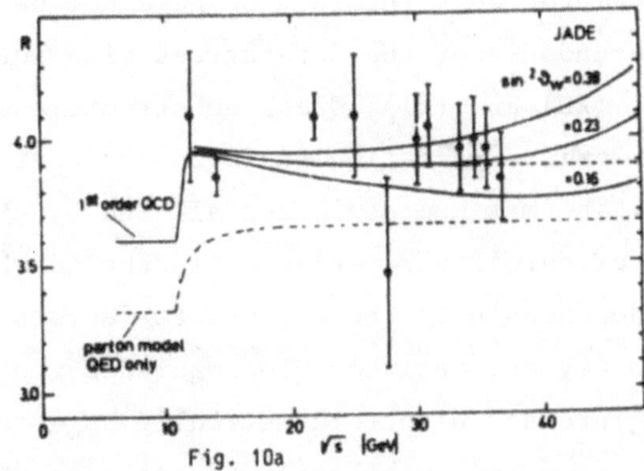

Fig. 19 Ratio R of the total cross section for $e^-e^+ \rightarrow$ hadrons, to the lowest order electromagnetic $e^-e^+ \rightarrow \mu^-\mu^+$ cross section, as a function of W. The dashed curve indicates the quark parton prediction; solid curves show QCD prediction with Λ = 0.3 GeV for different values of $\sin^2\theta_W$.

$$R = 3 \sum_{i=1}^{n_f} e_i^2 \left[1 + \frac{\alpha_s(W^2)}{\pi} + o(\alpha_s{}^2(W^2)) \right] \tag{20}$$

with the strong QCD coupling constant

$$\alpha_s(W^2) = \frac{12\pi}{(33 - 2n_f)\, \ell n \frac{W^2}{\Lambda^2} + o(\ell n\, \ell n\, \frac{W^2}{\Lambda^2})} \tag{21}$$

For simplicity the quark mass threshold factors of eq.(2) have been omitted from the above expression for R. The data show some evidence of this gluonic radiative correction (Fig. 19). A fit of eq.(20) to JADE data[12] gives $\alpha_s = 0.14 \pm 0.08$ for W in the 30 GeV region. One would like to push the accuracy of the measurement of R sufficiently such as to be able to directly test the $\ell n\ (W^2/\Lambda^2)$ dependence of $\alpha_s(W^2)$ because this is a fundamental property of QCD arising from the antiscreening of color charge connected with asymptotic freedom.

Effects on R from the weak interaction have not been observed until now. The W dependence of the data agrees with the one expected from equs. (20) and (21). Quark substructure would lead to a deviation from this expectation analogous to what we discussed for leptons. The maximum deviation allowed by the data is expressed in terms of cutoff parameters $\Lambda_\pm$ in the quark form factors. Assuming flavor-independence of the quark form factors lower bounds for $\Lambda_\pm$ of about 200 GeV are derived. One concludes that quarks, like leptons, appear to interact electromagnetically as structureless Fermi-Dirac objects down to a resolution of at least $1/\Lambda_\pm \approx 1 \cdot 10^{-16}$ cm.

We have seen that the presence of gluonic radiative corrections to the basic $q\bar{q}$ pair production process is indicated in the data on the total hadron production cross section but that the effect is small and lacks distinctive features. In order to isolate QCD effects

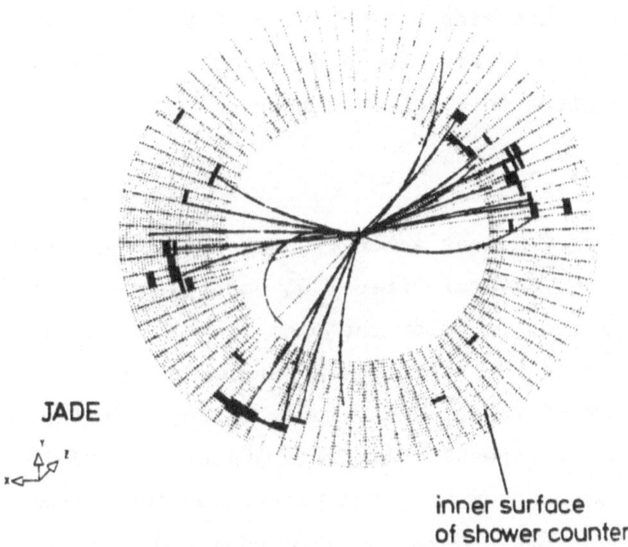

Fig. 20 Three-jet event observed at PETRA. The view is along the beam direction into the interior of the JADE detector, shown in perspective: the outer circle is the nearby edge, the inner circle the opposite distant edge of the cylindrical tracking region. This region has a diameter of 1.6 m and a length of 2.4 m. On the inner surface of the shower counter the actual segmentation is indicated; those segments in which energy is deposited are marked in black.

it is better to focus on specific final states which arise only due to these QCD radiative effects, such as three-jet production by the gluon bremsstrahlung process of Fig. 2a. Indeed this is the process in which gluon jets were first observed.[13] Of the order of 10 % of the hadron production events in e^-e^+ annihilation at PEP or PETRA above cm energies W of about 25 GeV are three-jet events. An example is shown in Fig. 20. A detailed discussion of three-jet events will be given in a later section.

Resonances

The behavior of R near a heavy quark threshold shows a striking pattern (Fig. 15). Several virtual $Q\bar{Q}$ bound states with vector quantum numbers can lie below the open flavor threshold. They are directly excited by the electroweak neutral current. These states have a number of different decay modes as shown in Fig. 21. For a ground state below threshold like $\psi(3.10)$ or $\Upsilon(9.46)$ the electroweak annihilation into lepton or quark pairs, and the strong annihilation into three gluons (in analogy to orthopositronium decay) are the leading modes. For heavier systems like the expected toponium $(t\bar{t})$ with a mass $m_{t\bar{t}} > 38$ GeV the weak processes (Fig. 21 a, c, d) are expected to become competitive. If a neutral Higgs scalar H^o with sufficiently low mass exists the radiative decay $(t\bar{t}) \rightarrow \gamma H^o$ (Fig. 21e) is predicted[14] to occur with a branching ratio of order 1 %. This may be the most promising process for the detection of the Higgs particle.

Excited $Q\bar{Q}$ states below open flavor threshold, like $\psi'(3.69)$ and $\Upsilon'(10.03)$, can decay by electromagnetic or gluonic de-exitation into a lower mass $(Q\bar{Q})$ bound state plus a photon or plus hadrons, respectively (Fig. 21 f,g). Resonant $(Q\bar{Q})$ states <u>above</u> open flavor threshold like $\psi''(3.77)$ and $\Upsilon'''(10.58)$ preferentially decay into flavored hadrons (Fig. 21 h), with partial widths of these decays exceeding those of the previously mentioned decays by orders of magnitude.

Ground states below open flavor threshold

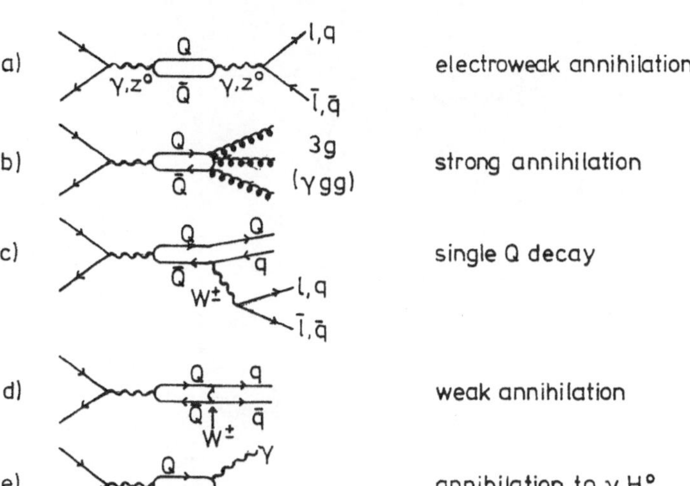

excited states below open flavor threshold

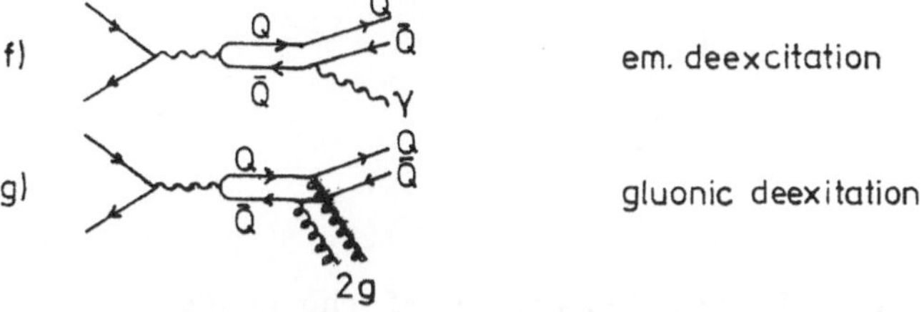

resonant $Q\bar{Q}$ states (above open flavor threshold)

Fig. 21 Production and decay of quarkonium vector states

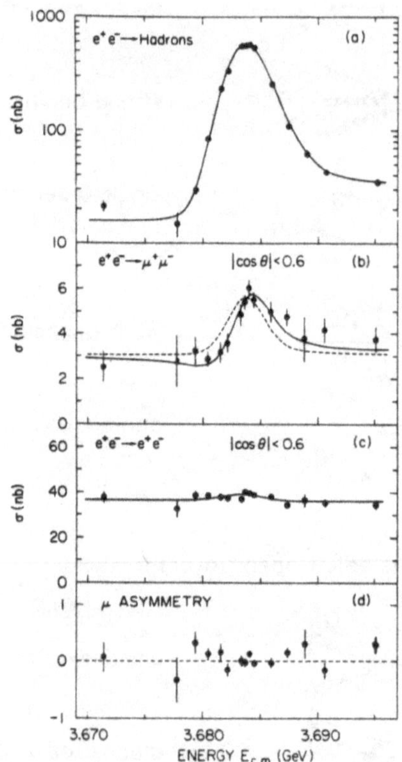

Fig 22 Energy dependence of the cross sections $e^-e^+ \to$ hadrons, $e^-e^+ \to \mu^-\mu^+$, $e^-e^+ \to e^-e^+$ and the forward-backward asymmetry in μ pair production in the vicinity of the $\psi'(3.69)$. From Ref. 15.

A detailed discussion of the properties of $Q\bar{Q}$ bound states and resonances has been given by Rosner in the previous volume of this series. Here we mention only one more practical point. The shape of the resonances as observed in the experiment is not a Breit-Wigner. First it is made asymmetric by bremsstrahlung from the initial state electrons (see Fig. 22a). These radiation effects also decrease the peak of the resonance. Second, there are interference terms between the resonant amplitude and the one-photon continuum, negative below and positive above the resonance (Fig. 22b). Third, quantum fluctuations in synchrotron radiation emission in an electron storage ring lead to a Gaussian distribution of beam energies. The resulting W smearing is typically several MeV, much larger than the width of the virtual bound states below the open flavor threshold.

e^-e^+ Storage Rings

In the modern high energy e^-e^+ colliders the electron and positron bunches are revolving in the same ring in opposite directions. The bunches are typically a few cm long and of transverse size ~0.05 mm(height) × 0.5 mm (width) at the intersection points. The luminosity of a storage ring is defined as the number of reactions per unit reaction cross section σ and unit time in an intersection point. Simple geometry gives for the number of reactions per unit time

$$\dot{N} = \sigma\ f \int n_1(x,y)\ n_2(x,y)\ dx\ dy = \sigma\ f\ n^2\ A \qquad (22)$$

where $n_i(x,y)$ = number of particles in the bunch i per area transverse to the direction of motion, A = effective transverse area of the bunches, f = number of colliding bunches per unit time = number n_b of stored bunches per beam times revolution frequency f_o in the ring. Each beam represents an electric current $I = efnA = ef_o n_b N$ where $N = nA$ is the number of particles per bunch and e the electron charge. For a two-dimensional gaussian transverse distribution of the particle density in the bunch integration yields $A = 4\pi\,\sigma_x\sigma_y$. Therefore, geometric considerations give the luminosity as

$$L = \frac{\dot{N}}{\sigma} = \frac{I^2}{4\pi\, e^2\, f_o\, n_b\, \sigma_x\, \sigma_y} = \frac{f_o\, n_b\, N^2}{4\pi\, \sigma_x\sigma_y} \qquad (23)$$

Values for the luminosity obtained in different storage rings are shown in Table 1.

The effects that limit the luminosity in a given electron storage ring are different ones, depending on the energy. At low and intermediate energies the limit is determined by electromagnetic beam beam interaction effects. A particle crossing a bunch of the opposite beam experiences the space charge of the other bunch in first approximation as a focussing quadrupole field. The focussing action changes the number Q of betatron oscillations per revolution for the particle, and this tune shift may bring the particle into a betatron or other type of resonance. A gaussian particle distribution in the bunch results in a continuous distribution of tune shifts, such that some particles will always become resonant and be lost unless the maximum tune shift is kept sufficiently small. The tune shift is given by

$$\Delta Q_x = \frac{r_e}{2\gamma}\,\frac{\beta_x^* N}{\pi\,\sigma_x(\sigma_x + \sigma_y)} \qquad (24)$$

and analogously for ΔQ_y, where $r_e = e^2/m_e c^2$, $\gamma = E_{beam}/m_e$ and β_x^*, β_y^* are the values of the horizontal and vertical amplitude functions at the intersection point.

It has been found empirically that ΔQ must be less than

Table 1 Some properties of electron positron storage rings

Name	Location	W range (GeV)	Maximum achieved luminosity $cm^{-2}s^{-1}$	Number of interaction regions
DCI	Orsay	1.0 - 3.8	$1.4 \cdot 10^{30}$	2
SPEAR	Stanford	2.5 - 9	$1 \cdot 10^{31}$	2
DORIS-II	Hamburg	3 - 11.2	$7 \cdot 10^{30}$	2
CESR	Cornell	8 - 16	$1.5 \cdot 10^{31}$	2
PEP	Stanford	10 - 30	$1.7 \cdot 10^{31}$	6
PETRA	Hamburg	10 - 36.8	$1.7 \cdot 10^{31}$	4

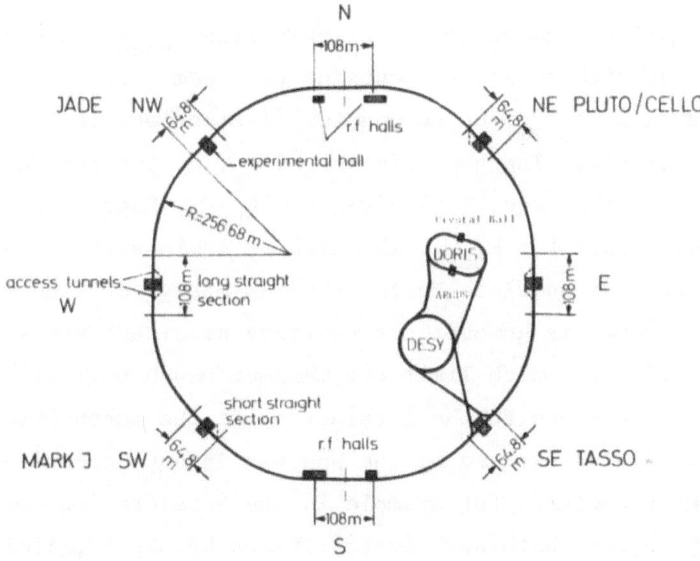

Fig. 23 General arrangement of storage rings and detectors at DESY.

about 0.04 in order to ensure stable operation of a high energy e^-e^+ collider. To maximize N the β^* have therefore to be minimized. Since β^* is the ratio of beam envelloppe to maximum slope on axis this means tight focussing at the intersection point. In the vicinity of the focus the β function behaves with distance ℓ from the focus like

$$\beta(\ell) = \beta^* + \ell^2/\beta^*.$$

In the focussing quadrupole the value of $\beta(\ell)$ will therefore be roughly proportional to the square of the distance of the lens from the intersection point. The amplitude function $\beta(\ell)$ at the lens must however not be made too large because of chromaticity effects which are proportional to $\beta(\ell)$. Consequently the quadrupole has to be positioned as close to the interaction point as compatible with the detectors. At PETRA the distance is $\ell = \pm 4.5$ m.

In the space charge limited region the luminosity for given ΔQ and β^* will depend on energy roughly like E^4_{beam}. This comes about because with constant focussing the beam emittance is growing with energy like E^2_{beam} due to quantum fluctuations in synchrotron radiation emission. The beam dimensions σ_x, σ_y thereby become $\propto E_{beam}$ such that equ. (24) gives $N \propto E^3_{beam}$. Putting this into eq. (23) we obtain $L \propto E^4_{beam}$. The highest luminosities should therefore be reached close to the maximum energy of a collider. The E^4_{beam} dependence is however not realized at higher energy due to beam instabilities which limit the maximum bunch current. These instabilities are due to the interaction of the bunch with electromagnetic fields excited by the bunches in any non-uniform surrounding structure, for example in the accelerating cavities.In the energy region where such instabilities become the limiting factor the luminosity will according to equ. (23) fall with energy like E^{-2}_{beam}

At the highest energies of a given collider the luminosity will eventually be limited by available rf power. The energy losses due to synchrotron radiation at given beam current I are $\propto E^4_{beam}$. The necessary rf accelerating voltage is therefore also $\propto E^4_{beam}$, which has the further unpleasant consequence that the electric losses in the cavities increase like E^8_{beam}. For given maximum rf power therefore $I \propto E^{-4}_{beam}$, and equ. (23) gives $L \propto E^{-10}_{beam}$ for this region.

Data on various storage rings are compiled in Table 1. Fig. 23 shows, as an example, the arrangement at DESY. The PETRA ring has a circumference of 2304 m and a bending radius in the arcs of 192 m. It is operated with two bunches per beam colliding in four interaction regions. The maximum energy so far reached is E_{beam} = 18.4 GeV; it will be raised to 23 GeV. Typical currents are 12 mA per beam. The vertical amplitude function at the interaction point is β^*_y = 9 cm, the maximum tune shift ΔQ_{max} = 0.04, the maximum measured luminosity $1.7 \cdot 10^{31}$ $cm^{-2}s^{-1}$.

An interesting feature of an electron storage ring is the buildup of transverse electron polarization due to radiative transitions between the two states of different spin energy in the magnetic guide field. This effect was first observed and exploited in an experiment at SPEAR.[16] The polarization depends on the imbalance between the spin flip transition rate in the two different directions of increasing or decreasing magnetic energy, as well as on various depolarizing effects due to resonances between spin and orbit motion, machine imperfections, and beam beam interaction. The theoretical maximum degree of polarization in an ideal storage ring is $P_{max} = 8\sqrt{3}/15 = 92$ %. At PETRA values of the polarization up to P = 80 % have been measured.

Detectors

A typical hadronic production event in the region of $W \gtrsim 30$ GeV cm energy has about 13 charged particles emerging in two jet cones of 17° half opening angle. The charged particles carry roughly 2/3 of the total energy W. The remaining energy goes mainly into neutral pions giving about 10 photons on average, and into neutral kaons and Λ's some of which decay in the detector. The neutrals are emitted into the same narrow jet cones as the charged particles are. On the other hand three-jet events, or decays of heavy quarkonia or pair production of heavy quarks, will give particles distributed over almost all solid angle.

To reconstruct the events and to interpret them properly one wants to cover a solid angle as large as possible, but also to be able to disentangle particles travelling closely together and to measure their momenta or energies and determine their identities. In particular electron and/or muon identification is important in order to recognize weak processes. For spectroscopic work on $Q\bar{Q}$ bound states good detection efficiency and resolution for low momentum photons is important. The investigation of photon-photon collisions requires special forward detectors for electrons and hadrons close to the beam direction. Since reaction rates are some 8 to 9 orders of magnitude smaller than the bunch collision rate in the storage ring, fast and efficient background rejection is a necessity.

As an example of how one tries to cope with these problems we will look at the JADE detector at PETRA[17] (Fig. 24). Like most large detectors at e^-e^+ storage rings it employs an axial magnetic field for momentum measurement. The field has a strength of 0.5 T and is produced by a thin aluminum solenoid of 2 m diameter and 3.6 m length. Tracks from charged particles are measured in a cylindrical high pressure drift chamber placed inside the solenoid and crystals of 16 radiation lengths thickness, arranged in the shape

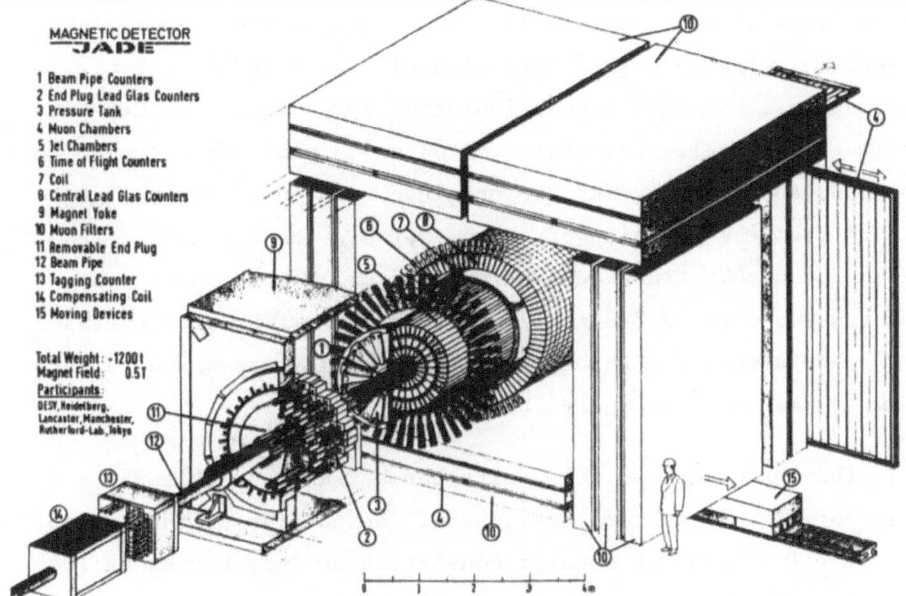

Fig. 24 The magnetic detector JADE at PETRA

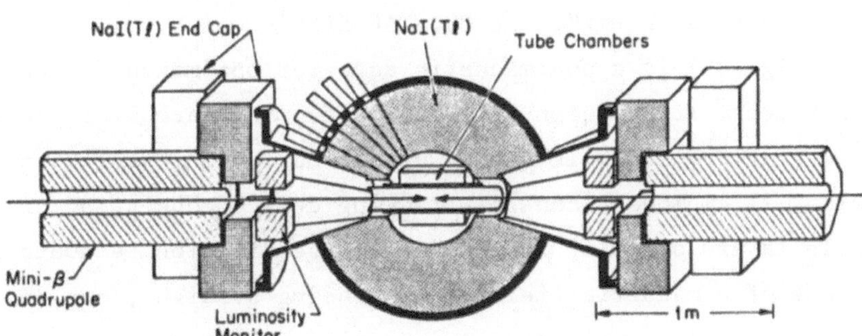

Fig. 25 The Crystal Ball detector at DORIS

surrounding the beam pipe; it covers 91 % of the total solid angle around the interaction point. The chamber employs axial sense wires, with 48 wires along a radius. The projection of the tracks onto the $r\Phi$ plane is measured with a precision of $\sigma_{r\phi} = 160\ \mu m$ per wire, resulting in a momentum resolution of $\sigma_{pT}/p_T = 0.022 \cdot p_T/\mathrm{GeV}$ and an azimuthal angular resolution of $\sigma_\phi = 1.3$ mrad for a track. The z coordinates of the tracks are obtained with an accuracy of $\sigma_z = 1.6$ cm from the division of charge between the two ends of the wires. The double track resolution is 7 mm. Tracks in jets are reconstructed with 99 % efficiency; no ghost tracks are generated. The chamber also has dE/dx measuring capability and would be sensitive to particles of charge $\pm e/3$.

In the space between this drift chamber and the solenoid a layer of time-of-flight counters is mounted. The outside of the solenoid is surrounded by a shower counter consisting of 2688 blocks of lead glass; the endcap regions are covered with 2×96 lead glass blocks. The whole detector is surrounded by layers of steel concrete absorber interleaved with drift chambers to measure muons. Close to the beam pipe on both sides of the central detector luminosity counters and a tagging hodoscope for photon-photon collisions are installed.

Whereas the JADE detector is an example of what may be called a "universal detector", the CRYSTAL BALL[18] which will be discussed next (Fig. 25) is a non-magnetic detector optimized for photon measurement. It has been very successful in charmonium spectroscopy work at SPEAR and is now in use at DORIS-II in the bottonium region. It employs a central tracking device consisting of three double layers of drift tubes with charge division readout. In the absence of a magnetic field tracks, being straight, can be recognized with this simple device over almost 80 % of 4π solid angle. The chambers are surrounded by the main component of the detector, an electromagnetic shower calorimeter consisting of 672 NaI(Tℓ)

of two half-spheres. This calorimeter covers 94 % of 4π solid angle and offers an energy resolution of $\sigma_E = 0.026\ E^{3/4}$ (energies in GeV) and an angular resolution of $\sigma_\theta = 2^o$ for showers. Further NaI crystals in the end cap region bring the total solid angle coverage to 98 % of 4π. In addition the detector employs a time-of-flight system.

Two recent devellopments in storage ring detector technology deserve particular mention. The first concerns vertex chambers. These are high precision drift chambers designed to measure tracks as accurately as possible in the vicinity of the interaction point. The chamber employed in the MARK-II experiment at PEP[19] is surrounding a Be beam pipe of only 0.006 radiation lengths thickness. The chamber has 7 axial layers, 4 of which are just beyond the beam pipe about 12 cm from the beam while the three other layers are at about 30 cm distance from the beam. The measurement accuracy is about 100 μm per layer. In practice, extrapolating tracks back to the interaction point the decay length of decaying τ leptons was measured in this chamber with an uncertainty of 120 μm(stat.)+ 150 μm (syst.).This lead to a determination of the τ lifetime of $(3.31 \pm 0.57\ (\text{stat.}) \pm 0.60\ (\text{syst.})) \times 10^{-13}$s, a value consistent with the expectation from $\mu\tau$ universality. In future experiments precision vertex chambers might allow us to recognize a certain fraction of those events in which charmed or bottom quarks are produced, such that e.g. the weak couplings of these quarks can be determined.

A second important new development is the time projection chamber (TPC)[20] at PEP. This is a high pressure cylindrical drift chamber surrounding the beam pipe, employing an axial magnetic field. Electrons produced along the tracks of ionizing particles are drifted along the magnetic field lines towards the end plates by an electric field which is parallel to the magnetic field. The readout planes at the chamber endcaps carry sense wires and cathode pads. The wires are used to measure the z coordinate via the drift time, while the pads give correlated $r\phi$ coordinates. In this way one is recording a three dimensional picture of the event. The signals are continously read into CCD's. The spatial resolution is $\sigma_{r\phi}$ = 195 μm in the bend plane and σ_z = 450 μm in the drift direction. A momentum resolution of $\sigma p/p$ = 0.07 at 1 GeV/c without, and .036 with vertex constraint, has been achieved. A track is sampled 183 times; the dE/dx distribution for pions near the mimimum has a resolution of $\sigma_{dE/dx}$ = 4.6 %, for high energy electrons and muons the corresponding numbers are 3.2 % and 2.8 %, respectively. The prospect of being able to identify essentially all the particles in a complicated event by ionization loss even in the relativistic region makes the TPC particularly attractive.

ELECTROWEAK INTERACTION AND NEW PARTICLES

Electroweak interaction of leptons

At the highest cm energies W = 36 GeV of PETRA the effective $Q^2 = W^2$ is already more than 10 % of the presumed mass of the Z^o. Consequently the standard model weak amplitude for fermion pair production (Fig. 1b) is of order 10 % of the electromagnetic amplitude (Fig. 1a). Large weak/electromagnetic interference effects are therefore predicted to occur if standard SU(2)×U(1) is correct. Testing the gauge theory of the weak and electromagnetic interaction by measuring the expected large interference between the electromagnetic and the weak neutral current indeed provided one of the principal original motivations for building PEP and PETRA.

From the diagrams of Fig. 1a and 1b one calculates the total fermion antifermion pair production cross section

$$\sigma_{tot}(e\bar{e}\to f\bar{f}) = \frac{4\pi\alpha^2}{3W^2}\left[1 + 2g_V^e\, g_V^f\, \mathrm{Re}\chi + (g_V^{e^2} + g_A^{e^2})(g_V^{f^2} + g_A^{f^2})|\chi|^2\right] \quad (25)$$

where

$$\chi = \frac{G_F\, m_Z^2}{8\sqrt{2}\pi\alpha}\ \frac{W^2}{W^2-m_Z^2+im_Z\Gamma_Z} \tag{26}$$

is the ratio of the weak to the electromagnetic amplitude, and g_V, g_A are the coupling constants of the weak neutral current. The lepton masses were assumed to be negligible; m_Z, Γ_Z are the mass and width of the Z^o, and G_F is the Fermi weak coupling constant. The standard $SU(2) \times U(1)$ model couplings are given by

$$g_V = 2\,T_3 - 4Q\,\sin^2\Theta_W,\quad g_A = 2T_3 \tag{27}$$

where Q = electric charge, T = weak isospin. With $\sin^2\Theta_W = 0.23$ the standard model predicts for leptons

$$g_V^\ell = -0.08,\quad g_A^\ell = -1. \tag{28}$$

At W = 35 GeV and with m_Z = 90 GeV we find from eq. (26) $|\chi| \approx \mathrm{Re}\chi = 0.06$. Due to the smallness of the vector coupling g_V^ℓ the expected weak effects in $\sigma_{tot}(e\bar{e}\to\ell\bar{\ell})$, $\ell = \mu,\tau$ are too small to be observable. In agreement with this expectation from the standard model the data on $\mu\bar{\mu}$ and $\tau\bar{\tau}$ production are found to follow the prediction from pure QED (i.e. the first term in eq. (25)), as discussed before.

A quantity that is sensitive to the weak axial coupling by virtue of an interference with the dominant electromagnetic (vector) amplitude, is the forward-backward angular asymmetry in the reaction $e\bar{e} \to f\bar{f}$,

$$A_{ff} = \frac{\sigma(\theta < 90^o) - \sigma(\theta > 90^o)}{\sigma(\theta < 90^o) + \sigma(\theta > 90^o)} \tag{29}$$

where θ is the cm angle between incoming e and outgoing f. From the diagrams of Fig. 1a, b one derives the angular distribution (for neglegible lepton masses)

$$\frac{d\sigma}{d\Omega}(e\bar{e} \to f\bar{f}) = \frac{\alpha^2}{4W^2}\left[1 + \cos^2\theta + \frac{8}{3}A_{ff}\cos\theta\right] \tag{30}$$

with

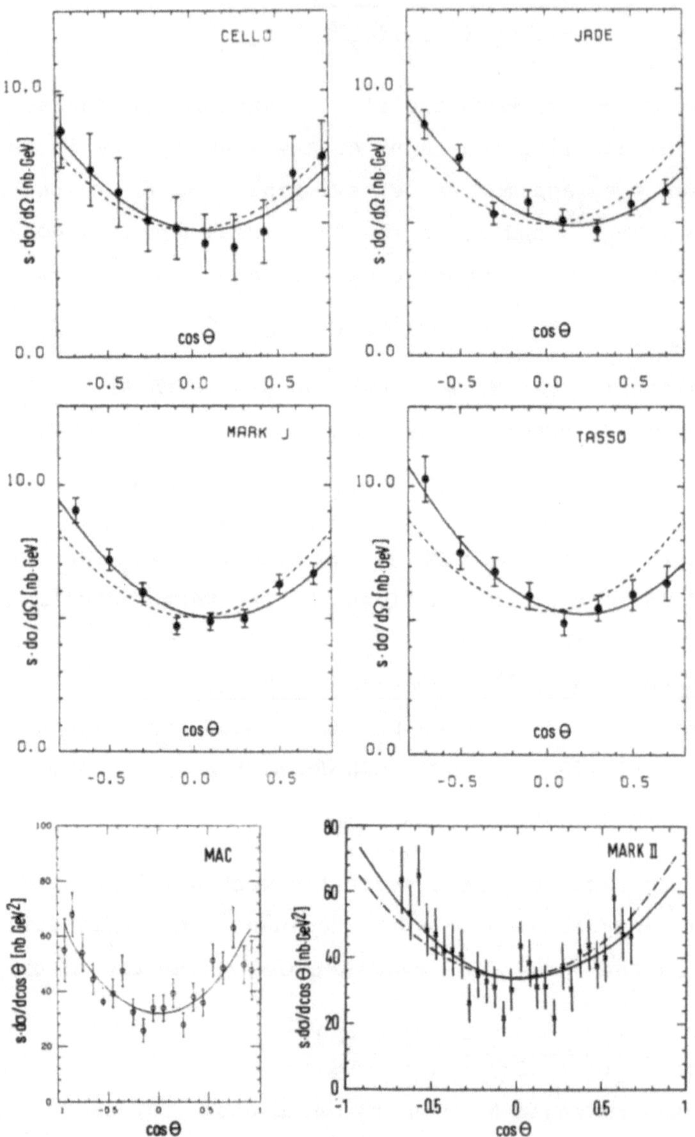

Fig. 26 Differential cross section for $e^-e^+ \rightarrow \mu^-\mu^+$ at cm energies W of 34 GeV from CELLO, JADE, MARK-J, TASSO, and at 29 GeV from MAC and MARK-II. The curves show the QED predictions (dashed) and the QED plus fitted weak neutral current contributions (solid lines), respectively.

$$A_{ff} = \frac{3}{2}\left[g_A^e\, g_A^f\, \mathrm{Re}\chi + 2\, g_V^e\, g_A^e\, g_V^f\, g_A^f |\chi|^2\right] \tag{31}$$

The asymmetry is due to parity mixing between the one photon vector and the Z^o axial vector state; it is not a parity violating effect.

Measurements of the angular distribution for $\mu\bar{\mu}$ production[4,21,22] are shown in Fig. 26. Similar measurements have been made for $\tau\bar{\tau}$ production. The possible systematic errors of the angular asymmetries have been carefully studied and were found to be $\leq$ 1 %. The radiative corrections which are as always applied directly to the data, produce asymmetry effects of purely electromagnetic origin due to C parity mixing of diagrams with even and odd numbers of photons; they contribute an asymmetry of about + 1.5 %. The angular coverage of the experiments for identifying the muon and tau pairs and for measuring their charge is typically about 80 %. In order to be able to compare results from different experiments with each other and with theory a small correction for the incomplete acceptance is usually made, using equ. (30). Asymmetries found in the various experiments are listed in Table 2. Averaged values for different energies are summarized and compared with the prediction from the standard SU(2) x U(1) model in Table 3. The predictions for A_{ff} from eqns. (27) and (31) are almost independent of the value of $\sin^2\theta_W$. The results support the standard model.

Independently of the standard model the data[22] yield with eq. (31)

$$g_A^e\, g_A^\mu = 1.11 \pm 0.13, \quad g_A^e\, g_A^\tau = 0.84 \pm 0.24 \tag{32}$$

Taking $|g_A^e| = 1.09 \pm 0.11$ from νe scattering one gets

$$|g_A^\mu| = 1.02 \pm 0.15, \; |g_A^\tau| = 0.77 \pm 0.25 \tag{33}$$

A combined analysis of the data on $\sigma_{tot}(e\bar{e} \to \ell\bar{\ell})$ and the angular asymmetries has also been made, using in addition to $\mu\bar{\mu}$ and $\tau\bar{\tau}$ production also Bhabha scattering ($e\bar{e} \to e\bar{e}$) data and assuming universality of the lepton couplings ($g_A^e=g_A^\mu=g_A^\tau$, $g_V^e=g_V^\mu=g_V^\tau$).

Table 2 Measured angular asymmetries of the reactions ee $\rightarrow \mu\bar{\mu}$, $\tau\bar{\tau}$, corrected for radiative and (partly) acceptance effects (Refs. 4, 21,22)

W(GeV)	$A_{\mu\mu}$ (%)	$A_{\tau\tau}$ (%)	Experiment	Accelerator
29	- 4.4 ± 2.4	-1.3 ± 2.4	MAC	PEP
29	- 9.6 ± 4.5	-3.5 ± 5.0	MARK-II	PEP
34	- 6.4 ± 6.4	-10.3 ± 5.2	CELLO	PETRA
35	-10.8 ± 2.2	-7.9 ± 3.9	JADE	PETRA
35	-10.4 ± 2.1	-7.4 ± 4.6 *	MARK-J	PETRA
35	-10.4 ± 2.3	-5.6 ± 4.4 *	TASSO	PETRA

* $|\cos\Theta| < 0.8$

Table 3 Angular asymmetries of the reactions ee $\rightarrow \mu\bar{\mu}$, $\tau\bar{\tau}$ averaged over experiments at different cm energies W, corrected for radiative and acceptance effects (Refs. 4,21,22)

W(GeV)	$A_{\mu\mu}$ (%)	$A_{\tau\tau}$ (%)	prediction from SU(2) x U(1)	Accelerator
14	+ 4.7 ± 3.2		-1.3	PETRA
22	- 8.1 ± 3.9		-3.6	PETRA
29	- 5.6 ± 2.1	-1.7 ± 2.2	-6.3	PEP
35	-10.4 ± 1.3	-7.9 ± 2.3	-9.3	PETRA

The result is[22]

$$g_V^2 = -0.06 \pm 0.13, \; g_A^2 = 1.13 \pm 0.14 \qquad (34)$$

The value of g_V^2 is equivalent (eq. 27) to $\sin^2\theta_W = 0.25 \pm 0.06$. Note that these results for g_V^2 and $\sin^2\Theta_W$ may have a systematic uncertainty exceeding the statistical error given above because g_V^2 is directly correlated with the absolute normalization of $\sigma_{tot}(e\bar{e} \to \ell\bar{\ell})$ (see eq. (25)). Fig. 27 shows results on g_A^2 and g_V^2 from different e^-e^+ experiments and compares them with results from νe scattering[22]. The neutrino results have a twofold ambiguity corresponding to $g_V^2 \leftrightarrow g_A^2$; one of the two ambiguous solutions agrees with the e^-e^+ results. It is the solution of the standard model. The data in Fig. 27 span a region of space-like momentum transfers of a few 10^{-2} GeV^2 for the neutrino data, to timelike values of more than 10^3 GeV^2 for the e^-e^+ annihilation data.

Up to this point in the discussion we have assumed a Z^o mass of $m_Z = 90$ GeV. From eqs (26) and (31) one finds that at $W = 35$ GeV the difference between the expected angular asymmetries for $m_Z = \infty$ and for $m_Z = 90$ GeV is about 15 %. This difference is of the same size as the present statistical uncertainty (see Table 3). Assuming $g_A^2 = 1$ (eq.27) the measured asymmetries require

$$m_Z = 77 \begin{array}{l} +45 \;\; (+\infty) \\ -13 \;\; (-20) \end{array} \text{GeV} \qquad (35)$$

where the values in parentheses give the two-standard deviation (95% confidence level) values. Thus, the data give a lower limit $m_Z > 57$ GeV. A smaller m_Z would have caused a significantly larger asymmetry than observed. The difference between $m_Z = \infty$, ie. a four-fermion (current current) interaction of the Fermi type, and a finite mass Z^o propagator is expected to become significant when PETRA reaches higher energies in the future.

Electroweak interaction of quarks

The weak neutral current also contributes to the total cross section for $e^+e^- \to q\bar{q} \to$ hadrons or, equivalently to the ratio

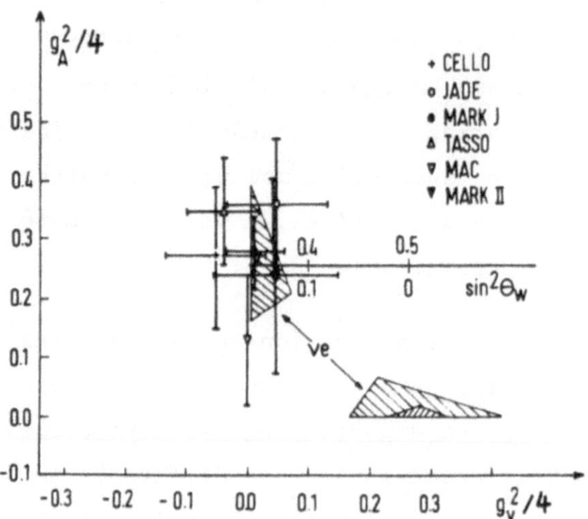

Fig. 27 Weak neutral current coupling constants g_A^2 vs. g_V^2. The shaded areas show the 1σ and 2σ regions determined in νe scattering experiments.

R (eq. 18). The contribution of each quark flavor and color to R and to the angular asymmetry A_{qq} is given by equs. (25), (30) and (31) with the replacement

$$\alpha \rightarrow \alpha e_e e_q = - \alpha e_q \tag{36}$$

where $e_q = 2/3, -1/3$ is the electric charge of the quark. The standard model couplings (eq. 27) are

$$\begin{aligned} g_V^u &= 1 - \frac{8}{3}\sin^2\theta_W, \quad g_A^u = +1 \\ g_V^d &= -1 + \frac{4}{3}\sin^2\theta_W, \quad g_A^d = -1 \end{aligned} \tag{37}$$

with analogous relations for the heavier generations. To these contributions the gluonic corrections from QCD have to be added.

Measurements of R were already shown in Figs. 15 and 19. R is sensitive mainly to the weak vector couplings g_V. In Fig. 19 the expected weak effects for different values of $\sin^2\theta_W$ are indicated, together with the QCD correction. The weak contribution is rising towards the Z^o pole. Sufficiently above $b\bar{b}$ threshold at W = 10 GeV the data can in principle be used to determine both α_s and $\sin^2\theta_W$ however, the energy range and accuracy reached until now are obviously not yet sufficient to independently measure both quantities. From the JADE data shown in Fig. 19 values of $\alpha_s = 0.14 \pm 0.08$ and $\sin^2\theta_W = 0.24 \pm 0.05$ are obtained, the errors neglecting correlations.[12] The main experimental problem is the overall normalization uncertainty of 3 to 6 %. It can be stated nevertheless that in these data, to which the c and b quarks contribute with almost one-half total weight, the weak vector couplings are consistent in strength with those found in lepton scattering which is dominated by light quarks, and which is characterized by a value of $\sin^2\theta_W = 0.23$.

First attempts to probe the axial couplings of the heavy quarks by measuring the forward-backward angular asymmetry A_{qq} in the reaction $e\bar{e} \to q\bar{q} \to$ 2jets have also been made. One method is to select a sample of events enriched with heavy quarks, by taking hadronic events with prompt leptons of relatively large transverse momentum relative to the jet axis. These leptons are likely to come from a weak decay with large Q value. The sign of the charge of the lepton serves to distinguish q from $\bar{q}$. Unfortunately, according to eqs. (30), (31), (36) with standard model couplings (37) the asymmetries in $c\bar{c}$ and $b\bar{b}$ production have identical sign while the leptons from the decays are oppositely charged, leading to a partial cancellation depending on the relative amounts of $c\bar{c}$ and $b\bar{b}$ in the selected sample. The asymmetry effects therefore become small and model dependent. They have nevertheless been observed in two experiments [23,24] with a significance of ~ 2 standard deviations each; the results are in agreement with the standard model couplings.

Another possibility consists in identifying c quarks by their fragmentation into D^{*+}. The $D^{*+} \to D^o\pi^+ \to K^-\pi^+\pi^+$ decay chain can be observed by virtue of a partial cancellation of mass resolution effects in the invariant masses of the $D^o\pi^+$ and the D^o systems[25]. A preliminary analysis[24] gave an asymmetry for $e\bar{e} \to c\bar{c}$ of A_{cc} $= -0.35 \pm 0.14$ at W = 35 GeV, to be compared with a standard model prediction of -0.14. Better statistics and the use of precision vertex detectors for direct observation of the weak decays shall allow a more precise determination of the quark weak couplings in the future.

Restrictions on a more general weak interaction scenario

We have seen that all the results on $ee \to \ell\bar{\ell}$ are in good agreement with the standard model. It is natural to ask to which extent the data force any alternative model to be "close" to the standard model. It is well known[26] that all the experimental data on weak interactions for $Q^2 << m_Z^2$ can be described by a local four-fermion phenomenological Lagrangian of the type

$$L_{eff} = \frac{4G_F}{\sqrt{2}} (J_\mu^{C^+} J_\mu^C + J_\mu^N J_\mu^N) - \frac{1}{2} \frac{e^2}{Q^2} J_\mu^{em} J_\mu^{em} + C \frac{4G_F}{\sqrt{2}} J_\mu^{em} J_\mu^{em} \quad (38)$$

where $\quad J_\mu^C = J_\mu^1 + iJ_\mu^2, \qquad J_\mu^N = J_\mu^3 - \sin^2\theta_W \, J_\mu^{em}$

with weak isospin current

$$\vec{J}_\mu = \sum_i \bar{\Psi}_L^i \gamma_\mu \frac{\vec{\tau}}{2} \Psi_L^i$$

and the usual left-handed fermion doublets

$$\Psi_L = \begin{pmatrix} \nu \\ e^- \end{pmatrix}_L, \begin{pmatrix} u \\ d' \end{pmatrix}_L, \ldots$$

J_μ^C, J_μ^N and J_μ^{em} are the weak charged and neutral and the electromagnetic currents, respectively. The first two terms of L_{eff} are the $Q^2 \to 0$ limit of the standard model. The last term, proportional to the constant C, is an additional parity-conserving term in the weak neutral-current Lagrangian which has the structure of the electromagnetic interaction. C must vanish if a single weak neutral boson (Z^o) dominates the low-Q^2 data. The low-energy phenomenology can be reproduced in more general frameworks, having e.g. more than one Z^o with masses of comparable size, or having a continuum of intermediate boson states as in models with composite fermions and bosons. Such schemes will in general give $C > 0$.

The C term in (38), being proportional to $(J_\mu^{em})^2$, is unobservable in neutrino scattering, and due to its parity-conserving nature it will be masked in charged lepton scattering by the much stronger electromagnetic contribution. In e^-e^+ interactions it will lead to a deviation of the lepton pair production cross sections $\sigma(e^-e^+ \to \ell^-\ell^+)$ from the QED predictions because of the additional electroweak interference effect. In particular, a lower bound for the QED cutoff parameter Λ_- can be converted[27] to an upper bound on C:

$$C < \frac{\pi\alpha}{\sqrt{2}G_F\Lambda_-^2} - \left(\frac{1}{4} - \sin^2\theta_W\right)^2$$

$$= \left(\frac{37.3 \text{ GeV}}{\Lambda_-}\right)^2 - \left(\frac{1}{4} - \sin^2\theta_W\right)^2 \tag{39}$$

Similarly, the C term will affect the total cross section for $e^-e^+ \to q\bar{q} \to$ hadrons or the ratio R. The quantity C can be interpreted[27] in terms of the relative deviation Δ of the total e^-e^+ cross section from the prediction of the standard theory (i.e. a theory with one Z^o), weighted by 1/s:

$$\Delta \equiv \frac{\int \frac{ds}{s}\, \sigma(e^-e^+ \to \text{all}) - \int \frac{ds}{s}\, \sigma(e^-e^+ \to \gamma,\, Z^o \to \text{all})}{\int \frac{ds}{s}\, \sigma(e^-e^+ \to \gamma,\, Z^o \to \text{all})} = 16\ C \tag{40}$$

The experimental 95 % confidence limit on C from the PETRA experiments [4,22] is

$$C < 0.015 \tag{41}$$

This has to be compared with the coefficient of $(J_\mu^{em})^2$ in the weak neutral current part of the standard theory, which according to (38) is $\sin^4\theta_W \approx 0.05$. Also note that from (40), $\Delta < 0.24$. Thus, the limit is already significant; any deviation from the standard model must be relatively small. In specific non-standard weak interaction models, as e.g. in those assuming several Z^o bosons[28], the limit on C is equivalent to a restriction on the masses of these bosons. If several weak neutral bosons contribute their masses cannot differ much from the mass of the standard Z^o.

Limits on pair production of scalar particles

If pointlike spin zero charged particles exist, they should be pair-produced through the one photon annihilation channel with a differential cross section as given by equs.(7) and (8). The

with different cross section as given by equs. (7) and (8). The angular distribution conveniently peaks at right angles to the beam. At energies high above threshold ($\beta \to 1$) the total production cross section slowly approaches a value of 1/4 times the muon pair production cross section. Possible scalars include charged Higgs particles or technipions, as well as the spin zero leptons and quarks required by supersymmetry.

Higgs particles and technipions (both denoted by H here) preferentially decay into the heaviest kinematically allowed lepton and quark pairs, e.g.

$$H^- \to \tau^- \bar{\nu}_\tau, \ \bar{c}\, s \qquad (42)$$

where the quarks subsequently fragment into hadrons. The Higgs mass is unknown. Technipions are expected to have masses in the range from about 5 to 14 GeV.

To find charged Higgs or technipion pairs one searches for unusual sources of τ leptons, in particular for $\tau\bar{\tau}$ pairs that are acollinear and/or non-coplanar with the beam, indicating a large missing momentum. Another signature is a τ recoiling against a jet of hadrons. Should the leptonic branching ratio happen to be very small then the signature would be a hadronic event with four (partly embryonic) jets that can be grouped into two pairs of equal effective mass.

The decay modes containing τ leptons,

$$e\bar{e} \to H^- H^+ \to (\tau\bar{\nu})(\bar{\tau}\nu), \ (\tau\bar{\nu}) + \text{hadrons} \qquad (43)$$

have been investigated in several experiments[29]. No candidates were found. The significance of these null results was assessed by Monte Carlo simulations of the production and decay process (43) in the detectors, applying the same selection criteria as for the real data. Thereby, 95 % confidence upper limits on the branching ratio BR($H \to \tau\nu$) and the product BR($H \to \tau\nu$) $\times$ BR($H \to$ hadrons) were obtained as shown by the shaded regions in Fig. 28. One

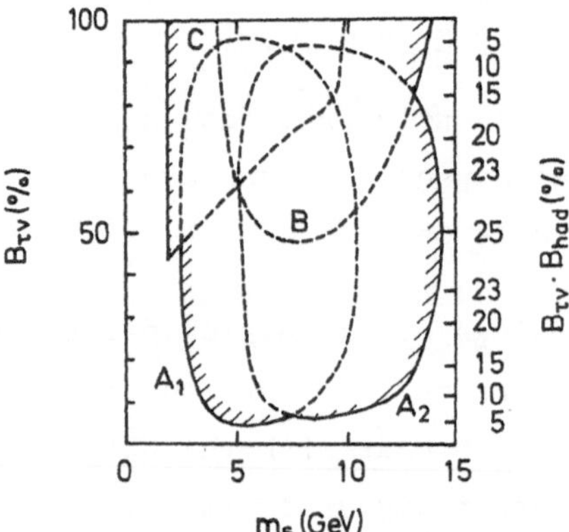

Fig. 28 The regions inside the contours show the domains of mass and branching ratios for charged Higgs or technipion particles which have been excluded by searches for

$e^-e^+ \rightarrow H^-H^+ \rightarrow (\tau\bar{\nu})\ (\bar{\tau}\nu)$: curves B,C

$\rightarrow (\tau\bar{\nu}) +$ hadrons: curves A_1, A_2

concludes that charge ±1 pointlike scalars $H^{\pm}$ with a branching fraction BR($H \to \tau\nu$) $\gtrsim$ 10% and masses in the range $5 < m_H < 13$ GeV are ruled out.

The scalar partners ℓ_o of the ordinary charged leptons, which are predicted in supersymmetry, would also have to be pair-produced with a cross section given by eq. (8); the rate for $e_o\bar{e}_o$ production would be further enhanced by t-channel photino or goldstino exchange. These scalars would decay into an ordinary lepton and a photino or goldstino. Since the photinos or goldstinos escape unobserved one would find a pair of leptons carrying on the average $\frac{1}{2}$ of the total energy W and having large non-coplanarity with the beam direction. Cuts have to be made to reduce background from photon-photon interactions and from bremsstrahlung in ordinary lepton pair production processes. Searches for such events at PEP and PETRA were unsuccessful[29]. The existence of e_o and μ_o of masses below about 16 GeV, and of τ_o below about 15 GeV, can therefore be excluded.

Besides looking for scalars, intensive searches for other new particles were also made at PEP and PETRA. The situation can be summarized as follows[30]: No new quarks, confined or free, with charge ±2/3 or ±1/3 have been detected up to W = 36.7 GeV. Neither have new sequential leptons, stable charged leptons, heavy unstable neutral leptons decaying into electrons, or excited states of the ordinary electron or muon been found. The existence of most of these particles is excluded for masses smaller than about 15 GeV. This value reflects the energy of about 35 GeV at which PETRA has been running.

JETS AND QCD TESTS

General properties of three-jet events

We have already discussed the dominant two jet character of the hadron production process

$$e^- e^+ \rightarrow \text{hadrons} \qquad (44)$$

at high energies and the evidence for an underlying electromagnetic quark pair production mechanism. We saw in particular that the cross section for (44), or equivalently the ratio R of hadron to muon production (equ. (18)), is in quantitative agreement with the quark parton model prediction for pointlike spin $\frac{1}{2}$ quarks including a small correction due to gluonic radiative corrections as predicted in quantum chromodynamics (QCD).

We now discuss in more detail those roughly 10 % of the hadron production events observed at cm energies W of about 25 GeV and above which show three jets. An event of this type is seen in Fig. 20. When studying such events one selects a kinematic region in which each of the jets has an energy in the cm of at least, say, 6 GeV such that they are easily recognized and comparable to the jets in two-jet events at $W \gtrsim 12$ GeV. It is then not difficult to convince oneself that three separate jets indeed exist, and that they tend to lie in (or nearly in) a plane. This has been demonstrated clearly in the PETRA and PEP experiments[31]. But what is the evidence that one of these jets comes from a gluon? Could it be a meson, a superposition of resonances, the product of a weak decay, or just a statistical fluctuation of the dominant two-jet configuration?

One of these possibilities can immediately be excluded, namely that the apparent jet arises from the decay of a single meson. This is because the three jets, apart from at most quite subtle differences, look very much alike. This is true for such gross

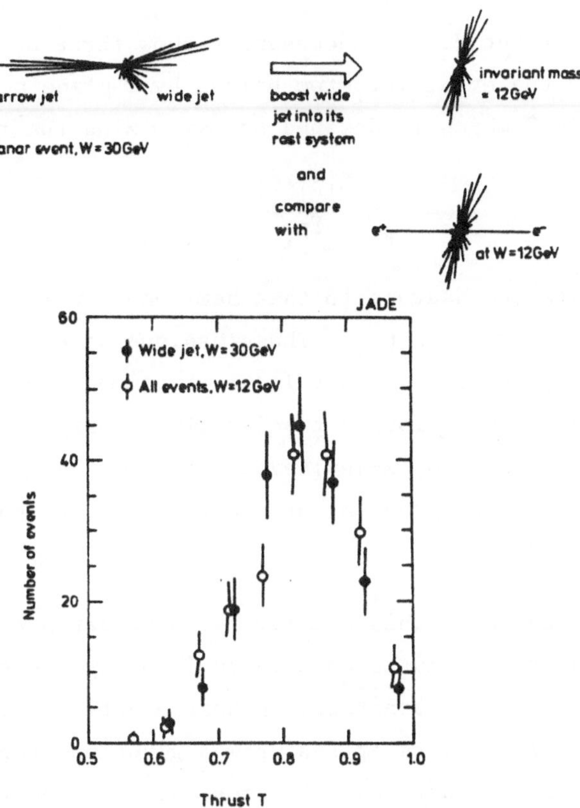

Fig. 29 On the top, a procedure of studying the structure of the wide jet in planar events from the reaction $e^+e^- \to$ hadrons is shown. The lower part shows the comparison of event shapes measured by thrust (T = 1/2 for spherical, T = 1 for collinear states), i) for the wide jet (transformed to its rest system) of planar events at 30 GeV, and ii) for complete events at 12 GeV (nearly all being 2-jet events). From ref. 32.

properties as the charged multiplicity $\langle n_{ch} \rangle$ at given energy E_{jet} of a jet, for the energy fraction E_{neut}/E_{jet} carried by neutrals, for the average transverse momentum $\langle p_T \rangle$ about the jet axis, and for the distribution of fractional longitudinal momentum x. As an example, Fig. 29 shows the result of a study by the JADE group[32] on a sample of $e^+e^- \to$ hadron events at W = 30 GeV which are planar but do not necessarily show three well resolved jets. One cuts the event into two halves by a plane perpendicular to the thrust axis $\hat{t}$ which is defined by maximizing the thrust

$$T = \operatorname*{Max}_{\hat{t}} \frac{\Sigma|\vec{p}_i \cdot \hat{t}|}{\Sigma p_i} \tag{45}$$

One then selects the hadrons in that hemisphere which has the larger transverse momentum sum $\Sigma\ p_{Ti}$. These hadrons have a total invariant mass of typically $\sim$12 GeV. In their commom center-of-mass frame they are found to form a two-jet like configuration that looks very similar in shape to (2-jet) events $e^+e^- \rightarrow$ hadrons at $W = 12$ GeV, as shown by the close agreement of the thrust distributions for the two cases (Fig. 29).

As a consequence of this similarity in appearance of the three jets one has up to now not been able to identify the gluon among the three jets by its characteristic hadronization properties. To prove that one of the jets arises from a gluon one therefore must rely on production properties like cross section, energy distributions, and angular correlations of the jets. All these properties are predicted[33] in perturbative QCD. In lowest order the predictions for gluon radiation are identical to those for electromagnetic radiation except that α is replaced by $4\alpha_s/3$.

For comparison with experiment the QCD predictions for partons have to be convoluted with hadronization distributions for which Monte Carlo models are used[34]. Different jet measures of course will have different degrees of model dependence. Thrust is a "bad" (strongly hadronization-dependent) measure while various angular correlations of jet axes, the opposite to same-side asymmetry of energy-weighted angular correlations between the hadrons, and the sphericity tensor are "good".

Energy flow in $e^+e^- \rightarrow$ hadrons

This section concerns one of the simplest and most straightforward measurements of the event shape[35]. Here, the energy of charged and neutral particles from reaction (44) is measured in a calorimeter surrounding the interaction region. Let the unit vector $\hat{k}_i$ describe the direction of the ith calorimeter element, and let E_i be the energy measured by it in a given event; we write $\vec{E}_i = E_i \hat{k}_i$. The thrust axis $\hat{t}$ is then given by maximizing the thrust

$$T = \underset{\hat{t}}{\mathrm{Max}} \frac{\sum_i |\vec{E}_i \cdot \hat{t}|}{\sum_i E_i} = \frac{\Sigma|\vec{E}_{i\parallel}|}{E_{tot}} \tag{46}$$

The event plane is the plane normal to which the total energy flow $\Sigma|\vec{E}_{Tout}|$ is a minimum, and the total energy flow in the event plane but perpendicular to $\hat{t}$ is denoted by $\Sigma|\vec{E}_{Tin}|$. Planar events at W ≈ 35 GeV are selected by requiring the "oblateness" of the events defined by

$$O = \frac{\Sigma|\vec{E}_{Tin}| - \Sigma|\vec{E}_{Tout}|}{E_{tot}} \tag{47}$$

to be larger than 0.3. The azimuthal energy flow in the event plane, superimposing many events all with the thrust axis $\hat{t}$ pointing to the left and the direction of maximum energy flow perpendicular to $\hat{t}$ pointing upwards, is shown in Fig. 30. The three-lobe pattern seen is partly an artifact of the construction of the energy flow diagram and does by itself not prove the existence of three distinct jets. Two-jet events or phase-space like events show a similar pattern (dashed and dashed-dotted curves). But it clearly appears that a QCD calculation for gluon bremsstrahlung in $o(\alpha_s)$ with $\alpha_s = 0.18$, using the leading order diagrams of Fig. 18 convoluted with a Monte Carlo simulation of quark and gluon hadronization[34], describes the data better (solid

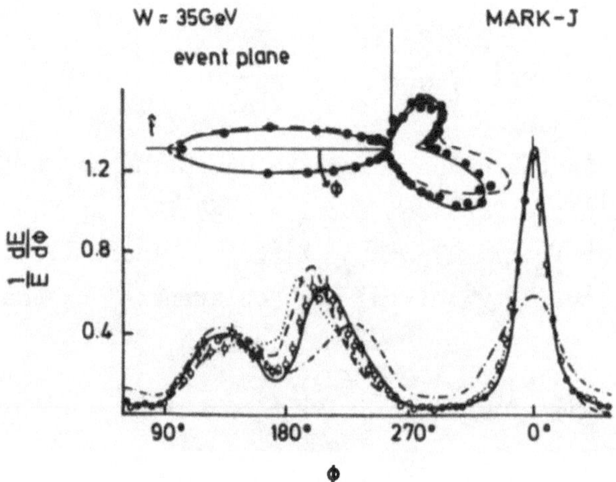

Fig. 30 Energy flow observed for planar events of the reaction $e^+e^- \to$ hadrons, as a function of the azimuthal angle in the event plane relative to the thrust axis $\hat{t}$, from ref. 35. The curves show fits by i) a QCD calculation (full curve) with $\alpha_s = 0.18$, ii) a two-jet $q\bar{q}$ model with a Gaussian (dashed) or exponential (dotted) p_T distribution, iii) a pure phase space distribution (dashed-dotted curve).

curve) than either the phase space distribution or a $q\bar{q}$ two-jet model without gluons, regardless of whether the hadronization is assumed to produce a gaussian or an exponential p_T distribution in the latter model. This conclusion does not rely on any assumptions about the mean p_T or the fragmentation functions in the $q\bar{q}$ models, since these parameters were independently fixed using the measured values of $\langle T \rangle$ and $\langle O \rangle$.

Energy-energy correlation in e^+e^- annihilation

A weakness of the energy flow analysis is that the dynamic effects from gluon emission show up only as relatively minor variations to the built-in three lobe pattern. The significance of the conclusions from this analysis relies strongly on the Monte Carlo simulation of the effects of hadronization. One has tried to improve on this situation by analysis methods which will be discussed in the following.

One of these methods uses the energy-energy correlation

$$\frac{1}{\sigma}\frac{d\Sigma_E(\Theta)}{d\Theta} = \lim_{\Delta\Theta\to 0}\frac{1}{\Delta\Theta}\sum_{j,k} x_j x_k \tag{48}$$

where

$$x_j = \frac{E_j}{E_{beam}} = \frac{E_j}{W/2} \tag{49}$$

is the fractional energy carried by the jth final state particle, and the sum runs over all those pairs j,k of an event whose momentum vectors span an angle Θ between them, falling into a given Θ bin of width $\Delta\Theta$. The result is then averaged over all events. Neglecting hadronization, i.e. considering q, $\bar{q}$ and g as the final state particles, this correlation has been calculated in perturbative QCD[36]. The result to $o(\alpha_s)$ is expected to be reliable for large angles Θ, say for $30^\circ \lesssim \Theta \lesssim 150^\circ$, such that regions of collinear multigluon emission are avoided. In this angular region the perturbative contribution will be

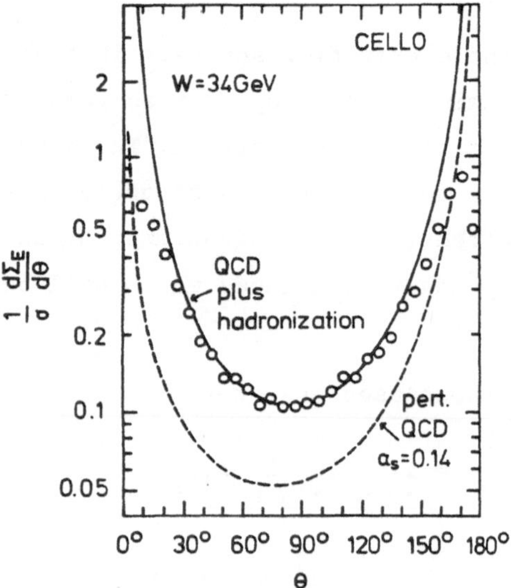

Fig. 31 Angular dependence of the energy-energy correlation function in the reaction $e^+e^- \to$ hadrons, corrected for detector acceptance (ref. 37). The curves show the parton-level $o(\alpha_s)$ QCD prediction (dashed) and the effect of adding a hadronization contribution (solid curve).

$$\frac{1}{\sigma}\frac{d\Sigma_E^{pert}(\Theta)}{d\Theta} \propto \alpha_s(W^2) \propto \frac{1}{\ln(W^2/\Lambda^2)} \qquad (50)$$

For comparison with experimental results the perturbative calculation has to be convoluted with the non-perturbative hadronization effects. A simplified method to deal with hadronization[36] consists in approximating its effect by an additive contribution

$$\frac{1}{\sigma}\frac{d\Sigma_E^{NP}(\Theta)}{d\Theta} \propto \frac{1}{W} \qquad (51)$$

which will vanish more strongly with increasing W than the perturbative contribution; however, in the W = 30 GeV region the two terms are still of similar magnitude. This is seen in Fig. 31 where the energy-energy correlation from the CELLO experiment is shown[37]. The agreement with the prediction adding (50) and (51) is good. The significance of this agreement is, however, difficult to judge due to the large size of the uncalculable non-perturbative contribution.

It is obviously preferable to consider quantities that are less affected by non-perturbative effects. In the simplest approximation the additive hadronization term (51) is expected to be symmetric under $\Theta \leftrightarrow \pi-\Theta$. On the other hand, the perturbative term (50) is asymmetric; this is easily understood from the fact that to inter-parton angles close to 180° both collinear and soft gluon emission contributes, while at $\theta=0$ only collinear emission can contribute. The forward-backward asymmetry

$$A(\Theta) = \frac{1}{\sigma}\left[\frac{d\Sigma_E}{d\Theta}(\pi-\Theta) - \frac{d\Sigma_E}{d\Theta}(\Theta)\right] \qquad (52)$$

of the energy-energy correlation function is therefore expected to be largely free from the non-perturbative effects. Results[37,38] on $A(\Theta)$ are shown in Fig. 32 and compared with an $o(\alpha_s)$ QCD prediction for $\alpha_s = 0.18$. One is tempted to conclude that the existence of this forward-backward asymmetry is a genuine manifestation of the presence of the perturbative QCD contribution in the energy-energy

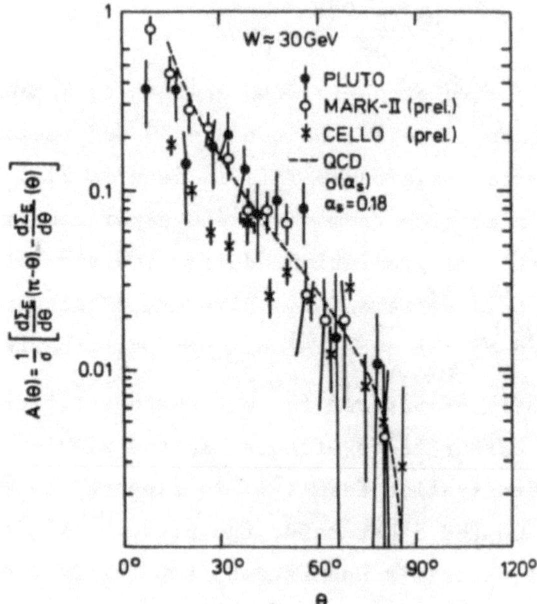

Fig. 32 Forward-backward asymmetry of the energy-energy correlation function in the reaction $e^+e^- \rightarrow$ hadrons, from refs. 37, 38. The curve indicates the QCD prediction to $o(\alpha_s)$ with $\alpha_s = 0.18$; precise predictions (ref. 38) involve small experiment-dependent corrections.

correlation function, and therefore of hard non-collinear gluon radiation. Detailed simulations of the hadronization and detector effects, using QCD calculations incorporating quark and gluon fragmentation in a Monte Carlo model, support this conclusion.

Jet angular correlations

The jet observables discussed so far depend heavily on measurements of particle energies, which in most experiments have relatively large uncertainties. Angles can in most storage ring detectors be measured much more precisely. Therefore, an approach more or less orthogonal to the ones described consists in using the particle energies only to divide the final state into "clusters" which can be associated with jets; once this is done, one uses only the directions of the jet axes.

To identify the jets as clusters of the particle momenta many different criteria and algorithms have been proposed and applied to the data[39]. Most of these methods have been found to reconstruct directions of jet axes to very good accuracy provided the energies and angular separations of the individual jets are not too small.

Let $x_j = E_j/E_{beam}$ $(j = 1,2,3)$ be the fractional energies of the three jets ordered such that

$$0 \leq x_3 \leq x_2 \leq x_1 \leq 1, \quad x_1 + x_2 + x_3 = 2 \qquad (53)$$

In the two-jet limit $x_1 = x_2 = 1$, $x_3 = 0$. The quantity x_1 is equal to the thrust of the event (eq. 45), calculated for the partons. Selecting events with $x_1 < 0.9$ limits x_3 to the range $0.2 < x_3 < 2/3$ which means that the lowest energy jet has, for $W = 35$ GeV, a cm energy between 3.5 and 12 GeV. Because of $M^2_{jet} << W^2$ the selection further implies

$$M_{k\ell} > 11 \text{ GeV}, \quad \Theta_{k\ell} > 70^o$$

for the invariant mass of any pair of jets and the angular separation between them. This angular separation is larger than the angular width of the jets.

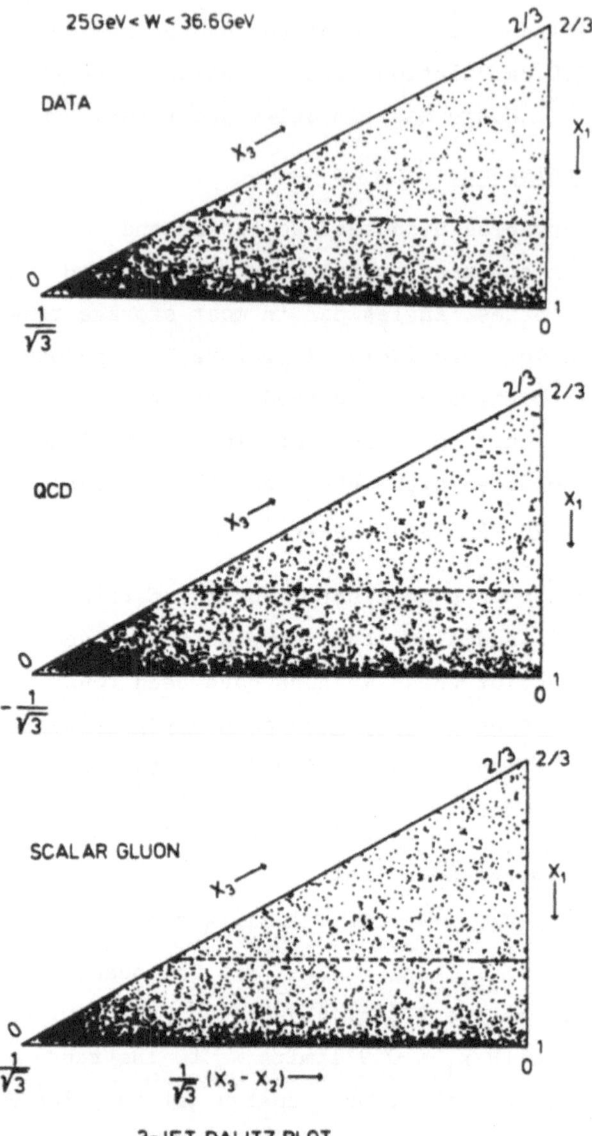

Fig. 33 Dalitz plot of the three reconstructed jet energies for the reaction $e^+e^- \to$ hadrons at $\langle W \rangle = 33$ GeV, from ref. 44 (top plot). For comparison, the QCD prediction to $o(\alpha_s)$ with $\alpha_s = 0.17$ and the prediction for scalar gluon radiation are shown using Monte Carlo simulations with the same number of events (middle and bottom plots). The scalar coupling is given by $\bar{\alpha}_s = 1.6$, choosen to optimize overall agreement with a number of event variables.

Instead of the angles $\Theta_{k\ell}$ between the jets one may equally well use, in the approximation $M^2_{jet} << W^2$, the fractional jet energies

$$x_j = \frac{E_j}{W/2} = \frac{2\sin\Theta_{k\ell}}{\sin\Theta_{12} + \sin\Theta_{23} + \sin\Theta_{31}} \tag{54}$$

calculated from the measured angles. The correlation between the $\Theta_{k\ell}$ can then equivalently be described by the Dalitz plot of the three x_j. This approach to obtain the x_j through jet angular measurements leads to values that agree closely with the fractional energies of the parent partons. This has been ascertained with Monte Carlo-generated events for various different cluster finding procedures and different detectors. The resolution in x_1 is typically 6 % and the systematic bias is found to be very small, of the order of 1 %.

A jet Dalitz plot determined in this manner from TASSO data at an average energy of W = 33 GeV is shown in Fig. 33. Such a plot contains the complete dynamical information on the gluon bremsstrahlung process $e^+e^- \to q\bar{q}g$ after integration over the beam and polarization directions. Only 1/6 of the full Dalitz plot needs to be shown because of the inability to identify the gluon among the three jets which suggested the ordering (53) of the x_j. Collinear two-jet configurations lie along the base line of the plot, symmetric three-star configurations at the top corner. The three-jet kinematic region discussed before is the sub-triangle above the dotted line at $x_1 = 0.9$; it contains approximately 1/10 of all events.

The QCD prediction in $o(\alpha_s)$ for this plot is also shown in Fig. 33, assuming $\alpha_s = 0.17$. It was produced with a Monte Carlo simulation[34] to describe hadronization, and was passed through the detector and the same analysis procedure as the data. It agrees perfectly with the data. We note that the QCD distribution varies slowly over the whole three-jet region, such that smearing

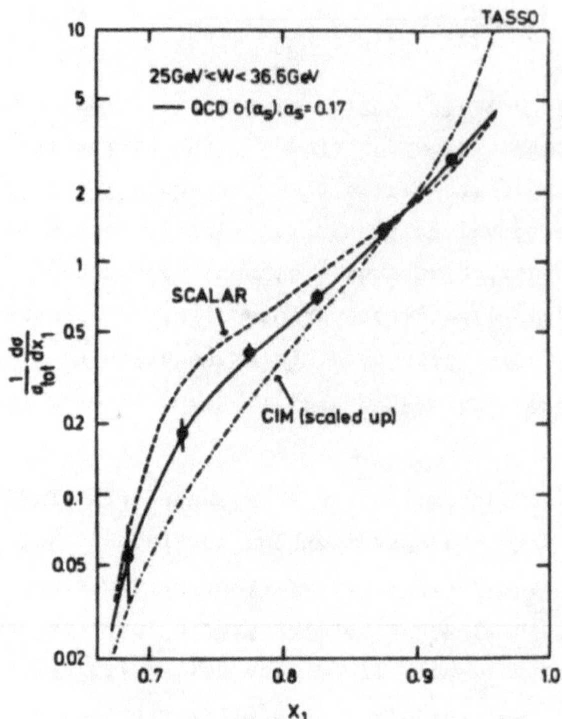

Fig. 34 Distribution of the maximum reconstructed jet energy x_1 (= parton thrust for three-jet events) in the reaction $e^+e^- \rightarrow$ hadrons at <W> = 33 GeV, from ref. 44. The full curve shows the QCD prediction to $o(\alpha_s)$ with α_s = 0.17, the dashed curve the prediction for scalar gluons (with the coupling $\bar{\alpha}_s$ = 1.6 optimized to furnish the best overall description of a number of event variables).The dashed-dotted curve is the prediction from the constituent interchange model with a scaled-up coupling constant such as to fit the p_T^2 distribution at the same energy (see Fig. 37 and the associated discussion).

effects from finite resolution are small. For comparison we also show the Dalitz plot expected if "gluons" were scalar[33]. It looks very similar to the other plots although a maximum likelihood analysis reveals significant differences between the scalar model and the data. The projection on parton thrust x_1, normalized to σ_{tot}, is shown in Fig. 34 and is compared there with the QCD and scalar gluon predictions. The agreement with QCD is quite impressive.

We now show results from a recent three-jet angular correlation analysis in the energy region $30 \leq W \leq 37$ GeV by JADE[41] in which comparison with the full $o(\alpha_s^2)$ QCD prediction,[40] ie. all the diagrams of Fig. 18, has been made. In this analysis, a "cluster" of final state hadrons with invariant mass below some maximum value given by $m_{jet}^2/W^2{}_{obs} \leq y^{max} = 0.04$ (where W_{obs} is the total observed energy of the final state) is called a jet. An analogous definition is used in the QCD calculation for parton jets[42]. The distribution of jet multiplicities obtained by such a cluster algorithm will depend on the jet defining parameter y^{max}. The chosen value is large enough to suppress sensitivity to fragmentation fluctuations yet small enough to resolve three-jet (and even four-jet) states from the two-jet dominated sample. The fractional jet energies $x_i = 2E_i^{jet}/W$ have, as discussed above, been calculated from the measured directions of the jet axes using massless jet kinematics. Three-jet events are described by 2 independent x_i; one may use $x_1 \equiv \mathrm{Max}\ x_i$ and $x_T \equiv x_2 \sin\Theta_{21} \equiv x_3 \sin\Theta_{31}$ (where Θ_{ij} is the angle between jet axes i and j). Measured distributions of x_1 and x_T are shown in Fig. 35. Second order QCD (shown by the full curves) describes the data well. Away from the two-jet region, e.g. for $x_1 < 0.9$, $x_T > 0.25$ the hadronization corrections are not larger than 20 %. The shapes of the distributions are therefore essentially parameter-free QCD predicitons. The rate of three-jet events fixes α_s;one obtains 0.16 ± 0.03 ($\overline{MS}$ scheme, $Q_{eff}^2 \approx 1200$ GeV^2)

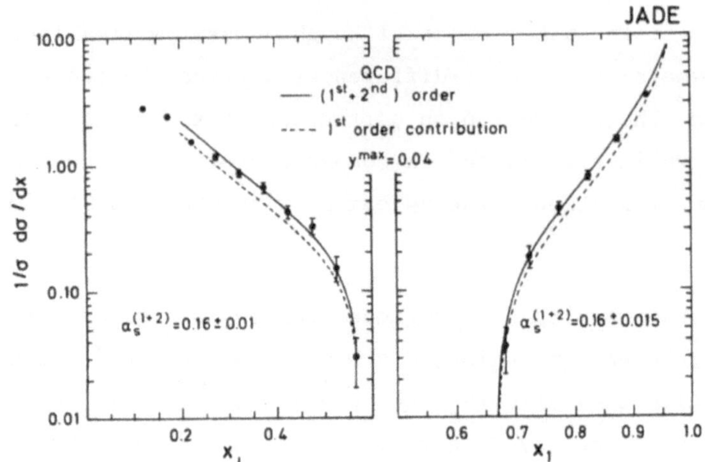

Fig. 35 Distributions of x_T and x_1 from JADE, corrected for detector effects, initial state bremsstrahlung, and hadronization. The full and dotted lines show the $o(\alpha_s^2)$ and $o(\alpha_s)$ QCD results, respectively.

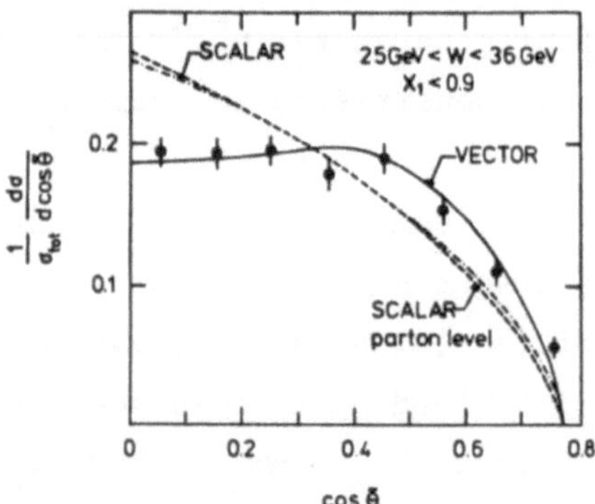

Fig. 36 Distribution of the cos of the Ellis-Karliner angle $\cos\tilde{\theta} = (x_2-x_3)/x_1$ for events in the three-jet region from TASSO. Full curve is QCD prediction with $\alpha_s = 0.17$ (vector gluons), while dashed curves show prediction for scalar gluons with/without hadronization effects. See refs. 45, 46

where the error includes systematic uncertainties.

Note that the $o(\alpha_s^2)$ effects in the distributions of Fig. 35 are small. To this order also four-jet events are expected, and some first evidence for them has been presented[43]. In contrast to QCD a QED-like Abelian vector theory[41] or a scalar gluon model[33] do not fit the data (Figs. 34 and 36).

Hadronization

It was indicated above that hadronization of quarks and gluons is a complicated process that cannot be treated by perturbative QCD. Instead, it has to be dealt with phenomenologically by Monte Carlo simulation[34]. The question therefore arises whether the apparent gluon jets can be "explained away" by assuming peculiar properties of quark hadronization, different from those usually put into the models.

When this is tried one finds that the frequency of occurrence of three-jet events is much larger than expected if these event shapes were the result of statistical fluctuations in the hadronization of $q\bar{q}$ events[31]. Even with extreme assumptions on quark hadronization (e.g. large exponential or power-law tails in the transverse momentum distributions of the hadrons relative to the direction of the quark,or peculiar fragmentation behaviour of c and b quarks), a successful description of the data in terms of a pure $e\bar{e} \to q\bar{q}$ process has not been found. The data clearly require a description in terms of three distinct jets[31,44]. A further problem for pure $q\bar{q}$ models is the strong increase of the average transverse momentum $<p_T>$ with cm energy W. This feature of the data also rules out the idea that the third jet resulted from the decay of a high mass meson emitted by the q or $\bar{q}$ in a large p_T (higher-twist or constituent interchange) process. This is apparent from Fig. 37 which shows that comparing such a process with the data one finds a discrepancy in W dependence of more than an order of magnitude over a W range from 12 to 33 GeV[44]. Evidently

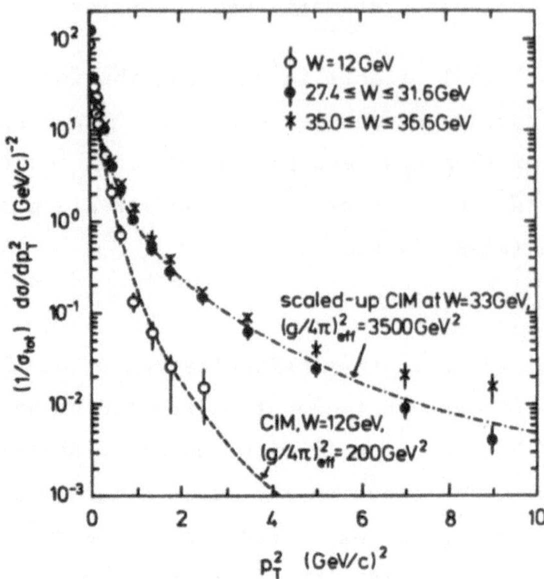

Fig. 37 Distribution of the square of the transverse momentum of charged particles with respect to the jet axis in the reaction $e^+e^- \rightarrow$ hadrons at different cm energies W (from TASSO, ref. 44). The dashed curve shows the prediction from the constituent interchange model (ref. 33) at W = 12 GeV; the dashed-dotted curve is for W = 33 GeV with a scaled-up $q\bar{q}M$ coupling so as to approximately fit the data.

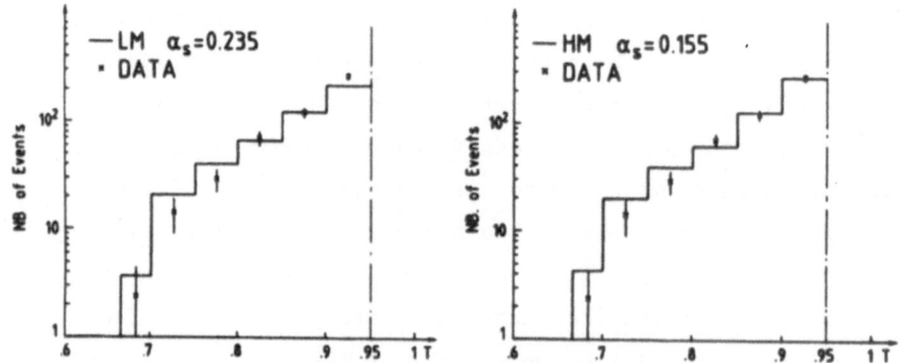

Fig. 38 Thrust x_1 of three-cluster events from CELLO, compared with QCD calculations assuming hadronization according to the Lund model (Ref. 47, left) or the Hoyer model (Ref. 34, right).

the W dependence of the p_T distribution indicates a pointlike process. Indeed QCD has provided the only successful description of the three-jet events, in terms of the emission of hard gluons by the quarks.

As discussed above direct identification of the gluon jet in a three-jet event has so far not been possible. It has however been reported that the jet of the lowest energy, which is the "best bet" for the gluon jet, has on average a somewhat larger hadron $\langle p_T \rangle$ with respect to the jet axis, than the jets from two-jet events at the same jet energy (i.e. at smaller W) have[12].

There is another interesting question concerning the way quarks and gluons fragment. The assumption jet direction = parent parton direction implies that each final state quark or gluon independently fragments in the overall cms. This is the physical basis of the Field-Feynman fragmentation model.[34] Alternatively in a color string model[47] each section of the string, extending from the q via the g to the $\bar{q}$, independently fragments in the rest frame of the separating colors. As these sections are boosted relative to the overall cms the jet directions, relative to which $\langle p_T^2 \rangle$ of the hadrons is minimum, will be slightly different from the parton directions. Indications for such a mechanism have been reported[48]. The two fragmentation schemes yield significantly different distributions of low momentum hadrons. While this does not affect the compatibility of the data with QCD it can cause significant uncertainty in the value of α_s deduced from the rate of events in the three-jet region[49] (Fig. 38).

Inclusive Hadron Distributions and Scaling Violation

Some inclusive particle yields for an average hadronic annihilation event at $W \gtrsim 35$ GeV are[50,51]

$10.2\ \pi^{\pm}$; $5.5\ \pi^{o}$; $1.8\ K^{\pm}$; $1.4\ K^{o},\bar{K}^{o}$; $0.8\ p,\bar{p}$; $0.3\ \Lambda,\bar{\Lambda}$;
$0.2\ \rho^{o}$; $\sim 0.8\ D^{*\pm}$

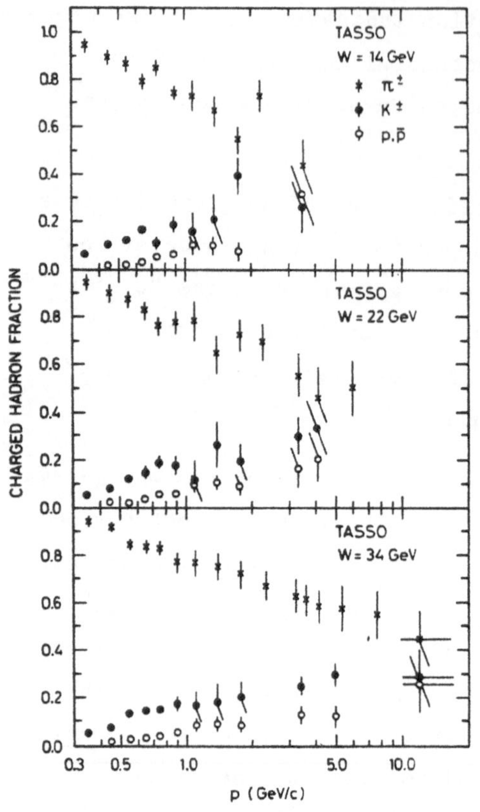

Fig. 39

Charged hadron fractions as a function of hadron momentum, from the reaction

$e^- e^+ \rightarrow$ hadrons

at cm energies W of 14, 22, and 34 GeV.

Fig. 40

K^O production cross sections. The lines show the origin of the strange quark in the K^O according to a cascade model. The curves labelled "primary s" and "b + c" show the contributions from primary s, c and b quarks; the curve labelled "s sea" shows the amount of K^O production with s quarks taken from the sea.

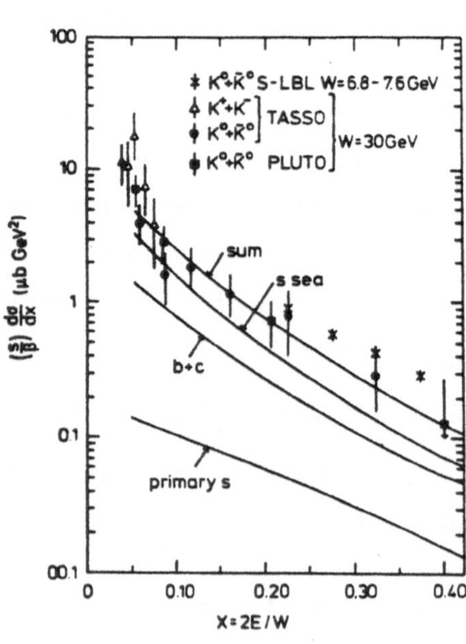

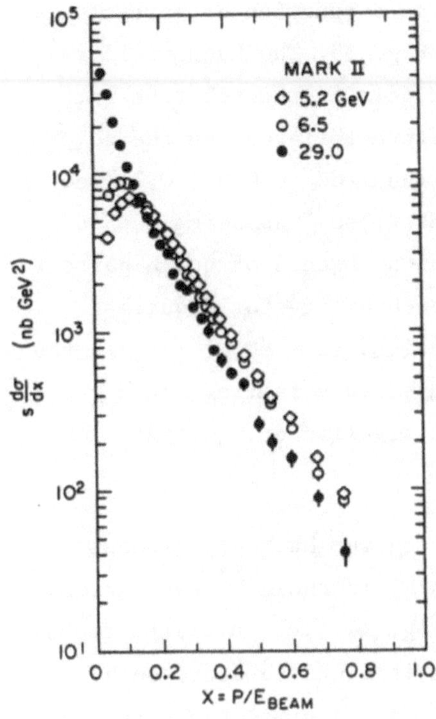

Fig. 41

The scaled cross section s dσ/dx at different values of $W = \sqrt{s}$, from MARK-II.

Fig. 42

The scaled cross section s dσ/dx versus $s = W^2$, from TASSO. Errors include systematics except 4.5% overall normalization uncertainty.

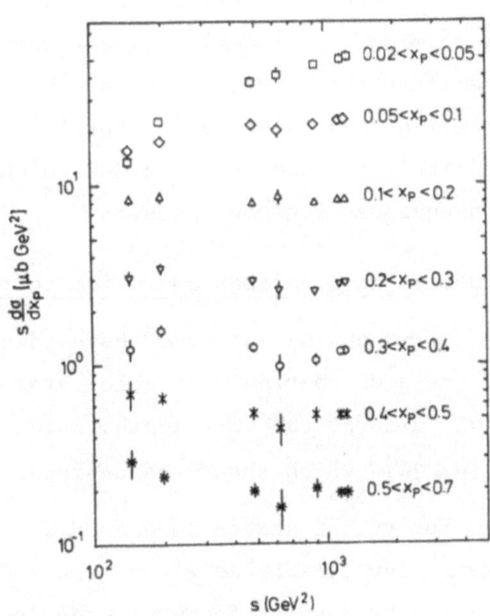

Relative fractions of $\pi^{\pm}$, $K^{\pm}$ and p,$\bar{p}$ as a function of particle momentum p at W ≈ 35 GeV are shown in Fig. 39. The kaon yield is reproduced in cascade fragmentation models if the relative rate for the production of $q\bar{q}$ pairs of different flavors in the color field is assumed to be given by $2s\bar{s}/(u\bar{u} + d\bar{d}) = 0.3 \pm 0.1$ (see Fig. 40). Baryon production can be described[52] as being due to the occasional creation of diquark pairs instead of quark pairs in the color field, with a relative probability for diquark creation of about 10%. The x distributions (x = 2E/W) for all the various particle species are similar, steeply falling for $x \gtrsim 0.15$ except for the $D^{*}(2010)$ which shows a significantly harder spectrum[51].

The overall inclusive x distribution for hadrons (without identification, approximating x by 2p/W) is shown [53] for various cm energies W from SPEAR and PEP in Fig. 41. The invariant cross section s dσ/dx in the large x region is found to decrease by about 30 % between W = 5.2 GeV and 29 GeV. A similar observation is made (Fig. 42) between the lowest and the highest PETRA energies (W = 14 GeV to 37 GeV)[54], i.e. above $b\bar{b}$ threshold, as well as for the inclusive distribution of charged pions (identified by time of flight and Cerenkov counters)[55]. The scale breaking is in qualitative agreement with QCD which predicts it to occur as a consequence of gluon emission.

Υ decays into hadrons and evidence for colored gluons

After having surveyed the evidence for gluon jets produced in the $e^{+}e^{-}$ continuum by color bremsstrahlung from quarks, we now shall discuss the self-annihilation of heavy $Q\bar{Q}$ bound states (Fig. 2b) which should be an ideal source of gluon jets[56].

For vector states such as J/Ψ or Υ the lowest order intermediate state involves three gluons since one gluon is forbidden by color and two by angular momentum conservation. The three-gluon Dalitz plot distribution has been calculated to leading order in QCD[57]. Unfortunately the most probable configuration is one where two of the gluons share almost all of the available energy; such configurations will be difficult to distinguish from two-jet events. However, 38 % of the decays have angular separations of more than 80° between the directions of any two gluons. Such states must, in spite of hadronization smearing, look obviously different from two-jet events even at cm energies as low as W = 10 GeV, i.e. at the Υ mass. In general, however, there is considerable overlap of the three jets at this energy, making detailed analyses somewhat model-dependent.

We consider first a simple measure of the event shape that does not require the identification of three jet axes. We use the sphericity

$$S = \operatorname*{Min}_{\hat{s}} \frac{3}{2} \frac{\sum(\vec{p}_i \times \hat{s})^2}{\sum p_i^2} = \operatorname*{Min} \frac{3}{2} \frac{\sum p_{Ti}^2}{\sum p_i{}^2} \tag{55}$$

which is 1 for ideally spherical and 0 for collinear states. Between W = 5 and 36 GeV the sphericity of e^+e^--produced hadronic final states (Fig. 43) decreases monotonically with increasing W as the particles become increasingly collimated, with the exception of a sudden strong jump at the Υ resonance. The two-jet structure of the nearby continuum states disappears on resonance, and the events take an average sphericity of $\langle S \rangle \approx 0.4$, a value close to that expected for a phase space like distribution (with given hadron multiplicity).

A detailed three-jet analysis of the final states from Υ decay has been carried out by the PLUTO collaboration[58]. The analysis concerns only the "direct" hadronic decays, after subtraction of the contribution from the electromagnetic decay via one intermediate photon. The QCD predictions with which the data were compared included the effects of hadronization. The gluon jets were assumed to fragment very much like the quark jets in the continuum.

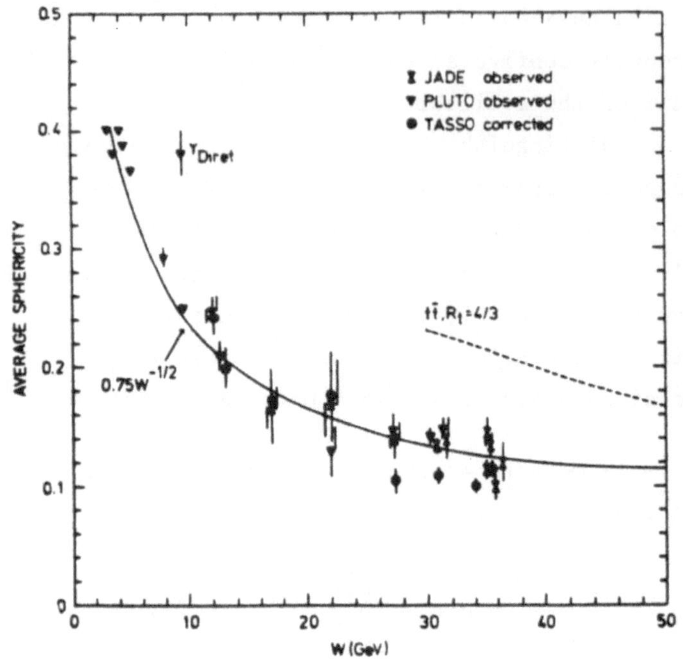

Fig. 43

The average sphericity in $e^-e^+ \to$ hadrons as measured by the JADE, PLUTO and TASSO experiments as a function of cm energy W.

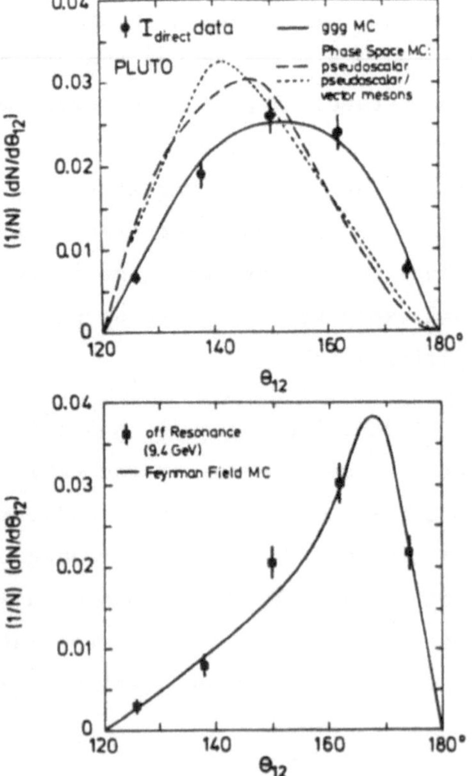

Fig. 44

Distribution of the reconstructed angle θ_{12} between the two most energetic jets, for the direct decay $\Upsilon \to$ hadrons and for off-resonance e^+e^- continuum events (from ref. 58). The curves show a three gluon QCD calculation, a phase-space Monte Carlo model for production of only pseudoscalar or of pseudoscalar and vector mesons, and a Field-Feynman $e^+e^- \to q\bar{q}$ calculation.

It is found that the characteristic Τ event shape, which is very different from that in the nearby continuum, is reflected in many different distributions e.g. those of thrust T, of the angles $\Theta_{k\ell}$ between the jets, and of the fractional jet energies x_i calculated from the $\Theta_{k\ell}$ according to eq. (54). One of many examples is shown in Fig. 44 where the distribution of the angle Θ_{12}, the angle between the two most energetic jets, is plotted. It is compared with the three-gluon Monte Carlo prediction as well as with two versions of a uniform phase space distribution, one with only pseudoscalars (π,K) and the other with pseudoscalars and vector resonances in equal proportion. The lower part of Fig. 44 shows the corresponding distribution off resonance in comparison with a Field-Feynman $q\bar{q}$ Monte Carlo model. The Τ final states are seen to be strongly acollinear ($<\Theta_{12}>$ very different from 180°); they are clearly distinguished from the more collinear $q\bar{q}$ events on the one hand, as well as from phase space like Monte Carlo events on the other. The three-gluon QCD prediction describes the data well. The same has been found for other relevant distributions. Similar conclusions have also been reached by others[59].

The inconsistency of the Τ decay states with two-jet $q\bar{q}$ states also proves that the gluons have color, since otherwise the decay $\Upsilon \rightarrow g \rightarrow q\bar{q}$ via a single gluon into a pair of light quarks would be allowed[60].

A consistency check on the interpretation of the decay Τ → hadrons as a three-gluon decay is provided by the total rate for this decay. Data from experiments both at DORIS and CESR give, after subtraction of the one-photon contribution, a value of Γ_{dir} = (27 ± 8) keV for the partial width corresponding to the "direct" hadronic Τ decays. With this measurement the QCD calculation[61]

$$\frac{\Gamma(\Upsilon \to 3g,\, 4g,\, q\bar{q}gg)}{\Gamma(\Upsilon \to \mu^+\mu^-)} = \frac{10(\pi^2 - 9)\,\alpha^3_{\overline{MS}}(m_\Upsilon)}{81\pi\, e_b^2\, \alpha^2}\left[1 + 9.1\,\frac{\alpha_{\overline{MS}}(m_\Upsilon)}{\pi}\right] \tag{56}$$

yields $\alpha_{\overline{MS}}(M_\Upsilon) = 0.14 \pm 0.01$ or $\Lambda_{\overline{MS}} \simeq 120$ MeV, a reasonable result in view of possibly substantial higher order corrections.

In summary, although the energies in Υ decay are too small to yield well-separated jets, it is clearly apparent that the final states are very different from the $q\bar{q}$ states in the nearby continuum. The QCD decay distribution folded with a hadronization model can describe the data while a phase space like decay will not do. All available evidence clearly points towards the existence of the $\Upsilon \to 3$ gluons decay process.

Jets in Photon-Photon Interactions

Since electrons and positrons are dressed with clouds of virtual photons, an electron-positron collider is at the same time a photon-photon collider. The virtual photons have an approximate $1/E_\gamma$ energy spectrum. Consequently the $\gamma\gamma$ collisions have an effective cm energy $W_{\gamma\gamma}$ which on average is only a fraction of $W = 2E_{beam}$.

The basic $\gamma\gamma$ diagram is shown in Fig. 45. The average angle of the scattered outgoing electron (positron) is

$$\Theta \approx \sqrt{m_e/E_{beam}} \sim 5 \text{ mrad} \tag{57}$$

for PEP or PETRA. The vast majority of these electrons will therefore be confined in the beam pipe. Special forward detectors cover angular regions between 20 and 100 mrad. Virtual photons associated with electrons detected in this angular region are called "small angle tagged photons". These virtual photons have squared four-momenta

$$Q^2 = 4E_{beam}\, E' \sin^2\frac{\Theta}{2} \tag{58}$$

of typically $Q^2 \lesssim 1$ GeV^2 (E' = energy of the scattered electron).

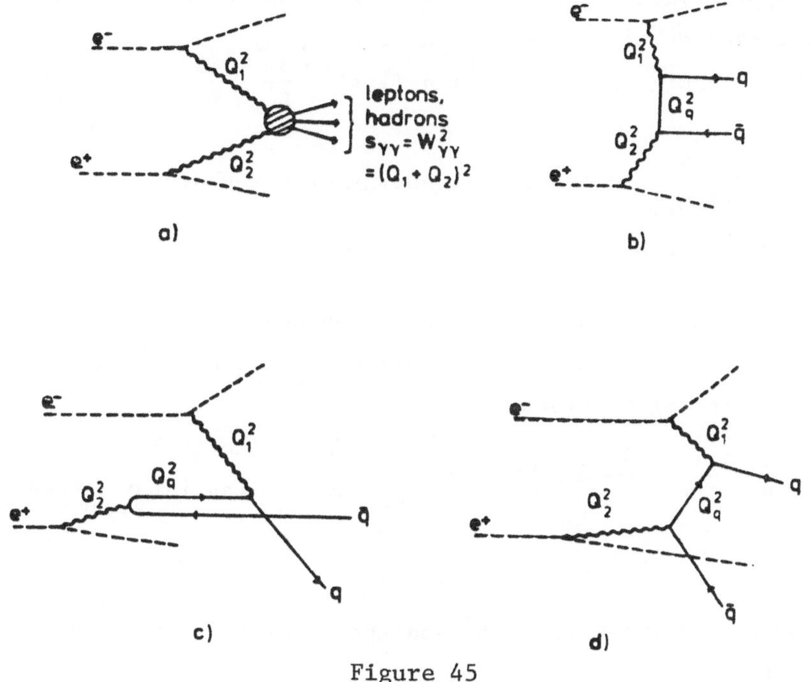

Figure 45

a) Basic photon-photon collision process $e^+e^- \to e^+e^-$ + anything.

b) Diagram for the quark parton process $e^+e^- \to e^+e^- q\bar{q}$

c) Same as b), for deep inelastic scattering of an electron on a nearly on-shell quark constituent of the quasi-real target photon ($Q_1^2 >> 1$ GeV^2, $Q_2^2 \approx 0$, $Q_q^2 \gtrsim 0$).

d) Same as b), for deep inelastic scattering of an electron on a highly off-shell quark that originated from dissociation of the quasi-real target photon at a point into a $q\bar{q}$ pair ($Q_1^2 >> 1$ GeV^2, $Q_2^2 \approx 0$, $Q_q^2 >> 1$ GeV^2).

There are, of course, additional diagrams where particles and antiparticles are interchanged.

For quasi-real photons, $Q_1^2, Q_2^2 \lesssim 1\ \text{GeV}^2$ one may use the equivalent photon approximation to relate a differential cross section $d\sigma(\gamma\gamma \rightarrow X)$ to the measured cross section $d\sigma(e^-e^+ \rightarrow e^-e^+X)$. In this approximation the number and distribution of virtual photons co-moving with an electron is given by

$$N(z,\theta)dzd\theta \simeq \frac{\alpha}{\pi}\left[1 + (1-z)^2\right]\frac{dz}{z}\ \frac{d\theta}{\theta} \qquad (59)$$

where $z = \dfrac{\nu}{E_{beam}}$ = fractional energy of the photon,

ν = photon energy.

The dz/z term is the usual bremsstrahlung distribution. The $d\theta/\theta$ distribution is cut off at small $\theta \approx m_e/E_{beam}$. Integration of (59) over θ therefore gives

$$N(z)dz \simeq \frac{\alpha}{\pi}\ \frac{1 + (1-z)^2}{z}\ \ell n\eta, \qquad \eta = \begin{cases} E_{beam}/m_e & \text{(no tag)} \\ \Theta_{max}/\Theta_{min} & \text{(tag in } \theta_{min}<\theta<\theta_{max}) \end{cases} \qquad (60)$$

The connection between the e^-e^+ and the $\gamma\gamma$ differential cross sections is

$$d\sigma(e^-e^+ \rightarrow e^-e^+X) \simeq \int N(z_1\theta_1)dz_1 d\theta_1 N(z_2\theta_2)dz_2 d\theta_2\ d\sigma(\gamma\gamma \rightarrow X) \qquad (61)$$

The $\gamma\gamma$ system has an invariant mass $W_{\gamma\gamma}$ that is given in terms of the photon energies and relative angle by

$$\begin{aligned} s_{\gamma\gamma} = W^2_{\gamma\gamma} = (Q_1 + Q_2)^2 \simeq 2Q_1 \cdot Q_2 \simeq 4\nu_1\nu_2 \sin^2\frac{\theta_{\gamma\gamma}}{2} \\ \approx 4\ \nu_1\nu_2 = 4\ E^2_{beam}\ z_1 z_2 \end{aligned} \qquad (62)$$

The product of the two photon flux factors in (61) is dominated by the 1/z terms,

$$N(z_1)N(z_2)dz_1dz_2 \sim \frac{dz_1\ dz_2}{z_1 z_2} \sim \frac{dW^2_{\gamma\gamma}}{W^2_{\gamma\gamma}} \qquad (63)$$

from which it is seen that the probability for the $\gamma\gamma$ system to have invariant mass $W_{\gamma\gamma}$ is approximately inversely proportional to $W^2_{\gamma\gamma}$ but for given $W_{\gamma\gamma}$ is independent of ν_1 and ν_2, ie. of the individual photon energies the $\gamma\gamma$ system is made up of. Introducing a variable y by

$$\nu_1 = W_{\gamma\gamma}\frac{e^y}{2}, \quad \nu_2 = W_{\gamma\gamma}\frac{e^{-y}}{2}, \quad y = \frac{1}{2}\ln\frac{\nu_1}{\nu_2} \tag{64}$$

we have for the total energy and momentum of the $\gamma\gamma$ system

$$\begin{aligned} E_{\gamma\gamma} &= \nu_1+\nu_2 = W_{\gamma\gamma}\cosh y \\ P_{\gamma\gamma} &\simeq \nu_1-\nu_2 = W_{\gamma\gamma}\sinh y \end{aligned} \tag{65}$$

from which it is evident that y is the rapidity of the $\gamma\gamma$ system. The $d\nu_1 d\nu_2/\nu_1\nu_2$ brems spectrum translates into a uniform y distribution of the $\gamma\gamma$ system for fixed $W_{\gamma\gamma}$:

$$\frac{d\nu_1 d\nu_2}{\nu_1\ \nu_2} = \frac{dz_1 dz_2}{z_1\ z_2} = \frac{dW^2_{\gamma\gamma}}{W^2_{\gamma\gamma}}\ dy = 2d(\ln W_{\gamma\gamma})\ dy \tag{66}$$

The $\gamma\gamma$ system is therefore in general moving with a high velocity approximately parallel to the direction of the incident beam. Integration of eq. (61) over the total rapidity range

$$-\ln\frac{2E_{beam}}{W_{\gamma\gamma}} < y < \ln\frac{2\,E_{beam}}{W_{\gamma\gamma}} \tag{67}$$

(which is limited by Max $\nu_1 = E_{beam}$, Max $\nu_2 = E_{beam}$) gives the $W_{\gamma\gamma}$ distribution

$$\frac{d\sigma_{e^-e^+\to e^-e^+X}}{d(\ln W_{\gamma\gamma})} \approx \left(\frac{2\alpha}{\pi}\ln\eta\right)^2\ 4\ln\frac{2E_{beam}}{W_{\gamma\gamma}}\quad \sigma_{\gamma\gamma\to X}(W_{\gamma\gamma}). \tag{68}$$

The states produced by the two interacting photons can have any J^P except $J = 1$ (a forbidden state for two photons). They have $C = +1$. The total cross section for two-photon interactions leading to hadrons in the domain of quasi-real photons is approximately[62]

$$\sigma_{\gamma\gamma\to\text{hadrons}} = 380\text{ nb} + \frac{520\text{ nb GeV}}{W_{\gamma\gamma}} \tag{69}$$

Inserting this into (68) and integrating over $W_{\gamma\gamma}$ (not requiring a tag) gives

$$\sigma_{e^-e^+\to e^-e^+ \text{ hadrons}} \approx 40 \text{ nb at } W \approx 30 \text{ GeV}.$$

This cross section is much larger than the annihilation cross section $\sigma_{e^-e^+\to \text{hadrons}}$ at the same W. However, most of the two photon interaction events have low values of $W_{\gamma\gamma}$ and show a considerable boost along the beam direction.The final state hadrons therefore often emerge at small angles relative to the beam and do not leave the beam pipe, such that most of these events remain undetected. If the event is detected some of the final state particles, emerging at angles close to the beam direction, are usually missed. This results in large corrections in determining the value of $W_{\gamma\gamma}$ of an event, necessary in order to measure $\sigma_{\gamma\gamma\to \text{hadrons}}$.

Among the experimental signatures used to identify $\gamma\gamma$ interactions there is in particular the"tagging" signal from a scattered electron. Single tagging is usually employed, the probability for double tagging being generally too small. One practical consequence is that $W_{\gamma\gamma}$ cannot be obtained from energy measurement of the scattered electrons but only from measuring the complete final state. With single tagging background in the tagging counter from a hard bremsstrahlung photon emitted by one of the beam particles has to be considered but can generally be handled. Untagged two photon events are distinguished from one-photon annihilation events primarily on the basis of their small final state energy (Fig. 17).

For small Q_1^2, Q_2^2 the leading contribution to the process

$$\gamma\gamma \to \text{hadrons} \tag{70}$$

is expected by the vector dominance model(VDM) to closely resemble the purely hadronic process

$$\rho^o\rho^o \to \text{hadrons}. \tag{71}$$

It will lead to low-p_T dominated hadron physics, with direct-channel resonances at low two-photon invariant masses $W_{\gamma\gamma}$ and

Reggeon and/or Pomeron exchange at high $W_{\gamma\gamma}$. A number of experimental results for such processes have been obtained [63], including resonance formation $\gamma\gamma \to \eta'$, S^*, A_2, f, f', exclusive channels like $\gamma\gamma \to \rho^o\rho^o$, $p\bar{p}$, 4π, and total cross section measurements for $\gamma\gamma \to$ hadrons. The results indicate that the size and the Q^2 behaviour of the total $\gamma\gamma$ cross section at $W_{\gamma\gamma} \gtrsim 3$ GeV are in reasonable agreement with the expectation from the hadronic part of the photon (VDM).

In the following we will not discuss this hadronlike aspect of $\gamma\gamma$ interactions. Instead we will concentrate on those <u>hard</u> $\gamma\gamma$ processes in which the parton nature of the participating quarks is revealed[64]. The simplest reaction involved is

$$e^+e^- \to e^+e^- q\bar{q}, \tag{72}$$

shown by the diagram of Fig. 45b. An analogous diagram contributes, of course, to charged lepton pair production. For large p_T of the produced $q\bar{q}$ pair the exchanged quark will be far off-shell, and perturbative pointlike couplings apply. One defines

$$d\sigma_{\gamma\gamma\to q\bar{q}} = R_{\gamma\gamma}\, d\sigma_{\gamma\gamma\to\mu\bar{\mu}}, \tag{73}$$

and the diagram Fig. 45b gives

$$R_{\gamma\gamma} = 3 \sum_{i=1}^{n_f} e_q^4 \left[1 + o(\alpha_s) \right], \tag{74}$$

where the factor 3 is the colour multiplicity and n_f is the number of produced flavours. In the $W_{\gamma\gamma}$ range relevant at PEP and PETRA, $n_f = 4$ and thus $3\,\Sigma\, e_q^4 = 34/27$. Finally, QED in leading order gives

$$\frac{d\sigma_{\gamma\gamma\to\mu\bar{\mu}}}{d\Omega} = \frac{\alpha^2}{W_{\gamma\gamma}^2}\,\frac{1+\cos^2\theta}{\sin^2\theta} \tag{75}$$

where θ is the angle of the muon (or quark) with respect to the incident $\gamma\gamma$ "beam" direction, which is close to the incident $e^\pm$ beam direction. The transverse momentum $p_T = (W_{\gamma\gamma}/2)\sin\theta$ of the produced lepton or quark is seen to have a power law distribution. Integrating over $W_{\gamma\gamma}$ (which is kinematically not well determined in the experiments) one obtains[65] distributions of the quark in the inclusive reaction $e^+e^- \rightarrow e^+e^- q\bar{q}$:

$$\frac{d\sigma}{(d^3\vec{p}/E)_q}(e^+e^- \rightarrow e^+e^- q\bar{q}) \simeq \left(\frac{\alpha}{2\pi}\ln\eta\right)^2 \frac{\alpha^2 R_{\gamma\gamma}}{p_{Tq}^4}(1 - x_q) \qquad (76)$$

with $x_q = 2p_q/\sqrt{s}$. The single hadron inclusive distributions from this process will scale in a similar manner,

$$\frac{d\sigma}{dp_{Th}^2}(e^+e^- \rightarrow e^+e^- h\ X) \sim p_{Th}^{-4} \qquad (77)$$

giving a much slower decrease with transverse momentum than that exhibited by soft hadronic reactions. This feature will therefore distinguish the hard parton process of Fig. 45b from the VDM-type processes. Evidence for a p_{Th}^{-4} tail in the inclusive hadron distribution from single-tagged two photon interactions is shown in Fig. 46.

An even more characteristic signature of the hard process(72) is the occurrence of a pair of jets that are typically acollinear and of total energy smaller than $\sqrt{s}$. An example is shown in Fig. 47. The same event shapes occur, however, as a consequence of hard photon bremsstrahlung from the initial e^+ or e^- in one-photon annihilation reactions. Tagging helps to distinguish the two processes, and detailed Monte Carlo studies show that the background to reaction (72) from one photon processes is well manageable.

Transverse momentum distributions[66,67] of single jets from the reaction

$$e^-e^+ \rightarrow e^-e^+ + \text{jet} + \text{anything} \qquad (78)$$

are shown in Fig. 48. For $p_T^2 \gtrsim 6\ \text{GeV}^2$ the data are compatible with a p_T^{-4} behavior. The dotted line in Fig. 48b shows the expectation from vector dominance (VDM). It is quite sizable for transverse momenta $p_T^2 \lesssim 6\ \text{GeV}^2$. Also shown is the result of calculations

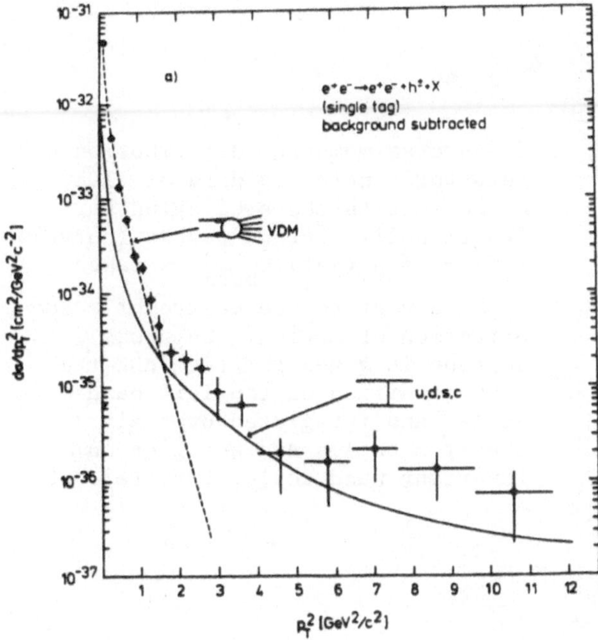

Fig. 46

Inclusive hadron differential cross section $d\sigma/dp_T^2$ plotted against p_T^2. Also shown are the predictions from the vector dominance model (VDM) and from the process $e^+e^- \to e^+e^-q\bar{q}$ with subsequent quark fragmentation (full line). From ref. 67.

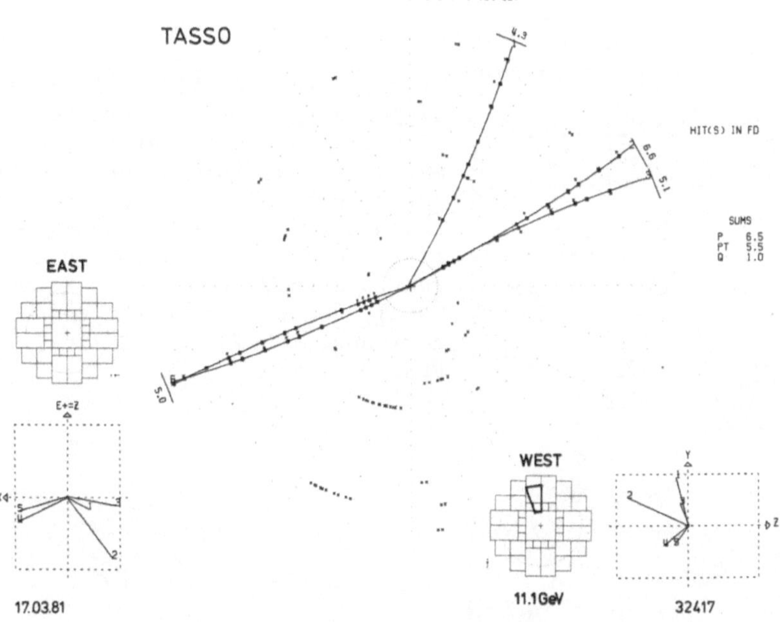

Fig. 47 An event of the reaction $\gamma\gamma \to 2$ jets observed at PETRA at W = 34.5 GeV. The energy of the single tagged e^- is 11 GeV.

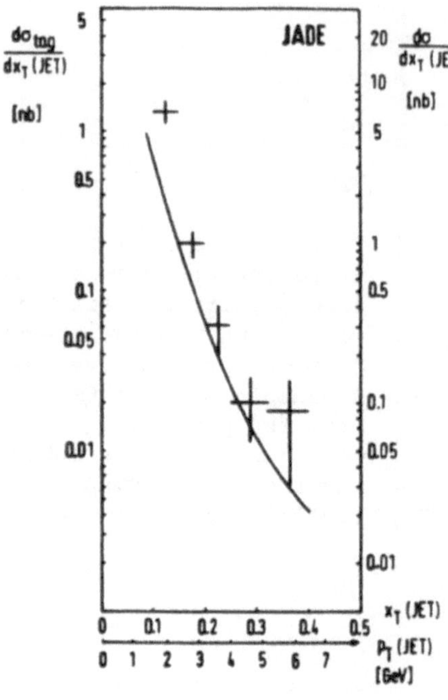

Fig. 48

a)
Transverse momentum distribution of jets for single tag data of JADE, compared with the prediction for fractionally charged quarks (curve). $x_T(\text{jet}) = p_T(\text{jet})/E_{beam}$ is taken with respect to the centre-of mass direction of motion. The cross section is given for the JADE tagging condition on the left hand scale, and integrated over all electron angles and energies on the right hand scale. From ref. 66.

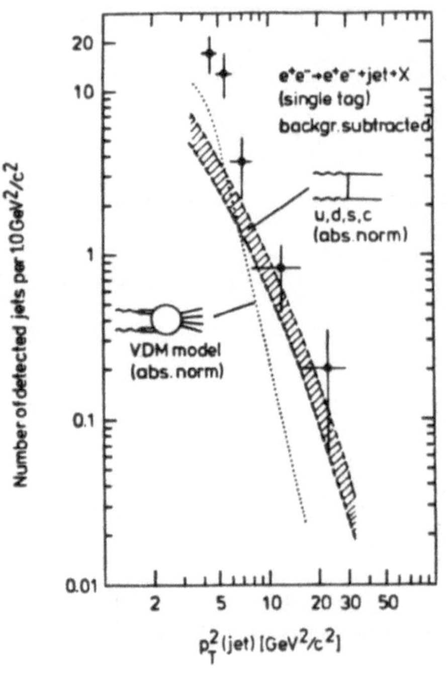

b)

Distribution of p_T^2(jet) from single tagged data of TASSO for the reaction $e^+e^- \to e^+e^- + \text{jet} + X$, compared with a calculation for the parton process $e^+e^- \to e^+e^-q\bar{q}$ with fragmentation of the final quark pair according to various different hadronization models (shaded band), and to the vector meson dominance model (dotted line). From ref. 67.

of the diagram of Fig. 45b with a Monte Carlo program[68] using the relations (73), (74) and employing different models to describe the fragmentation of the quarks. These evaluations of the hard (pointlike) $\gamma\gamma \to q\bar{q}$ process lie below the measured distributions for $p_T^2 \lesssim 6\ GeV^2$, but for higher p_T^2 the observed cross section approaches the pointlike one. However, a o(50%) contribution in this p_T^2 region from other hard processes like $\gamma q \to qg$ or $\gamma q \to qM$ leading to three-jet final states, or from processes of higher order in QCD like $\gamma\gamma \to gg$, cannot be excluded. Therefore $R_{\gamma\gamma}$ is experimentally not yet well determined. The data do, however, disfavour integral quark charges[69] since these would make $R_{\gamma\gamma}$ much larger, such that the data points would then be nearly a factor 2 below the prediction.

Structure Function of the Photon

If one of the outgoing electrons in a $\gamma\gamma$ reaction is tagged at a large angle while the other remains untagged we have

$$Q_1^2 \gg 1\ GeV^2, \quad Q_2^2 \ll 1\ GeV^2. \tag{79}$$

Note that in the reaction discussed before, $e^+e^- \to e^+e^- q\bar{q}$ with quasi-real photons $Q_1^2 \simeq Q_2^2 \ll 1\ GeV^2$, the transverse momentum p_T of the q had to be essentially balanced by the p_T of the $\bar{q}$. In the large angle single tagging condition (79) to be discussed now, however, one of the electrons in the final state has high transverse momentum p_T and two limiting cases arise, depending on the four-momentum transfer Q_q of the virtual exchanged quark (Fig. 45b):

i) $Q_q^2 \ll 1\ GeV^2$. The exchanged quark can be considered a nearly on-shell (longlived) constituent of the quasi-real "target photon" Q_2. It is struck by a highly virtual photon Q_1 from deep-inelastic scattering of the tagged electron. The situation is exactly analogous to the familiar one in hadron structure function measurements. In fact, one measures here the hadronic portion of the photon structure function for which the vector dominance model(VDM) is expected to apply. In the final state one high-p_T jet occurs, balancing the p_T of the large-angle scattered electron, while the other "target jet" remains at low p_T (Fig. 45c).

ii) $Q_q^2 >> 1\ \text{GeV}^2$. The exchanged quark is highly virtual. Now also the "target photon" → quark antiquark pair transition is pointlike and can be treated perturbatively. One measures a portion of the target photon structure function that has no direct analogy in hadrons, arising from the pointlike nature of the target.

The most interesting case is of course the second one, because this hard quark component of the structure function can be explicitly calculated in perturbation theory[70]. In lowest order (ie. in the quark parton model) we consider the diagram of Fig. 45d which gives for the photon structure function (neglecting longitudinal target photons as well as quark masses)

$$F_2^\gamma(x,Q^2) \simeq x \sum_i e_i^2\ (q_i + \bar{q}_i) \simeq 2x\ F_1(x,Q^2)$$
$$F_1^\gamma(x,Q^2) \simeq \frac{1}{2} \sum_i e_i^2\ (q_i + \bar{q}_i) \tag{80}$$

as for a hadron target, where

$$q_i(x,Q^2) = \bar{q}_i(x,Q^2) = 3\ \frac{\alpha}{2\pi}\ e_i^2 \left[x^2 + (1 - x)^2\right] \ell n\ \frac{W_{\gamma\gamma}^2}{m_q^2} \tag{81}$$

is the quark (antiquark) distribution function for flavor i inside the target photon. Here x is defined as usual by $x = Q^2/2(Q \cdot Q_2)$, and we have written $Q_1^2 = Q^2$. The factor 3 in (81) is the color multiplicity, and the x dependence is due to the spin (as familiar eg. from the well known double-humped energy distribution observed in electromagnetic pair production). The shape of the structure function $F_2^\gamma(x,Q^2)$ at fixed Q^2 is then given by

$$x \left[x^2 + (1 - x)^2\right], \tag{82}$$

peaking at x = 1 in sharp contrast to (soft) hadronic structure functions.

Even more remarkable is the effect of the QCD modifications to the photon structure function[70]. The factor $\ln(W_{\gamma\gamma}^2/m_q^2)$ in (81) is replaced in QCD by a factor $\ln(Q^2/\Lambda^2)$, which arises from the dressing of the recoil quark by gluons giving the quark an effective mass of order Λ, the scale parameter of QCD. Due to this factor $\ln(Q^2/\Lambda^2) \sim 1/\alpha_s(Q^2)$ the hard perturbative portion of the structure function will dominate at large Q^2 and x over the VDM contribution. The latter is expected to behave like meson structure functions

$$F_2^{\gamma,\mathrm{VDM}} \simeq \frac{\alpha}{f_\rho^2/4\pi} \, \frac{1-x}{4} \, , \tag{83}$$

falling with x and, in addition, shrinking towards small x with rising Q^2. In higher orders of perturbative QCD hard gluons are radiated by the quarks; this will cause the x dependence to become somewhat softer than (82) but the important $\ln Q^2$ behavior remains.

The connection between the structure function and the total $\gamma\gamma$ interaction cross section is given (in the equivalent photon approximation for the low-Q_2^2 target photon) by

$$d\sigma(ee \to eeX) = \int dz_2 \left(\frac{\alpha}{\pi} \ln \eta_2\right) \frac{1 + (1-z_2)^2}{z_2} \, d\sigma(e\gamma \to eX) \tag{84}$$

with

$$\frac{d\sigma(e\gamma \to eX)}{dxdy} \simeq \frac{4\pi\,\alpha^2}{Q^4} \, s\, z_2(1-y)\, F_2^\gamma(x, Q^2) \tag{85}$$

where $s = W^2 = 4\,E_{\mathrm{beam}}^2$, $\eta_2 = E_{\mathrm{beam}}/m_e$ (for an untagged target photon), z_2 = energy of target photon/E_{beam}, and

$$x = \frac{Q^2}{2(Q \cdot Q_2)} \approx \frac{Q^2}{Q^2 + W_{\gamma\gamma}^2} \, , \qquad y = \frac{(Q \cdot Q_2)}{(e \cdot Q_2)} = 1 - \frac{E_{\mathrm{tag}}}{E_{\mathrm{beam}}} \cos^2\left(\frac{\theta_{\mathrm{tag}}}{2}\right) \tag{86}$$

$$Q^2 = 4E_{beam}E_{tag}\ \sin^2\left(\frac{\theta_{tag}}{2}\right) \quad (E_{tag},\ \theta_{tag} = \text{energy and angle of the tagged electron})$$

This holds for the large angle single tag condition (79). Longitudinal polarization states of the target photon have been neglected which is justified for a quasi-real (untagged or anti-tagged) target photon.

Measurements of the total cross section for single-tagged events from the process

$$e^-e^+ \rightarrow e^-e^+ + \text{hadrons}$$

have been made by the PLUTO, CELLO and JADE groups[71]. Electrons were detected at angles between 100 and 430 mrad. A major experimental problem is the determination of $W_{\gamma\gamma}$ and therefore of x, since it can never be excluded that final state particles are lost in the beam pipe. Monte Carlo simulations indicate that the "visible" portion, $W_{\gamma\gamma}^{vis}$, is on average about 30 % smaller than the complete $W_{\gamma\gamma}$. Figure 49a shows an event distribution at $\langle Q^2\rangle = 23\ GeV^2$ from JADE, plotted against $x_{vis} = Q^2/(Q^2+W^2_{\gamma\gamma,vis})$, compared with theoretical predictions from a simulation of the process in the detector using QCD predictions for the photon structure function. In Fig. 49b the result of an attempt by PLUTO to extract the photon structure function itself at $\langle Q^2\rangle = 5\ GeV^2$ is shown. Here, corrections for the invisible portion of $W_{\gamma\gamma}$ have been applied to the data. In both cases comparison with leading order (full curves) as well as with higher order QCD predictions (dashed-dotted) are shown. These calculations agree well with the data for $x \gtrsim 0.3$ (note that $x_{vis} > x$), the values of $\Lambda_{\overline{MS}}$ being in the 100 to 300 MeV range. There is room for some VDM contribution at lower x.

From the measurements shown it clearly appears that in deep-

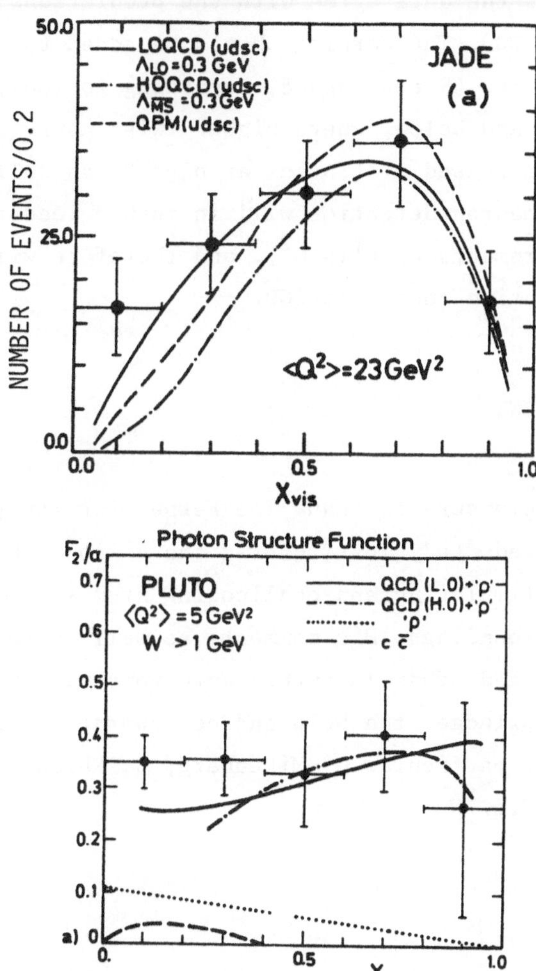

Fig. 49

Top: x_{vis} distribution in deep-inelastic electron photon scattering from JADE. Curves show leading order and higher order QCD, and quark parton model predictions.

Bottom: Photon structure function $F_2^{\gamma}(x)/\alpha$ extracted from PLUTO data. Curves show leading order and higher order QCD contribution, ρ dominance contribution, and effect from $c\bar{c}$ production.

inelastic $e\gamma$ scattering the pointlike component of the photon structure function is dominant. Within the still rather large uncertainties the data agree with the predictions from perturbative QCD. The warning should be added that the high x points in Fig. 49 correspond, according to (86), to values of $W_{\gamma\gamma}$ of 2 GeV and below, where higher twist contributions may be important. Increased statistics at high Q^2 as well as improved small angle hadron detection will in future lead to much improved measurements of $F_2^{\gamma}(x,Q^2)$, and therefore will provide important quantitative checks of QCD.

ACKNOWLEDGEMENT

It is a pleasure to thank Tom Ferbel for inviting me to the Advanced Study Institute in the Adirondacks, which provided a stimulating and challenging athmosphere in very pleasant surroundings. His comments as well as those from the students and other lecturers were very valuable. I also wish to acknowledge the help and collaboration of many collegues, in particular G. Mikenberg, U. Karshon, G. Rudolph, G. Wolf and S.L. Wu.

References

1. For a review see P. Dittmann and V. Hepp, Z. Physik C10, 283 (1981)

2. See, for example, F. Berends and R.Kleiss, Nucl.Phys. 178, 141 (1981);
F. Berends, K.F.J. Gaemers, and R. Gastmans, Nucl. Phys. B 63, 381 (1973)

3. R. Brandelik et al., Phys. Letters 117 B, 365 (1982)

4. D.P. Barber et al., Phys.Rev. Lett 48, 1701 (1982); Phys. Lett. 95B, 149 (1980)
W. Bartel et al., Phys. Lett. 108B, 140 (1982); 99B, 281 (1981)
H.-J. Behrend et al., Phys. Lett. 114B, 282 (1982); Z. Physik C 14, 283 (1982)
C. Berger et al., Phys. Lett. 99B, 489 (1981); Z. Physik C7, 289 (1981)
R. Brandelik et al., Phys. Lett. 110B, 173 (1982); DESY 82-032 to be published

5. G. Wolf, Selected Topics on e^+e^- Physics, Lectures given at the 1979 JINR-CERN School of Physics, Dobogokö, Hungary; DESY 80/13 (1980)

6. W. Bartel et al., Phys. Letters 92B, 206 (1980);
R. Brandelik et al., Phys. Lett. 94B, 259 (1980);
B. Adeva et al., Phys. Rev.Letters 48, 967 (1982);
H.-J. Behrend et al., DESY 81-021 (1981); DESY 82-080 (1982)

7. G.G. Hanson et al., Phys. Rev. Lett. 35, 1609 (1975);
R.F. Schwitters et al., Phys. Rev. Lett. 35, 1320 (1975)

8. D.P. Barber et al., Phys. Rev. Lett. 46, 1663 (1981)
W. Bartel et al., Phys. Lett. 88B, 171 (1981)
H.-J. Behrend et al., DESY 81-029 (1981)
C. Berger et al., Phys. Lett. 86B, 413 (1979)
R. Brandelik et al., Phys. Lett. 113B, 499 (1982)
See also P. Söding and G. Wolf, Ann.Rev. Nucl.Part. Sci. 31, 231 (1981)

9. W. Braunschweig, Recent results from PETRA on jet formation; Proc. 1981 Int. Symp. on Lepton and Photon Interactions at High Energies, Bonn 1981, p. 68

10. R. Brandelik et al., Phys. Lett. 100B, 357 (1981);
Ch. Berger et al., DESY 82-058 (1982)

11. M. Dine and J. Sapirstein, Phys. Rev. Lett. 43, 668 (1979);
K.G. Chetyrkin et al., Phys. Lett. 85B, 277 (1979);
W. Celmaster and R.J. Gonsalves, Phys. Rev. Lett. 44, 560 (1979)

12. G. Heinzelmann, Recent results from JADE at PETRA; Proc. 21st Int. Conf. on High Energy Physics, Paris 1982, p. 59

13. R. Brandelik et al., Phys. Lett. 86B, 243 (1979);
D.P. Barber et al., Phys. Rev. Lett. 43, 830 (1979);
C. Berger et al., Phys. Lett. 86B, 418 (1979);
W. Bartel et al., Phys. Lett. 91B, 142 (1980);

14. F. Wilczek, Phys. Rev. Letters 39, 1304 (1977)

15. G.S. Abrams et al., Phys. Rev. Letters 33, 1453 (1974)

16. A.A. Sokolov and I.M. Ternov, Sov.Phys. Dokl. 8, 1203 (1964)
V.N. Baier, Sov. Phys. Uspekhi 14, 695 (1972)
J.D. Jackson, Rev. Mod. Phys. 48, 417 (1976)
R.F. Schwitters et al., Phys. Rev. Lett. 35, 1320 (1975)

17. W. Bartel et al., Phys. Letters 88B, 171 (1979)

18. M. Oreglia et al., Phys. Rev. D 25, 2259 (1982)

19. J.A. Jaros, in Proc. 21th Int. Conf. on High Energy Physics, Paris (1982), p. 106

20. D.R. Nygren and J.N. Marx, Physics Today 31, No. 10, 46 (1978)

21. J.M. Dorfan, Proc. of Summer Inst. on Part. Phys., The Strong Interaction (Stanford 1981) p. 569 (1981) and private communication.
J.G. Smith, SLAC-PUB-2921 (1982)

22. In addition to Refs. 4 and 21, results submitted to the XXI Int. Conf. on High Energy Physics 1982 in Paris, and summaries by M. Davier at the Paris conference and by B. Naroska at the 2nd Internat. Conf. on Physics in Collisions at Stockholm (1982), were also taken into account.

23. J. Burger, Int. Conf. on High Energy Physics, Paris 1982, p.63

24. D. Lüke, Int. Conf. on High Energy Physics, Paris 1982, p.67

25. G.J. Feldman et al., Phys. Rev. Lett. 38, 1313 (1977)

26. J.D. Bjorken, Phys. Rev. D19, 335 (1979)
P.Q. Hung and J.J. Sakurai, Nucl. Phys. B143, 81 (1978)
E.H. DeGroot, D. Schildknecht and G.J. Gounaris, Phys. Lett. 90B, 427 (1980)

27. E.H. DeGroot and D. Schildknecht, Phys. Lett. 95B, 128 (1980);
G.J. Gounaris and D. Schildknecht, preprint BI-TP 81/09 (1981)

28. H. Georgi and S. Weinberg, Phys.Rev. D17, 275 (1978);
V. Barger, W.Y. Keung and E. Ma, Phys. Rev. Lett. 44, 1169 (1980);
E.H. DeGroot, D. Schildknecht and G.J. Gounaris, Phys. Lett. 90b, 427 (1980)

29. B. Adeva et al., Phys. Lett. 115B, 345 (1982)
W. Bartel et al., Phys. Lett. 114B, 211 (1982)
H.-J. Behrend et al., Phys. Lett. 114B, 287 (1982)
C. A. Blocker et al., Phys. Rev. Lett. 49, 517 (1982)

30. For a summary see eg. A. Böhm, DESY 82-027 (1982) and Proc. of the Rencontre de Moriond, Les Arcs 1982, (ed. Tran Thanh Van) to be publ.

31. D.P. Barber et al., Phys. Reports 63, 337 (1980); Phys. Lett. 108B, 63 (1982)
W. Bartel et al., Z. Physik C9, 315 (1981); Phys. Lett. 91B, 142 (1980)
H.-J. Behrend et al., Phys. Lett. 110B, 329 (1982); Phys. Lett. 113B, 427 (1982); Z.Physik C14, 95 (1982)
C. Berger et al., Phys. Lett. 97B, 459 (1980)
R. Brandelik et al., Phys. Lett. 94B, 437 (1980)
W.D. Schlatter et al., Phys. Rev. Lett. 49, 521 (1982)
P. Söding and G. Wolf, Ann. Rev. Nucl. Part. Sci. 31, 231 (1981)

32. A. Petersen, DESY report 80-46 (1980); W. Bartel et al., Phys. Lett. 91B, 142 (1980)

33. J. Ellis, M.K. Gaillard, and G.G. Ross, Nucl. Phys. B111, 253 (1976)
T.A. DeGrand, Y.J. Ng, and S.H. Tye, Phys.Rev. D16, 3251 (1977)

34. A. Ali et al., Phys. Lett. 93B, 155 (1982)
R.D. Field and R.P. Feynman, Nucl. Phys. B136, 1 (1978)
P. Hoyer et al., Nucl. Phys. B161, 349 (1979)

35. D.P. Barber et al., Phys. Lett. 108B, 63 (1982)

36. C.L. Basham, L.S. Brown, S.D. Ellis, and S.T. Love, Phys.Rev. Lett. 41, 1585 (1978); Phys. Rev. D17, 2298 (1978); Phys. Rev. D19, 2018 (1979)

37. H.-J. Behrend et al., Phys. Lett. 110B, 329 (1982)

38. C. Berger et al., Phys.Lett. 99B, 292 (1981)
R. Hollebeek, Proc. 1981 Int. Symp. on Lepton and Photon Int. at High Energies, Bonn (1981), p. 1

39. S. Brandt and H.D. Dahmen, Z. Physik C (Particles and Fields) 1, 61 (1979)
S.L. Wu and G. Zobernig, Z. Physik C (Particles and Fields) 2, 107 (1979)
S.L. Wu, Z. Physik C (Particles and Fields) 9, 329 (1981)
J. Dorfan, Z. Physik C (Particles and Fields) 7, 349 (1981) and private communication.
K. Lanius, H.E. Roloff and H. Schiller, Z. Physik C (Particles and Fields) 8, 251 (1981)

H.J. Daum, H. Meyer and H. Bürger, Z. Physik C (Particles and Fields) 8, 167 (1981).

40. R.K. Ellis et al., Phys.Rev. Lett. 45, 1226 (1980); K. Fabricius et al., Phys. Lett. 97B, 431 (1980)

41. W. Bartel et al., DESY 82-060 (to be published)

42. K. Fabricius et al., Z. Physik C11, 315 (1982); G. Kramer, DESY 82-029 (to be published)

43. W. Bartel et al., Phys. Lett. 115B, 338 (1982)

44. P. Söding, in Particles and Fields 1981: Testing the Standard Model (C.A. Heusch Ed.), AIP Conference Proceedings No. 81, p. 107 (1982)

45. R. Brandelik et al., Phys. Lett. 97B,453 (1980)

46. J. Ellis and I. Karliner, Nucl. Phys. B148, 141 (1979)

47. B. Andersson et al., Phys. Lett. 94B, 211 (1980)

48. W. Bartel et al. Phys. Lett. 101B, 129 (1981)

49. H.-J. Behrend et al., DESY 82-061 (to be published)

50. W. Bartel et al., Phys. Lett. 104B, 325 (1981)
H.-J. Behrend et al. DESY 82-018 (to be published)
C. Berger et al., Phys. Lett. 104B, 79 (1981)
R. Brandelik et al., Phys. Lett. 94B, 444 (1980); 94B 91 (1980); 108B, 71 (1982); 105B, 75 (1981); 113B, 98 (1982); DESY 82-046 (1981)

51. J.M. Yelton et al., Phys. Rev. Lett. 49, 430 (1982) (see also Ref. 24)

52. T. Meyer, Z. Physik C12, 77 (1982)

53. J.F. Patrick et al., SLAC-PUB 2936 (1982)

54. R. Brandelik et al., Phys. Lett. 114B, 65 (1982)

55. R. Brandelik et al., Phys. Lett. 113B, 98 (1982)

56. T. Appelquist and H.D. Politzer, Phys. Rev. Lett. 34, 43 (1975); Phys. Rev. D12, 1404 (1975)

57. K. Koller and T.F. Walsh, Phys.Lett. 72B, 227 (1977), (E: 73B, 504); Nucl.Phys. B140, 449 (1978);
S.J. Brodsky, T.A. DeGrand, R.R. Horgan, and D.G.Coyne, Phys. Lett. 73B, 203 (1978);
H. Fritzsch and K.H. Streng, Phys. Lett. 74B, 90 (1978);
A. deRujula et al., Nucl.Phys. B138, 387 (1978);
K. Koller, H. Krasemann, and T.F. Walsh, Z. Physik C (Particles and Fields) 1, 71 (1979)

58. C. Berger et al., Z. Physik C 8, 101 (1981)

59. B. Niczyporuk et al. (LENA collaboration), Z Physik C (Particles and Fields) 9, 1 (1981);
D. Schamberger (Cornell), Proc. 1981 Int. Symp. on Lepton and Photon Interactions at High Energies (Bonn 1981); A. Silverman, ibid.

60. T.F. Walsh and P.M. Zerwas, Phys.Lett. 93B, 53 (1980)

61. P.B. Mackenzie and G.P. Lepage, Phys. Rev. Letters 47, 1244 (1981)

62. C. Berger et al., Phys. Lett. 99B, 287 (1981)

63. C. Berger et al., Phys. Lett. 94B, 254 (1980); 99B, 287 (1981); 89B, 120 (1980)
R. Brandelik et al., Phys. Lett. 97B, 448 (1980); Z. Physik C 10, 117 (1981);
W. Bartel et al., Phys. Lett. 113B, 190 (1982)
G.S. Abrams et al., Phys. Rev. Letters 43, 477 (1979)
D.L. Burke et al., Phys. Lett. 103B, 153 (1981)
H.-J. Behrend et al., DESY 82-008 (1982)

64. S.M. Berman, J.D. Bjorken, and J.B. Kogut, Phys.Rev. D4, 3388 (1971)

65. S.J. Brodsky, T. DeGrand, J. Gunion and J. Weis, Phys.Rev. D19, 1418 (1979); Phys.Rev. Lett. 41, 672 (1978)

66. W. Bartel et al., Phys. Lett. 107B, 163 (1981)

67. R. Brandelik et al., Phys. Lett. 107B, 290 (1981)

68. J.A.M. Vermaseren, in $\gamma\gamma$ Collisions, Proc. Int. Workshop at Amiens, April 1980; Vol. 134 of Lecture Notes in Physics, Springer 1980, p. 35

69. Y. Nambu and M.Y. Han, Phys.Rev. D10, 674 (1974);
J.C. Pati, Proc.Int. Summer Inst. on Theor. Particles Physics, Hamburg 1975, p. 384

70. E. Witten, Nucl.Phys. B120, 189 (1977);
C. Llewellyn Smith,Phys.Lett. 79B, 83 (1978);
W.R. Frazer and J.F. Gunion, Phys.Rev. D20, 147 (1979);
C. Peterson, T.F. Walsh, and P.M. Zerwas, Nucl.Phys. B174, 424 (1980)

71. C. Berger et al., Phys.Lett. 107B, 168 (1981)
W. Bartel et al., DESY 82-064 (1982), to be published
H.-J. Behrend et al., DESY 82-065 (1982), to be published

PROBABILITY, STATISTICS, AND ASSOCIATED COMPUTING TECHNIQUES

F. James

Data Handling Division CERN, Geneva

1. INTRODUCTION

The experimental physicist, having already devoted considerable effort to designing his experiment and collecting his raw data, is usually faced with the problem of using considerable statistical analysis of this data in order to produce statements about the system under study. He naturally wants and expects to find optimal statistical techniques to provide a unique and unambiguous quantitative measure of the significance of the data. The object of these lectures is to explore the extent to which this is possible. In some cases we will be able to find techniques which are optimal, but in others we will be faced with a kind of complementarity reminiscent of Heisenberg's principle of quantum mechanics which prevents us from attaining all our statistical goals simultaneously.

1.1 STATISTICS AS THE INVERSE OF PROBABILITY

Probability theory is a branch of pure mathematics, and although physicists often use it just as they use algebra, it has nothing to say directly about the physical world. It was first developed systematically in the second half of the seventeenth century, mostly in France.

The classical problem in probability is: 'Given the probabilities of certain elementary events, find the probability of a more complicated combination of such events.' For example, given that events occur randomly in time, with equal probability per unit time, and given that this probability per unit time is known to be, for example, one per second, what is the probability of observing exactly three events in an interval of one second? As is well-known, this problem was solved centuries ago by Poisson, and the answer is found by evaluating the Poisson function:

$$P(n) = k^n e^{-k} / n!$$

where k=1.0, the expected number of events in the time interval, and n=3, so the probability is 0.0613.

Now let us consider the inverse problem: 'Given that a certain number of events have been observed to occur in a given time interval, and given that events are known to occur randomly with constant, but unknown probability k per unit time, what can we say about the value of the constant k?' Here we leave the domain of pure probability and enter the realm of statistics, also a branch of mathematics, but considerably more recent and much closer to experimental science. The systematic development of statistics did not begin until this century, and was clearly linked to the development of experimental techniques, not only in physics but especially in biological sciences.

Already the statistical question is more vague than we would expect for a proper mathematical problem, and as we shall see below, the answer furnished by statistics is open to at least two different interpretations according to whether one adopts the Bayesian or non-Bayesian point of view. All this is not really surprising since the question we are asking (and therefore the mathematics we are going to use to solve it) is basically _inductive_ whereas we are used to mathematics being _deductive_.

2. NORMAL THEORY OF PARAMETER ESTIMATION

In this chapter we address ourselves to parameter measurement (for which we use the statistical term _estimation_), and in particular to the determination of measurement uncertainties, known to physicists as _error calculation_ and known to statisticians as _interval estimation_.

It is assumed here that the reader is familiar with the usual least-squares fitting procedure, which is an optimal statistical method in many cases, and the discussion here will be centered on the assignment of statistical uncertainties, or confidence intervals, to the fit parameters. Later on we will consider cases where least-squares is not the optimal method.

2.1 NORMAL THEORY (GAUSSIAN MEASUREMENTS)

A large class of phenomena yield variables which are distributed according to a normal or Gaussian distribution.[1] A discussion of where and why we should expect to meet such a distribution is given in the subsection below; let us assume for the moment that we are dealing with a random variable such as the measurement of the length of a table, where the variable is distributed according to:

$$P(x<x'<x+dx) = c \exp[-(x-e)^2/2s^2]\ dx$$

where e is the true length of the table and s is the standard deviation of the measurement.

The above formula should be read:"The probability of observing (measuring) a value between x and x+dx is proportional to dx and to the exponential of $-(x-e)^2/2s^2$." [The normalisation constant c does not interest us here, but it is important as the total probability for observing any x′ must be one.] In order to calculate this probability one must know the true value e, which in practice we can only determine by using measurements x, so the whole process appears to be circular. In fact, although we don't know e, we do know a value of x, our measurement, and interestingly enough we notice that the formula is symmetric upon interchange of x and e, so that we can consider that it also gives the distribution of e for a given x.

As we will see in more detail below, the precise interpretation of the Gaussian formula will depend on which of the two schools of statistics we choose to follow:

1. The classical approach is that e is unknown but fixed, and we can talk only about probability distributions for observing x.

2. The Bayesian approach allows for distributions of degrees of belief in the value of e.

In any case the symmetry of the Gaussian distribution in e and x causes the numerical determination of the statistical significance of the measurement to be exactly the same in the two cases, namely

---

[1] As used in these lectures, the words normal and Gaussian are completely equivalent.

the area under the relevant part of the Gaussian probability density curve. For example, the probability that e and x are less than s apart (one standard deviation) is about 68%; less than 2s apart is about 95%, etc.

2.1.1 <u>The universality of the Gaussian distribution</u>

Physicists are so used to the usual rules-of-thumb connecting standard deviations and significance levels (for example, "two standard deviations is a 5% effect.") that they often do not realize that the usual correspondence is only true for variables (measurements) which are Gaussian distributed. The assumption is that everything is Gaussian distributed, and indeed simple measurements of continuous physical phenomena usually do appear to assume a Gaussian shape, when a series of independent identical measurements are made and the results plotted as a histogram. On the other hand it is easy to see that this cannot be an exact and universal effect, since if it were true for some variable y, it would in general not be true for any function of y such as y^2.

There is in fact some mathematical grounds for "universal normality", as expressed by the Central Limit Theorem. This theorem states that the sum of n random variables approaches a Gaussian distribution for large n, no matter how the individual variables are distributed. This amazing but well-known theorem, which apparently is contradicted by the reasoning given directly above,[2] would explain universal normality if a complex measurement could be considered as the sum of a large number of component measurements. For example, if the inaccuracy in measuring some parameter results from the additive combination of a large number of smaller elementary inaccuracies, the total inaccuracy should be approximately Gaussian, for identical but independent measurements.

[2] The resolution of this paradox resides in the nature of convergence to limits for distributions of random variables. The convergence will be different for different functions of the random variable, even if the limit is the same.

2.1.2 Real-life resolution functions

If the sum of random variables is asymptotically Gaussian distributed, even when the individual random variables are not, the sum of probability distributions is in general not at all Gaussian, even if all the component distributions are Gaussian, as long as they have different widths.

Let us assume, for example, that the process of measuring the length of a table with some measuring instrument provides Gaussian distributed values of the length, with a given standard deviation s. We make a series of measurements with this instrument, but then continue with another instrument having a different accuracy and therefore a different value of s. Now if we look at the combined histogram of all measurements together, we will see a superposition of two Gaussians, with an overall shape which is not at all Gaussian, even though the basic process is Gaussian.

This situation is familiar in high energy physics where the resolution function is in general a superposition of Gaussians of different widths lying in a continuous range corresponding to the fact that the accuracy with which one can measure the momentum and angles of a particle track are continuous functions of the momentum and angles themselves. The resulting resolution function is always more sharply peaked than a Gaussian (due to the measurements with exceptionally small errors) and also has longer tails than a Gaussian (due to measurements with the largest errors).

2.2 COMBINATION AND PROPAGATION OF UNCERTAINTIES

It often happens that one wants to know the uncertainty of a certain value which is not measured directly but is a known function of other quantities whose uncertainties are known. For example, we may estimate the mass of the μ lepton using the mass of the π meson (and its uncertainty) and the mass difference π-μ, with its uncertainty. Physicists call this propagation of error, since the error in the final result arises from combining the errors of the two component values. It can be considered simply as a transformation of variable rather than a true statistical problem, but it so often arises in a statistical context that it is appropriate to discuss it here.

2.2.1 The sum or difference of two variables

It follows directly from the definition of expectation and variance that the expectation of the sum of two independent random variables is the sum of the expectations of the variables, and the variance of the sum is the sum of the variances. (It is important to note that this is true only when the two component variables (measurements) are independent.) This means that the standard deviation of the sum (or difference) is the square root of the sum of the squares of the individual standard deviations. As long as the individual variables are independent and their variances are finite, this is an exact result for any number of measurements and for any distribution of deviations. If, in addition, the individual measurements are Gaussian, the sum or difference will also be Gaussian, and in fact -- by the Central Limit Theorem -- the sum or difference will always be "more Gaussian" than the individual distributions.

2.2.2 Local theory, or the propagation of small errors

Apart from the simplest case of sums or differences, the exact calculation of uncertainties of transformed variables can be extremely complicated and highly dependent on the exact distributions involved. For this reason one usually uses only the local properties of the distributions around the estimated parameter values, which is usually an excellent approximation, especially when errors are indeed small. Thus one linearizes the transformations by using only the first derivatives of the new variables with respect to the old variables:

$$(\Delta R)^2 = \sum_{ij} \frac{\partial R}{\partial \theta_i} V_{ij} \frac{\partial R}{\partial \theta_j} + \ldots$$

where $R=R(\theta_1,\theta_2,\ldots)$ is the new variable and V_{ij} is the variance-covariance matrix of the component variables θ. The three dots indicate that there are higher order terms containing higher derivatives of R with respect to θ, but the usual linear approximation is to neglect the higher terms and keep only the sum given here.

2.2.3 Error on the ratio of two continuous variables

If x and y are two independent variables with variances σ_x^2 and

σ_y^2, then tstraightforward application of the linear approximation above gives for the variance of the ratio $R=x/y$:

$$(\Delta R/R)^2 = \sigma_x^2/x^2 + \sigma_y^2/y^2$$

This is a well-known rule-of-thumb, often used by physicists. It is interesting to see how close it is to the exact answer, assuming x and y to be Gaussian-distributed variables. The shocking answer is that the exact variance of R in this case is infinite! At first glance, hardly a good approximation. In fact, as long as $\sigma_x \ll |x|$ and $\sigma_y \ll |y|$ the approximation is good locally, and gives about the right width to the distribution of R, but it underestimates the extreme tails of the distribution which cause the integrated square deviation of R to diverge. The approximation is in some sense closer to what the physicist really wants to know than the exact answer, and indeed the exact 68% confidence interval for R (which is of course finite) is close to that given by the linear approximation. The physicist should however be aware of the fact that the distribution of R deviates strongly from a Gaussian in the tails, especially when the errors in x and y are not small compared with the values of x and y (especially y).

2.2.4 Error on the ratio of two discrete variables

An important case where the local approximation does not work is where small samples of discrete data are involved. Fortunately, exact methods are often relatively easy to apply for these cases, so good solutions can be found, but the usual approximations must be avoided.

Consider the measurement of a branching ratio for the decay of an unstable particle with only two decay modes. We observe, in a given time interval, n decays of one mode and N-n decays of the other. The branching ratio may be estimated as $R \simeq n/N$, and it is tempting to estimate the uncertainty on R by combining uncertainties of $\sqrt{n}$ and $\sqrt{N}$ using the formula given above for the error of a ratio. Two difficulties arise: (1) n and N are not independent, and (2) the linear approximation is very poor when n and N are small (say less than 10). The exact solution of this problem follows from the fact that the distribution involved is really binomial (since there are only two outcomes for a decay). A complete treatment of this problem including examples of how it arises in physics experiments (and how some experimenters have published incorrect error analyses by using the approximation when it was not justified) is given in James and Roos (1980).

2.2.5 The propagation of large errors

The question now arises of what to do in the general case for continuous variables when the linear approximation for error propagation is suspected of being poor. Straightforward calculation of the distribution of the new variable R involves complicated integrals over the component distributions which, even if they are independent Gaussians, quickly become intractable, and one must resort to numerical calculations even in relatively simple cases.

One such case came up recently in the analysis of an experiment by Reines, Sobel, and Pasierb which gives evidence for the instability of the neutrino. This result is of the greatest importance in high energy physics since it had generally been believed that all neutrinos were massless and could not decay. In view especially of the cosmological consequences of massive neutrinos, it is necessary to determine the significance of these results accurately. The final result of the experiment is the measurement of the ratio of two cross sections, let us call this R. Expressed in terms of the elementary quantities measured in the experiment, it can be written as:

$$R = \frac{a}{\dfrac{d}{k^2 e}(b-c) - 2\left(1 - \dfrac{k^2 d}{ke}\right) a}$$

where

$$\begin{aligned} a &= 3.84 \pm 1.33 \\ b &= 74 \pm 4 \\ c &= 9.5 \pm 3 \\ d &= 0.112 \pm 0.009 \\ e &= 0.32 \pm 0.002 \\ k &= 0.89 \end{aligned}$$

Straightforward application of the linear approximation gives:

$$R \simeq 0.191 \pm 0.073$$

But theoretical calculations show that the neutrino is unstable if R is less than about 0.42. Therefore, based on approximate error analysis, the result appears to be very significant: 3.2 standard deviations or about one chance in a thousand that the neutrino is stable.

However, two of the elementary quantities have large errors, and two quantities enter into the formula twice, producing correlations. In addition, there are several fractions, which we have seen cause non-Gaussian distributions, so let us try to calculate the exact confidence intervals for R. The easiest (and perhaps the only) way to do this is by Monte Carlo. Choose values of a,b,c,d,e randomly according to the appropriate Gaussian distributions (we will be optimistic and assume that at least the elementary measurements are Gaussian with known variances), and plot the resulting values of R. The FORTRAN program to do this is so simple that I include it here (Calls to subroutines beginning with H are for the HBOOK histogramming package; NORRAN is a Gaussian random number generator; all subroutines called here are from the CERN Program Library):

```
      PROGRAM REINES(INPUT,OUTPUT)
C          CALCULATION OF ERROR ON NEUTRAL TO CHARGED CURRENT
C          NEUTRINO INTERACTIONS, D'APRES REINES AND ROOS.
C
C     SET UP HISTOGRAM OF R
      CALL HBOOK1(1,10H N OVER D  , 50 ,0.,0.5,0.)
C
C  FILL HISTOGRAM BY LOOPING OVER RANDOM SAMPLES OF R
      DO 100 I= 1, 10000
      CALL NORRAN(XN)
      XN = XN*1.33 + 3.84
      CALL NORRAN(X112)
      X112 = X112 * .009 + 0.112
      CALL NORRAN(X74)
      X74 = X74 * 4. + 74.
      CALL NORRAN(X95)
      X95 = X95 * 3. + 9.5
      CALL NORRAN(X32)
      X32 = X32 * 0.02 + 0.32
      X89 = 0.89
      D1 = X112*(X74-X95)/(X89*X32)
      D2 = 2.0 * XN * (1.0 - (X89*X112/X32))
      XXX = XN/(D1-D2)
      CALL HFILL (1,XXX)
  100 CONTINUE
C
C  ASK FOR PRINTING OF HISTOGRAM, WITH INTEGRATED CONTENTS
      CALL HINTEG(1, 3HYES)
      CALL HISTDO
C
      STOP
      END
```

The histogram showing the distribution of the 10000 Monte Carlo values of R is shown in Figure 1. Those of you familiar with the reading of HBOOK output will quickly find the significant number,

```
     500                        -
     480                      - I-
     460                      I-II--
     440                     -I    I
     420                   --I     I
     400                   I       I-
     380                   I        I
     360                   I        I
     340                 --I        I-
     320                 I           I-
     300                -I            I -
     280              -I              I-I
     260               I                I-
     240               I                 I-
     220               I                  I-
     200              -I                   I-
     180              I                     I--
     160              I                       I-
     140            --I                        I
     120            I                          I---
     100           -I                             I--
      80           I                                I-
      60          -I                                 I----
      40         -I                                      I------
      20       --I                                             I-

CHANNELS  10   0         1         2         3         4         5
           1   12345678901234567890123456789012345678901234567890

CONTENTS 100        1112233444444444333222221111111
          10   12358238693201265965493168531876401089754543322221
           1.  90992724101391557441788476794339467366374772477661

INTEGRAT1000            11122334445566677778888899999999999999999
         100      12346925826150594826924792357890123455666777888
          10   13731476214678173294982409580852779988515159257034
           1.  99879682334767274823086073093698285840304180418401

LOW-EDGE  1.             1111111111222222222233333333334444444444
*10**  1   0   01234567890123456789012345678901234567890123456789

 ENTRIES = 10000   UNDERFLOW = 23   OVERFLOW = 136
 MEAN VALUE = .2009E+00      R . M . S =.9219E-01
```

Figure 1. Distribution of R for experiment of Reines et al. namely the number of entries falling above 0.42. This is almost 4%, so that the true significance of the result is only 4% instead of the apparent 0.1%. Notice also the skew, non-Gaussian distribution of R.

3. CONFIDENCE INTERVALS

3.1 CLASSICAL THEORY

The classical theory of confidence intervals allows us to find an interval (a range of parameter values) which, with a given probability, will contain the true value of the parameter being measured. We show here how such intervals can be constructed, at least in theory, for the most general case.

Let t be an estimate (measurement) of a parameter whose true value (unknown) is θ. The value of t for a given experiment will be a function of the measured data x:

$$t = t(x)$$

For any given value of θ, we can in principle always find the expected distribution (probability density) of t, which we denote by $f(t|\theta)$. This function describes the behaviour of the measurement apparatus. For the simplest case, it would simply be a Gaussian centered at $t=\theta$, with width given by the measurement accuracy of the apparatus. For the most complicated experiments it is necessary to find f numerically by Monte Carlo simulation of the measuring system.

Now consider a range of values of t from t_1 to t_2, denoted by $\langle t_1,t_2\rangle$. Since $f(t|\theta)$ is known, we can calculate, for any given value of θ, the probability of obtaining a value of t in the range $\langle t_1,t_2\rangle$:

$$\beta = P(t_1 \leq t \leq t_2|\theta) = \int_{t_1}^{t_2} f(t'|\theta)dt'$$

Similarly, for a given value of β, $0<\beta<1$, it is possible to find values of t_1 and t_2 satisfying the above relation, namely that the probability of t lying in the range $\langle t_1,t_2\rangle$ is β. In fact, there are in general many different ranges which will satisfy this relation; it can be made unique by requiring in addition that the range be <u>central</u>:

$$\int_{-\infty}^{t_1} f(t'|\theta)\, dt' = \int_{t_2}^{+\infty} f(t'|\theta)\, dt' = \alpha$$

and therefore $\alpha = (1-\beta)/2$

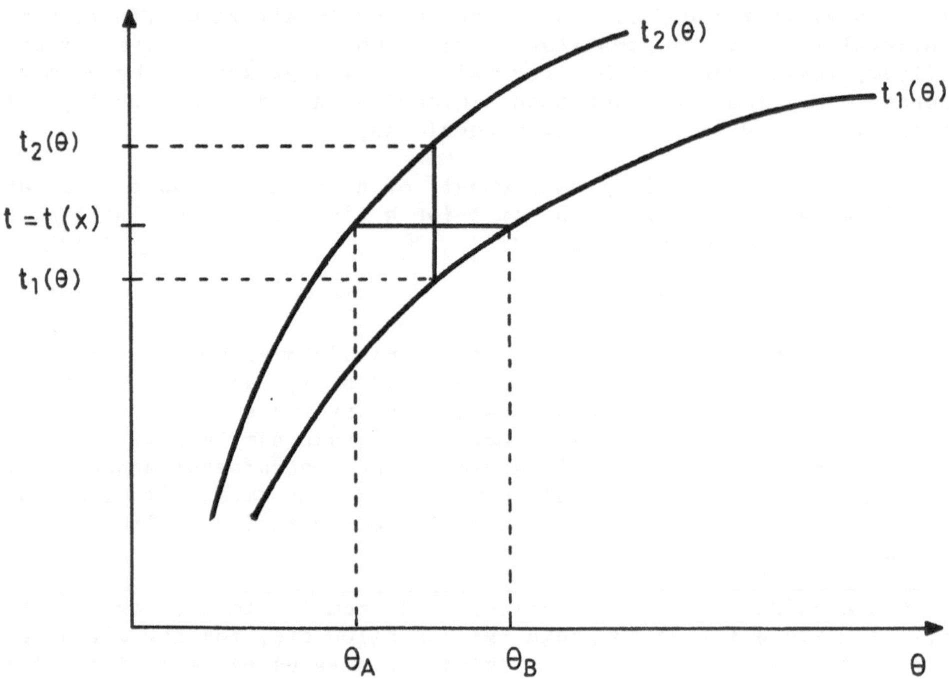

Figure 2: Constructing confidence intervals.

The above relations, for a given confidence level β, uniquely define two functions $t_1(\theta)$ and $t_2(\theta)$. Since the functions are only defined implicitly, the actual calculation of function values would in general have to be performed numerically by iteration, but it is in principle always possible, and may even be quite easy in certain simple but important cases. Thus we can draw two curves in (t,θ)-space, representing $t_1(\theta)$ and $t_2(\theta)$. Let us take the θ-axis is horizontal, and the t-axis vertical. Then along any vertical line (constant θ), the distance between the two curves represents by construction a probability content β (that is, the probability of obtaining an estimate t in that range is β for that value of θ.)

The usefulness of the diagram comes from the fact that this is also true along a horizontal line: namely, for any given value of t, the probability content of the range of θ lying between the two curves is also β. The diagram is constructed vertically, but can also be read horizontally. To convince oneself that this is true requires some mental gymnastics and a good understanding of the meaning of the probability content of a range of parameter values in the classical sense, but it is rigorously true. Since the classical interpretation of confidence intervals can best be understood by opposition to the Bayesian interpretation, this should become clearer after the discussion of Bayesian theory below.

3.1.1 Example: confidence intervals for the binomial parameter

The technique outlined above was used, for example, by Clopper and Pearson to find the confidence limits for the parameter of a binomial distribution. This is the distribution which describes processes where only two kinds of events are possible, and a given event has a constant probability of being of one kind, independent of the other events. [Examples: measuring the branching ratio for a particle with only two decay modes; measuring a forward-backward or left-right asymmetry] Then, if the probability of one event being of one type is p, the probability of n events out of a total sample of N being of that type is given by the well-kmown binomial formula:

$$P(n) = \begin{vmatrix} N \\ n \end{vmatrix} p^{n} (1-p)^{N-n}$$

where $\begin{vmatrix} N \\ n \end{vmatrix}$ is the binomial coefficient $N!/(N-n)!n!$

This was used to construct (vertically) the diagram of Figure 3, which was in turn used horizontally to solve the inverse problem: Given observed values of N and n, what confidence limits can be placed on the value of p?

Clopper and Pearson give elaborate diagrams which allow the reader to construct confidence intervals for essentially any N, n, and β. Later it was recognized that these values are in fact just those of Fisher's F-distribution with appropriately transformed ar-

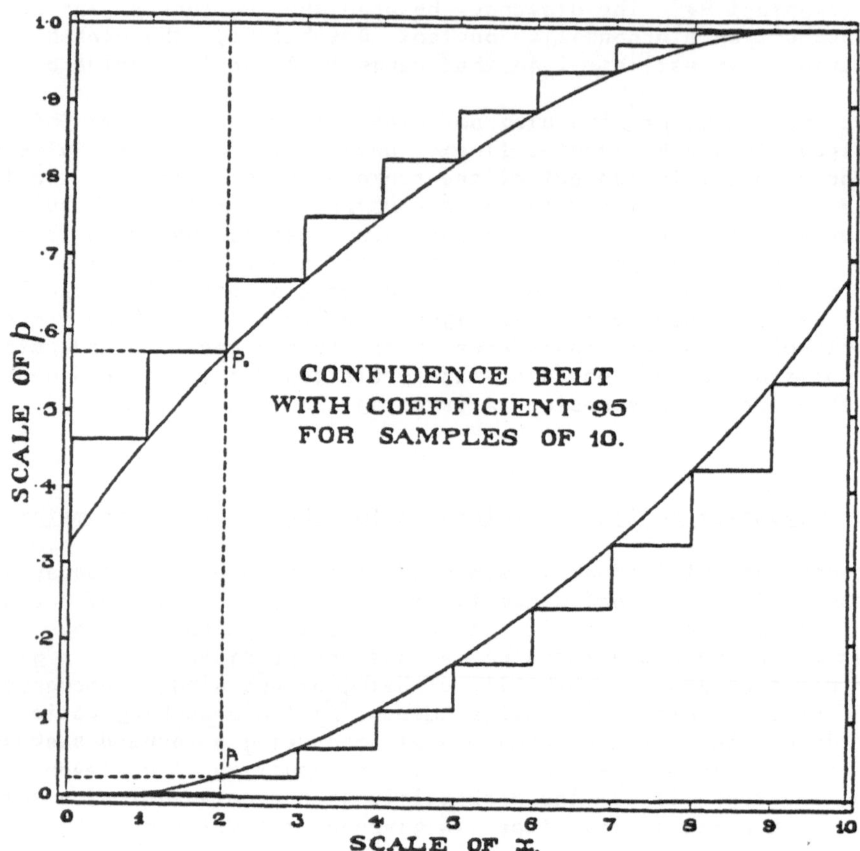

Figure 3: Clopper-Pearson confidence intervals for the binomial distribution

guments, so that existing tables or subroutines of F can be used to determine exact confidence intervals for p. Additional details concerning this problem and the way in which it arises in the context of physics experiments, may be found in James and Roos (1980).

3.2 BAYESIAN THEORY

The Bayesian theory of parameter interval estimation, which is not accepted by all statisticians, is based on an extension of the domain of validity of Bayes' theorem beyond what a "classical" statistician would deem proper. We therefore begin this section by recalling Bayes' theorem, a fundamental theorem of probability theory.

3.2.1 Bayes' Theorem

Consider a set of elements, each of which may belong to set A, or to set B, or to neither or to both A and B. Then the probability of a random element belonging to both A and B is the probability of its belonging to A given that it already belongs to B, multiplied by the probability of its belonging to B. The same is clearly true with A and B inverted, so that we can write:

$$P(A \text{ and } B) = P(A|B)P(B) = P(B|A)P(A)$$

This is Bayes' Theorem. It gives a relationship between a <u>conditional probability</u> $P(A|B)$, and the inverse conditional probability $P(B|A)$.

$$P(B|A) = P(A|B)P(B)/P(A)$$

3.2.2 The Bayesian use of Bayes' Theorem

We have assumed known the conditional probability density $f(t|\theta)$, namely the distribution of estimates t we would obtain if the true value of the parameter were known. Our experiment however gives us just the inverse: it provides a value of t and we wish to make a probability statement about θ. Apparently Bayes' Theorem is just what we need since it allows us to express a conditional probability in terms of its inverse. Straightforward application of the theorem to the case of interest gives:

$$P(\theta|t) = P(t|\theta)\ P(\theta)\ /\ P(t)$$

(Each of the above "probabilities" is in fact a probability density and should be followed by a dθ or a dx, but these differentials clearly cancel, so that we can work directly with the probability densities.)

Now let us examine each of the factors in the above expression:

1. $P(\theta|t)$ is just what we want (says the Bayesian), the probability density for the true value of the parameter θ, given the measured value of t. The "classical" statistician says that although θ is unknown, it has only one true fixed value, and it does not make sense to talk about the probability of its taking on any value. The Bayesian counters by calling $P(\theta|t)$ the "degree of belief" in the value θ, and says that as long as the true value is not known, this quantity behaves like a probability.

2. $P(t|\theta)$ is nothing but our old friend $f(t|\theta)$.

3. $P(\theta)$ presents a serious problem both in interpretation and in practical calculation. This quantity is known as the prior knowledge of the parameter θ. The Bayesian claims that we nearly always have prior knowledge about the phenomena we study, and that the experimental results will always be interpreted in terms of this knowledge anyway, so why not build it into the results from the beginning? Even the Bayesian admits however to practical difficulties in expressing vague knowledge about θ, and especially in expressing complete ignorance, a subject to which statisticians have devoted considerable effort. Moreover, the non-Bayesian insists that it must be possible to express the results of an experiment independently of other outside knowledge, but this is apparently impossible in the Bayesian formulation.

4. $P(t)$, the a priori probability of observing t, is apparently even more intractable than $P(\theta)$, but in fact can be reduced to the same problem since we can express it as:

$$P(t) = \int_{-\infty}^{+\infty} P(t|\theta)\, P(\theta)\, d\theta$$

3.3 USE OF THE LIKELIHOOD FUNCTION

In practice, Bayesian theory of confidence intervals is not used by physicists, probably because of the problem of subjectivity involved in the prior knowledge, and also because of practical difficulties arising in computation. Similarly, the classical technique of construction of exact confidence intervals is applied primarily to the solution of rather general problems (such as the example of 1.3.1) and is rarely used to estimate errors for particular experiments. The reason why these fundamental exact techniques are not used is that there exists a much simpler approximate method for obtaining interval estimates by using the likelihood function directly.

Consider the log-likelihood function for the simplest possible experiment: the direct measurement of a quantity (θ) using a measuring engine which produces Gaussian-distributed errors of zero bias and known variance σ^2:

$$\ln L = \ln f(t|\theta) = \ln\left| \frac{1}{\sigma\sqrt{2\pi}} \exp[-(t-\theta)^2/2\sigma^2] \right|$$

$$= -(t-\theta)^2/2\sigma^2 + \text{const.}$$

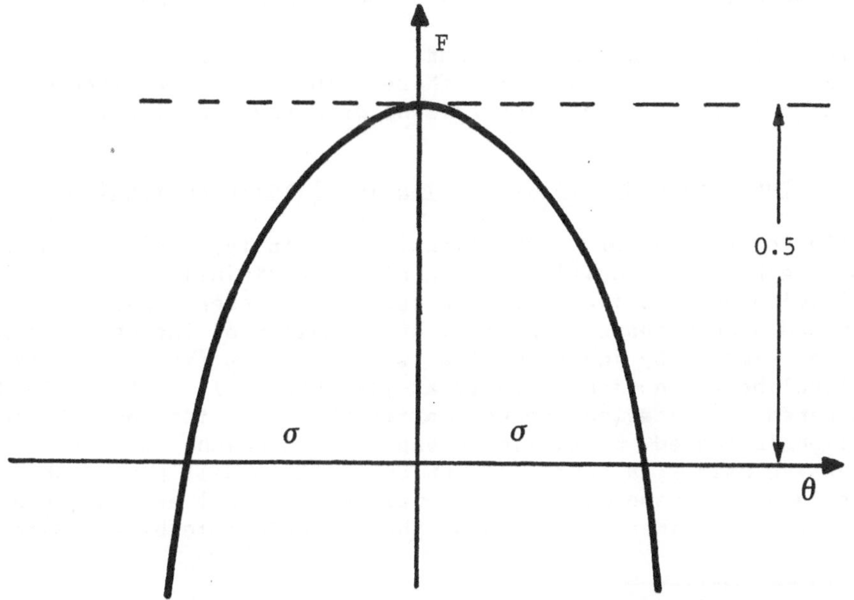

Figure 4. Parabolic log-likelihood for Gaussian measurement.

This is the parabola shown in Figure 4. We know by construction that this represents a measurement with standard deviation $= \sigma$. That is, if θ_0 is the value of θ for which ln L has its maximum, then:

$$P(\theta_0-\sigma \leq \theta_{true} \leq \theta_0+\sigma) \simeq 0.68$$

$$P(\theta_0-2\sigma \leq \theta_{true} \leq \theta_0+2\sigma) \simeq 0.95$$

etc.

Now if the value of σ is not known, it can be measured from the shape of the parabola in either of two ways:

1. By measuring the second derivative of the log-likelihood function at its maximum (or in fact anywhere, since it is constant). Then:

$$\sigma^2 = \left|\partial^2 \ln L/\partial\theta^2\right|^{-1}$$

2. Or by taking $\sigma=|\theta_1-\theta_0|$, where θ_1 is either of the two points where

$$\ln L(\theta_1) = \ln L(\theta_0) - 1/2.$$

   We will call this second way the method of MINOS.[3]

For the case at hand, both methods give the same value of σ. We consider below application of these methods to the more general case of a likelihood function arising from any experiment.

3.3.1 The second derivative of the log-likelihood function

Largely for reasons of computational simplicity, this is the most common method of calculating parameter uncertainties directly from the likelihood function. In the case where several parameters are estimated simultaneously the full error matrix of the parameters is obtained simply by inverting the second derivative matrix of the log-likelihood function (usually at its maximum). This method is considered an excellent approximation whenever the log-likelihood function is indeed parabolic in shape, (in which case the second derivative matrix is constant) which in turn will happen whenever there is a large amount of data, small measurement errors, or a model which is a linear function of the parameters to be estimated.

---

[3] MINOS is the name of the subroutine in MINUIT which implements this method. Other references to MINUIT will be made in these lectures for the benefit of MINUIT users. MINUIT is available from the CERN Program Library and is described in James and Roos (1975).

3.3.2 The method of MINOS

This method is much more recent than the second-derivative method; it was probably originated by Sheppey (CERN, unpublished) around 1965 and first implemented in a general program MINROS (now obsolete, superceded by MINUIT) shortly afterward. It is believed to be valid (at least to a good approximation) also in the general non-linear case where the log-likelihood function is not parabolic.

The justification for its validity for non-parabolic cases lies in the fact that it is invariant to any (even non-linear) transformation of the parameter θ. That is, if the 68% confidence interval as determined by MINOS for parameter θ is $\langle\theta_1,\theta_2\rangle$, and we make a transformation of variables to $\phi=\phi(\theta)$, then the 68% confidence interval in ϕ will be $\langle\phi_1,\phi_2\rangle$, where $\phi_1=\phi(\theta_1)$ and $\phi_2=\phi(\theta_2)$. Now there must be some (in general non-linear) transformation of θ which would transform the log-likelihood function to an exact parabola, for which the method is believed to give a very good approximation to the exact interval. And since the method is invariant under such a transformation, it must also give a good approximate answer in the general case, unlike the second-derivative method.

Note that this method will in general yield asymmetric intervals, with different parameter uncertainties in the positive and negative directions. Intervals may also be non-linear in the sense that two-standard-deviation errors [given by $\ln L(\theta_1)=\ln L(\theta_0)-2.0$] are not necessarily twice as big as one-standard-deviation errors [given by $\ln L(\theta_1)=\ln L(\theta_0)-0.5$].

3.4 MULTIPARAMETER CONFIDENCE INTERVALS

When more than one parameter is estimated simultaneously in an experiment, some additional problems in interpretation may arise. The correct procedure in such cases is straightforward, although it may be somewhat complicated, and is discussed in some detail in James (1980).

4. LEAST SQUARES OR NOT LEAST SQUARES

The basic method of parameter estimation is based on the maximization of the likelihood function, which in many cases reduces to the method of least squares. The traditional method of fitting particle trajectories to position measurements is that of least squares. This method is certainly justified for detectors such as bubble chambers for which measurements of points on tracks are approximately Gaussian-distributed and independent. The same method is however also used for quite different detectors such as multiwire proportional chambers (MWPC) in which position measurements are neither Gaussian nor independent. The resulting difficulties have long been recognised, namely:

1. The track-parameter uncertainties as calculated by standard least-squares techniques are often incorrect (generally too small).

2. The overall distribution of the chisquare statistic does not follow the expected law.

3. The value of the sum of squares of residuals after fit is not a reliable measure of the goodness of the fit. (This is related to the preceeding point.)

Typically various "corrections" are applied to compensate for these shortcomings. For example, parameter errors may have to be scaled up, or detector planes may be omitted in the fit, and chisquare cutoffs for acceptance of fits may have to be determined empirically since the chisquare distribution obtained may be very far from that expected ideally.

A more fundamental approach has been taken by Drijard, Ekelof, and Grote 1980 who show that the effects of correlations between wire planes can largely explain the observed behaviour of least-squares fits, particularly the slower-than-expected convergence as more wire planes are added.

We present here a quite different approach to the fitting of tracks in any detectors which are "quantized", that is, where a signal in one element (or in several adjacent elements) assures that the track trajectory passed within a certain "window", but where the probability of having passed through any given point within this window is *a priori* uniform. The best example of such a detector is the MWPC (for which we give actual results below) but these techniques can probably be applied as well to silicon strip detectors and other devices.

Although this study was originally motivated by a desire to understand and correct the anomalous behaviour of least-squares track fitting in wire chambers, another major goal is to find a more efficient estimator as well.

4.1 LEAST SQUARES

The justification for using least squares is that it is equivalent to the maximum likelihood method when the residuals (measurement errors) are Gaussian-distributed. This can be seen very easily, since the (log-) likelihood function is just the sum of logarithms of Gaussian terms:

$$L(\underline{\alpha}) = \prod_i e^{-(t_i(\underline{\alpha})-y_i)^2/2\sigma_i^2}$$

$$\ln L(\underline{\alpha}) = \sum_i \ln e^{-(t_i(\underline{\alpha})-y_i)^2/2\sigma_i^2}$$

$$= \sum_i -(t_i(\underline{\alpha})-y_i)^2/2\sigma_i^2$$

which is, apart from a factor of -1/2, the well-known quadratic form of chisquare. The best fit is then obtained by maximizing (or dividing by -2 and minimizing) this expression with respect to the vector of track parameters $\underline{\alpha}$. Notice first-of-all that if the measurements y_i are correlated, then the quadratic form must include off-diagonal terms:

$$\chi^2 = \sum_{ij}\sum (t_i(\underline{\alpha})-y_i)\ V_{ij}\ (t_j(\underline{\alpha})-y_j)$$

But notice also that even if one takes account of such correlations, the method still corresponds to maximum likelihood if and only if the residuals t_i-y_i are Gaussian-distributed. That is, the quadratic form of chisquare is a direct result of the assumption of Gaussian measurement errors, and we should expect that it would no longer be optimal when residuals are distributed in some other way.

4.1.1 The Gauss-Markov Theorem

Justification for the use of least squares is often based on the well-known Gauss-Markov theorem which states that under the usual regularity conditions and for linear models, the linear least-squares estimator has the minimum variance among all estimators which are unbiased and linear functions of the data. Restricting oneself to linear combinations of the data may have been an important consideration fifty years ago, but in the age of the computer it begins to appear ridiculous. Indeed we will give a counterexample later in this paper, in which the best estimator is considerably better than least-squares, and in addition it is just as easy to compute, although it is not strictly speaking a linear function of the observations.

I prefer to state the Gauss-Markov Theorem as follows: Any estimator which is more efficient than least-squares cannot be reduced to a linear function of the observables. Consider for example the class of estimators which minimize deviations between models and observations using the L_p norm, that is minimizing the sum of the pth powers of the absolute values of the residuals. Then for all values of p from one to infinity, only p=2 (least squares) yields an estimator that can be reduced to a linear function of the observables. In recent years, especially due to the interest in robust estimation, there has been considerable study of at least two other members of this infinite class of estimators, namely for p=1 and p=∞, which both possess very interesting statistical properties and for both of which reasonably fast algorithms exist. Since they are not linear functions of the data, the Gauss-Markov theorem has nothing to say about these estimators.

4.2 MAXIMUM LIKELIHOOD FOR UNIFORM ERROR DISTRIBUTION

It is natural to seek a solution to our problem by considering the likelihood function for the case of a uniform (rectangular) error distribution. (It is now convenient to consider the likelihood rather than the log-likelihood.) The individual factors in the likelihood function are functions which take on either the value one (if the trajectory passes through the corresponding window) or zero (if it does not). The overall value of the likelihood is therefore either one (if the trajectory passes through all the windows) or zero (if it misses at least one).

The maximum value of this likelihood function will in general be taken on not at a point, but over some region in the space of the parameters describing the track trajectory. Inside this region, where L=1, all trajectories are equally "good" since they would all give rise to the same wire hits as were observed. The entire space outside this region corresponds to trajectories which are impossible since they would have resulted in different wire hits. If the track is described by only two parameters, we can visualize this likelihood function as resembling a mesa rising out of the Arizona desert, where L=0 in the desert and L=1 on the top of the mesa. Real tracks are usually described by five parameters, but five-dimensional mesas are more difficult to visualize.

Although all trajectories corresponding to points in the allowed region are equally good, it is natural to choose the point in the middle of this region as the solution of the fit. The parameter uncertainties are completely defined by specifying the allowed region (the walls of the mesa) since all combinations of parameters lying on the top of the mesa are equally good fits. The walls of the mesa correspond then to the error ellipse of least-squares theory, except that they must be understood in the sense of an absolute bound rather than a statistical uncertainty.

The above exercise demonstrates the basic principles of a complete solution to the problem of fitting with uniform error distribution, in the sense that it allows us to understand the meaning of a best fit, a good fit, and the uncertainties of the fitted track parameters. Unfortunately it does not allow us to find these solutions numerically since one cannot maximize the likelihood function described here with standard maximization techniques, all of its derivatives being almost everywhere zero.

4.3 THE CHEBYSHEV FIT

The numerical technique used to obtain the solutions described above is to minimize the <u>maximum deviation</u> between hypothetical and measured track positions instead of minimizing the sum of the squares of the residuals. In mathematical terms we use the Chebyshev norm[4] rather than the L_2 norm. The quantity to be minimized is then:

$$C(\underline{\alpha}) = \max_i \left|(t_i(\underline{\alpha}) - y_i)/\sigma_i\right|$$

$$= \max_i \left|r_i(\underline{\alpha})\right|$$

where $t_i(\underline{\alpha})$ is the position where the trajectory described by parameters $\underline{\alpha}$ would cross the ith plane, y_i is the middle of the window established by the detector at the ith plane (i.e. the position of the wire which was hit), and σ_i is the distance from the middle of the window to the edge (i.e. half the wire spacing). Notice that σ is no longer a standard deviation, it is here an absolute bound.

It is clear that for any set of parameters $\underline{\alpha}$:

---

[4] The Chebyshev norm is the L_∞ norm since minimizing the maximum residual is equivalent to minimizing the sum of the pth powers of the residuals in the limit as $p \to \infty$.

$$\text{if } C(\alpha)<1, \quad L(\alpha)=1$$

$$\text{and if } C(\alpha)>1, \quad L(\alpha)=0.$$

Therefore the Chebyshev statistic C_{min} is the goodness-of-fit statistic with critical value 1, and if the Chebyshev solution has a value of $C_{min}>1$, then there is no allowed region (the track fit is not good; no trajectory will pass through all the windows).

Furthermore, the Chebyshev solution, if $C_{min}<1$, will in some sense be in the middle of the allowed region, so that we can consider that both the problem of finding the best fit and that of determining the goodness of fit will be solved if we can perform a Chebyshev fit. Fortunately such techniques have long been used in numerical analysis, particularly in approximation of functions, where it is desired to minimize the maximum deviation between a function and its approximation. As with least squares, only the <u>linear</u> Chebyshev problem has a straightforward algorithm, so we proceed by linearization and iterations. The algorithm used here for the linear Chebyshev solution is that of Barrodale and Phillips 1975, based on the simplex method of linear programming.

The Barrodale-Phillips algorithm exists in the form of a FORTRAN subroutine which accepts as input the vector of residuals $\underline{r}$ and the matrix of first derivatives of each r_i with respect to each element of $\underline{\alpha}$. It returns as output the solution of the linear Chebyshev problem, the point $\underline{\alpha}$ for which $C(\underline{\alpha})$ takes on its smallest value, assuming each r_i is just a linear function of $\underline{\alpha}$. It also returns the vector of residuals $\underline{r}$ at the solution and the value of C_{min}.

In practice the Chebyshev fit is computationally more expensive than a least-squares fit. For the fits I have made, the linear Chebyshev solving was considerably slower than the matrix inversion required for the usual least-squares analysis, but for reasons not yet understood by the author, the linearity hypothesis seems to be obeyed better so that fewer iterations are required for the complete Chebyshev fit. Depending on the track parameterization used, this may mean fewer trackings through the magnetic field which may compensate the longer time per iteration.

4.4 THE PARAMETER UNCERTAINTIES

The parameter errors are simply the distances from the solution point to the extremities of the allowed region in the different directions, and the boundary of the allowed region is the set of all $\underline{\alpha}$ for which $C(\underline{\alpha})=1$. Thus the positive error bound on the first track parameter α_1 is the largest value of α_1 for which $C(\underline{\alpha})=1$. Similarly the negative error bound is the smallest value of α_1 for which $C(\underline{\alpha})=1$, and so on for the other components of $\underline{\alpha}$. Geometrically this is very similar to the MINOS error analysis described above.

The numerical technique used here to find the error bounds defined above is as follows: In order to find the positive error bound on the kth parameter, find the linear Chebyshev solution to the following situation.

1. Let the parameters of the best Chebyshev fit be $\underline{\alpha}=\underline{\alpha}_0$. The residuals $\underline{r}$ and their first derivatives are known at this point.

2. Add a fictitious residual r_{n+1} with value slightly greater than one and all derivatives zero except for the kth which is slightly negative.

3. Find the linear Chebyshev solution for this modified problem. The value of α_k at this solution is the upper error bound of the kth parameter.

The above procedure is followed for each parameter in turn. For negative error bounds, the derivative mentioned in step 2 is set slightly positive instead of slightly negative.

Let us see how this method works. Since the solution to the original fit gave $C<1$, the additional residual which is everywhere almost equal to one will assure that the solution to the modified problem will occur where $C\simeq1$. If it turns out to be too far from one, additional iterations may be performed with the original value of r_{n+1} and its derivative adjusted to force the solution to be as close as desired to $C=1$. The fact that the kth derivative is slightly negative (positive) will assure that the solution gives the largest (smallest) value of α_k for which $C=1$.

4.5 THE EFFICIENCY OF THE CHEBYSHEV ESTIMATOR

Whereas it is clear that the hypothesis-testing properties of the Chebyshev method are always superior to those of the least-squares method for the situations treated here, we have not yet any measure of the relative efficiency of the two methods for the estimation of parameters. Since the Chebyshev method always finds a maximum likelihood solution (if one exists), whereas the least-squares does not necessarily, we expect the former to be asymptotically superior and hopefully even optimal, but we would like to have a more precise idea of what to expect on theoretical grounds.

A somewhat simpler but closely related situation in which the behaviour of different estimators has been studied in some detail is that of the estimation of the center of a distribution [Rice and White 1964, summarized in Eadie et al 1971, p. 187]. The exact problem is: Given a sample of N values $x_i, i=1,N$, drawn randomly and independently from a uniform distribution of unknown central value, what are the variances of the least-squares and Chebyshev estimators of the center of the distribution? The answer[5] is:

Least-squares: μ = sample mean = $(\Sigma x_i)/N$
$V(\mu) = 1/12N$

Chebyshev: μ = sample midrange = $(x_{max} + x_{min})/2$
$V(\mu) = 1/(2N^2+6N+4)$

For very large N, the variance of the Chebyshev estimate is smaller by a factor of N/6, so that the Chebyshev error is smaller by a factor $\sqrt{(N/6)}$. For small N, the difference is less striking, but the Chebyshev estimate remains better for all N right down to N=3. For N=1 and N=2, the two estimators are of course the same.

For situations as complicated as the track-fitting problem, the statistical properties of Chebyshev estimators are apparently unknown, and we can only speculate that their behaviour might be similar to that of the location estimators given above. Then we would expect the least-squares estimator to be as good as Chebyshev only for six wire planes (since there are now five free parameters instead of one), with the Chebyshev estimator becoming progressively superior as the number of wire planes increases. If the analogy is exact, the Chebyshev estimator should become twice as good as least-squares (i.e. one-fourth the variance) when the number of wire planes $N+5 \simeq 25$. The effect of the correlations observed by Drijard et. al. 1980 is that some of the planes do not add any

[5] The variance for the Chebyshev estimate is given in its asymptotic limit $1/2N^2$ in Rice and White 1964. The expression given here and valid for all N was calculated by the author.

information, so that for track-fitting one would expect the factor-of-two improvement to be obtained only for a number of planes somewhat larger than 25.

4.6 ERROR SYMMETRIZATION

The parameter errors can be made symmetric if the fitted parameter values are chosen to be the middle of the allowed region rather than the Chebyshev solution. Such a procedure is somewhat arbitrary since it is not invariant under non-linear transformations of the parameters α, but this particular choice facilitates later stages in the event analysis.

On the other hand, if we adopt the "unsymmetrized" Chebyshev solution, we will have to carry around asymmetric errors for the rest of the analysis, but then the solution is invariant under even non-linear transformations of the parameters, that is, it is independent of the way the tracks are parameterized.

Fortunately, the numerical experiments I have made indicate that the symmetrized estimates are slightly more efficient than the unsymmetrized, so we have used symmetrized estimates here. Of course asymptotically the errors should become symmetric so it should not matter which solution is taken.

4.7 ROBUSTNESS VS. EFFICIENCY

In the above sections we have outlined a method of track fitting and error analysis which is in some sense statistically optimal for the situation assumed. Unfortunately, optimal statistical efficiency is generally coupled with extreme sensitivity to the correctness of the assumptions made, for the simple reason that statistical optimality is achieved by taking advantage of all the information about the problem. If some of this "information" is not quite true, the results may be catastrophic.

It is therefore necessary in practice to use extreme care in applying the Cebyshev method to real wire chambers. The behavior of the chambers must be well understood, including the phenomenon of "clustering". And the method must be protected against all kinds of noise such as pattern-recognition errors. As a bonus, however, one obtains a rather powerful tool to detect such noise, since the Chebyshev goodness-of-fit criterion is very precise.

5. TESTING OF HYPOTHESES

In this chapter, we consider choosing between two or more well-defined hypotheses, for example measuring the parity of an elementary particle, which can only be positive or negative.

5.1 THE TEST STATISTIC AND OTHER DEFINITIONS

In order to discuss the theory of hypothesis testing, a few basic concepts need to be defined, with the corresponding notation. The notation, although rather arbitrary, is fairly well standardized throughout the statistical literature, and we adopt here the most commonly used symbols.

The two hypotheses under consideration may have free parameters whose values must also be estimated (composite hypotheses), but we assume for the moment that they are completely defined (simple hypotheses). The simplest or most important of the two hypotheses is denoted H_0 and called the <u>null hypothesis</u>, probably because it often corresponds to "no effect" or "zero dependence". The other hypothesis, H_1, is called the <u>alternative hypothesis</u>. Our aim is to choose between the two hypotheses, based on some observations, and especially to be able to express the significance of those observations in distinguishing between the two hypotheses.

We denote the space of all observations by W. This includes all possible outcomes of our experiment. We will want to divide this space into two regions: <u>the critical region</u>, ω is the set of all observations for which we would reject H_0, and the <u>acceptance region</u>, $W-\omega$, for which we would accept H_0. This will be done with the help of a function X(W) called the <u>test statistic</u>. The study of hypothesis-testing is thus reduced to the study of the properties of different test statistics and the resulting regions of rejection and acceptance.

We can now define the important quantities in terms of probabilities of obtaining different values of the test statistic:

1. The <u>level of significance</u> of the test, α, is the probability of X falling in the critical region ω when H_0 is in fact true.

$$P(X\epsilon\omega|H_0) = \alpha$$

2. The <u>power</u> of the test, $1-\beta$, is the probability of X falling in the critical region when H_1 is true:

$$P(X \in \omega | H_1) = 1-\beta$$

or, alternatively:

$$P(X \in (W-\omega) | H_1) = \beta$$

3. The error of the first kind, or loss, occurs when H_0 is rejected even though it is true. The probability for this happening is clearly α.

4. The error of the second kind, or contamination, occurs when H_0 is accepted even though the alternative H_1 is true. The probability for this is from the above definitions β.

A good statistical procedure for testing hypotheses will therefore be one which chooses the function X and the region ω so as to minimize both α and β.

5.1.1 Example: Separation of two types of events

As an example let us suppose we are studying elastic proton-proton scattering, and we have a sample of events which contains both true elastic events and those with additional π^0 production. On the basis of the measured momenta of the protons alone, we have to decide, for each event, whether an unseen π^0 was also produced, in order to obtain finally the biggest and purest possible sample of elastic events. For each event, the two hypotheses are as follows:

H_0: $p+p \rightarrow p+p$

H_1: $p+p \rightarrow p+p+\pi^0$

We could choose as test statistic the missing mass for the event, a function of the measured momenta which, if the momenta were measured exactly, would be equal to the π^0 mass for unwanted events and zero for elastic events. Because of measurement errors, the expected distributions of missing mass under the two hypotheses will have some width, and in practice may appear as in Figure 5. It is clearly not possible to choose ω so as to make both α and β zero at the same time, although either one could be made arbitrarily small at the expense of increasing the other. Physically, this means that if we set the acceptance level so as to lose very few true elastic events, we will also have to accept a large number of background events; if, on the other hand, we require a very pure sample of elastic events, we will have to settle for a big loss of number of events. Some compromise is necessary, based on the physics to be done. In order to reduce both α and β simultaneously,

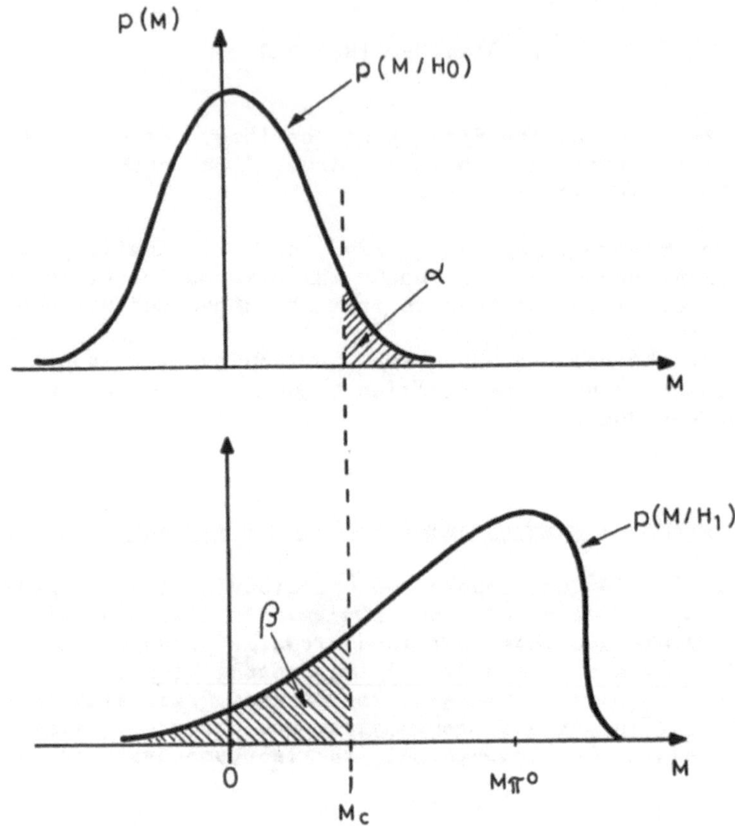

Figure 5. Test statistic distributions for H_0 and H_1.

we would have to find a better test statistic. For example, if it is known that the π^0 are usually very fast, it may be better to use missing energy or missing momentum to discriminate between the two types of events. The choice of test will be discussed in the next subsection.

5.2 CHOOSING A TEST

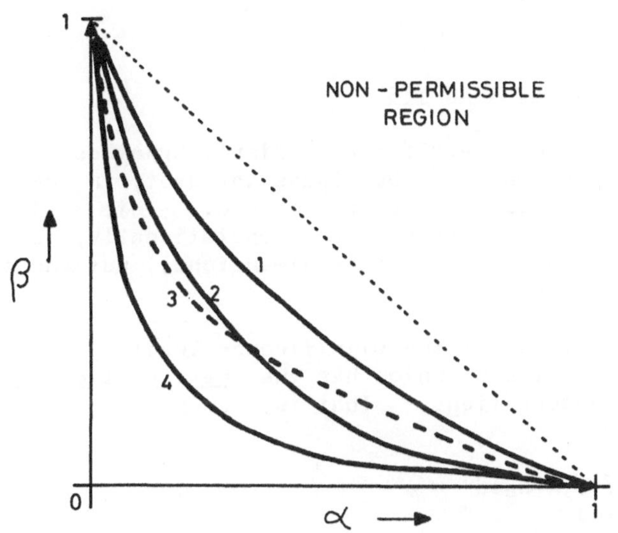

Figure 6. Basic properties of four different tests.

The properties of a test between two simple hypotheses can be seen from the diagram of Figure 6, which shows β as a function of α for four different tests. Such a curve for any admissible test must lie below the diagonal indicated, since for any point above the diagonal the probability of accepting the wrong hypothesis would be greater than the probability of accepting the right one, and surely it must be possible to do better than that.

The best test will correspond to the lowest (α,β) curve, since that will have the lowest value of β for a given value of α. Thus test one is always worse than the other three, test two is sometimes better than test three (namely if we are interested in small values of β rather than small values of α), and test four is always better than the other three.

If the test is being applied in order to select data for later analysis, additional criteria may be important, namely that the selection procedure not introduce a bias in the distributions of interest for the data selected, or at least that the bias so introduced be calculable.

5.3 THE NEYMAN-PEARSON TEST

In the case of completely defined simple hypotheses, there is one test which can be shown to be always the best, in the sense of giving always the smallest β for a given α. This test, the Neyman-Pearson test, may be very complex computationally, especially when the space of observables is many-dimensional, but can always be defined in principle.

For a given value of the significance level α, the most powerful test will be that one which has the best critical region in the space of the observations. That is, among all those regions $\omega(\alpha)$ which satisfy

$$\int_{\omega(\alpha)} f(X|H_0)dX = \alpha$$

we wish to find that region which minimizes β, or maximizes the power

$$1-\beta = \int_{\omega(\alpha)} f(X|H_1)dX$$

$$= \int_{\omega} \frac{f(X|H_1)}{f(X|H_0)} f(X|H_0)dX$$

But the last expression above is just the expectation of the likelihood ratio for hypothesis one divided by hypothesis zero, assuming hypothesis zero to be true. This means that the best critical region is such that this likelihood ratio is larger for all points inside ω than outside ω, with the constraint that the "size" of the region is α.

In the general case, the observable space of X is many-dimensional and the Neyman-Pearson test consists of finding the hypercontour of constant likelihood ratio which divides this space into two parts such that the part corresponding to larger values of the likelihood ratio has an integrated probability (test size) of α. This can be a very lengthy calculation. The usual simplification is to consider not the whole observable space X, but a one-dimensional test statistic t(X) as introduced earlier. Then the test is no longer Neyman-Pearson and not necessarily optimal, depending on how good a choice of test statistic was made.

5.4 COMPOSITE HYPOTHESES

In practice, real experiments often give rise to situations where both parameter estimation and hypothesis testing must be performed simultaneously. That is, the hypotheses to be tested are not completely defined in advance, but have free parameters whose values must be estimated. Such a hypothesis is called a composite hypothesis. The mathematical theory of composite hypothesis testing is not as well developed as that of simple hypothesis testing. The general techniques which can be found are valid only asymptotically (and the asymptotic limit may be very high!) or for certain related "families" of hypotheses. In many real cases, the only way to obtain realistic confidence levels is by resorting to Monte Carlo simulation.

5.4.1 The maximum likelihood ratio

The most important general tool for studying composite hypotheses is again based on the likelihood function. It is the maximum likelihood ratio, defined as the ratio of the likelihood functions for the two hypotheses, each one maximized with respect to all the free parameters. If there were no free parameters, this would correspond to the Neyman-Pearson test, which is optimal and calculable. Unfortunately, the effect of the free parameters makes the test not necessarily optimal, and in general one does not even know how to calculate α exactly.

Let us therefore restrict ourselves to the relatively common case where H_0 and H_1 have the same free parameters except that H_1 has some parameters free whose values are fixed for H_0. for example:

$$H_0: \theta_1 \text{ free}, \quad \theta_2=c$$
$$H_1: \theta_1 \text{ free}, \quad \theta_2 \text{ free}$$

The maximum likelihood ratio is:

$$\lambda = \frac{\max f(X|\theta_1,\theta_2=c)}{\max f(X|\theta_1,\theta_2)}$$

where X is all the data, and the maximization is with respect to θ_1 in the numerator and with respect to both θ_1 and θ_2 in the denominator. Since the denominator is maximized over the same parameter as the numerator and in addition one more, clearly $0 \leq \lambda \leq 1$. The difficulty in knowing the expected distribution of λ arises from the fact that since the value of θ_2 will in general be different for the two hypotheses, the value of θ_1 will also be different due to the maximization, and the effect of the maximization will depend on the correlation between θ_1 and θ_2.

Asymptotically (for large amounts of data, or small errors on θ) the distribution of λ is known, so this is normally used in evaluating its significance. Namely, the quantity $-2 \ln \lambda$ is distributed as $\chi^2(r)$ where there are r more free parameters in the denominator of λ than in the numerator. Thus one evaluates the maximum likelihood ratio for the data, and looks up the value of $-2 \ln \lambda$ in a table of chisquare with r degrees of freedom, which gives the significance level α. A small value of α is evidence against the more restrictive hypothesis H_0 and in favor of the more general hypothesis H_1.

5.5 SMALL-SAMPLE PROBLEMS

The maximum likelihood ratio test and significance level calculation as defined above is widely used since it is nearly the only real statistical tool for this situation. It is however notoriously unreliable for small samples of data. Unfortunately, it is hard to know exactly when a data sample is small, since it depends strongly on the hypotheses and distributions involved in ways which are not obvious.

A famous example of this problem is the attempt to determine whether a peak in a distribution of effective mass is a single peak or a split peak. In several examples it has been crucial to establish the significance of the evidence for split peaks because simple peaks would correspond to elementary particle states easily accomodated by the existing quark theory whereas split peaks would have implied resonances with exotic properties requiring a quite different approach to the whole theory. The importance of the problem triggered extensive statistical studies which revealed a marked tendency for data generated with a simple-peak hypothesis to fit the split-peak hypothesis better than the simple-peak model. This bias in the testing procedure can only be evaluated by Monte Carlo simulation, drawing samples from a known hypothesis and fitting them in the same way as the experimental data.

6. GOODNESS-OF-FIT

In this chapter we consider the significance with which we can accept or reject a hypothesis, without specifying any alternative hypotheses. As for hypothesis testing, we will have a critical region ω such that for all data $X \epsilon \omega$ we will reject the null hypothesis. As before, we can find the probability α of rejecting H_0 when it is true:

$$\int_{\omega(\alpha)} f(X|H_0)dX = \alpha$$

However we can no longer evaluate the probability of accepting H_0 when it is false, since this would depend on the alternative hypothesis which is not specified. We have therefore a measure of the significance of evidence against H_0 (if the fit is bad) but the significance of the evidence in favor of H_0 coming from a good fit cannot be quantified.

6.1 CHOOSING A GOODNESS-OF-FIT TEST

Since there is no alternative hypothesis, we cannot calculate β and cannot know the power of a test. This in principle deprives us of any means of comparing goodness-of-fit tests since a test is better than another only if it is more powerful at rejecting unwanted hypotheses. Since the unwanted hypothesis is not specified, it appears that we cannot have a realistic basis for choosing a test. Indeed we can expect some tests to be sensitive to certain kinds of deviations from the null hypothesis, and other tests to be sensitive to others. With this in mind, there are at least two approaches which will help in choosing a test.

The first approach is rather intuitive, but can be made more rigorous with some use of statistical information theory (which we do not discuss explicitly in these notes). The idea is to make sure somehow that the test makes use of all the information relative to the hypothesis being tested, and that it does not have any arbitrary procedures or parameters which would affect the value of α independently of the data and the hypothesis. We will see examples of how this is used below.

The second approach is to invoke a class of alternative hypotheses which allow us to estimate the power without being too specific about the alternatives. For example one can define the local power of a test as its power against infinitesimal deviations from the null hypothesis. Still this is not as general as it may seem since even infinitesimal deviations may take different forms, but it is usually possible in this way to get a good comparison of tests under rather general conditions.

6.2 DISTRIBUTION-FREE TESTS

The calculation of the confidence level for a given test involves an integration of a probability density function over a region which may be many-dimensional. Although such numerical calculations may not be too difficult when performed on modern computers, it has traditionally been necessary, and even today is still desirable, to avoid this calculation by means of a distribution-free test. Such a test involves a test statistic $t(X)$ whose distribution is known (under the null hypothesis) independent of the distribution of the X. Many distribution-free tests have been found, and the appropriate distributions of their test statistics are either known analytically or, more frequently, tabulated so that the user can simply calculate t and read the corresponding significance level from a table.[6]

A well-known example of a distribution-free test is the chisquare test of goodness-of-fit of a probability density $g(x)$ to density estimated experimentally in the form of a histogram. If the number of events observed in the ith histogram bin is n_i, the value of x in the middle of the bin is x_i, and the density $g(x)$ is normalized to the total number of events observed, then the test statistic for goodness-of-fit is:

$$t = \sum_i \frac{(g(x_i)-n_i)^2}{n_i}$$

This is often called the chisquare function, because under the null hypothesis (that the histogram really comes from $g(x)$), the quantity t should be distributed like a χ^2 variable with n degrees of freedom if the number of bins is n. This is distribution-free

[6] Traditionally all tests were distribution-free, so the term was not even used, it being assumed. Very recently we see a new kind of test being used (see 6.5), in which the confidence level must be recomputed for each case, something that would not have been thinkable without modern computers.

since the expected distribution of t does not depend on g(x). If we fit another function with different data we would still use the same table of χ^2 to obtain the level of significance α.

6.3 COMPARING TWO ONE-DIMENSIONAL DISTRIBUTIONS

In this section we will show what criteria may be used to choose a goodness-of-fit test by comparing two such tests for compatibility of one-dimensional distributions. Both tests, the chisquare test mentioned above and the Kolmogorov test, may be used either to compare two experimental samples of events or to compare an experimental sample with a completely-defined theoretical probability density.

6.3.1 The Chisquare test

This test is defined just above for the fit of a one-dimensional experimental sample to a known curve g(x). It is easily generalised to compare two experimental distributions:

$$t = \sum_i \frac{(m_i - n_i)^2}{m_i + n_i}$$

We notice that this test requires the grouping of observations into bins (called by statisticians 'data classes') and does not prescribe exactly how this is to be done. We are in principle free to make as many bins as we want, and place the boundaries where we want. This arbitrariness, which clearly will affect the value of t and probably also its properties as a test statistic, is one of the undesireable features pointed out above.

In view of the popularity of the chisquare test, the subject of optimal binning has been the object of considerable study. We summarize here some results of these studies:

1. In order to increase the local power of the test, there should be as many bins as possible.

2. The upper limit on the number of bins comes from the requirement that t follow a χ^2 distribution under the null hypothesis, which is only true for a 'large' number of events per bin, where the Poisson distribution becomes approximately Gaussian. Opinions vary as to how many events are needed for this, but most studies indicate that there should be very few bins of less than ten events and no bins of less than five events.

3. The distribution of bin boundaries is usually chosen equally spaced for practical reasons, but all studies indicate that it is better statistically if bins are chosen to be equally probable, that is, approximately the same number of events should fall in each bin.

4. Additional experiment-dependent considerations, such as the accuracy with which x can be measured, may be important.

Another apparent source of arbitrariness is the exponent 2 in the expression for t. Any other positive non-zero exponent would give rise to an admissible test, although the expected distributions for χ^3, χ^4, etc. would have to be recalculated. In fact it can be shown using the theory of information that χ^2 is indeed optimal whenever the deviations in each bin are Gaussian (here they are in fact Poisson which is approximately Gaussian). The 'square' in 'chisquare' is therefore not at all arbitrary, and is optimal for the usual case. One could, however, imagine cases where the measurements were not Gaussian-distributed and where a different test statistic would be better.

6.3.2 The Kolmogorov test

We now turn to a somewhat different kind of test, based on what the statisticians call the order statistics, which are nothing but the experimental observations ordered by increasing x-value. This allows us to form the cumulative distribution of the data S(x) as follows: The distribution starts at zero for $x=-\infty$, and increases by an amount 1/N at each point x where an experimental point x_i has been observed. [N is the total number of points observed, so $S(+\infty)=1$.] We can use the Kolmogorov test either to compare two experimental distributions $S_1(x)$ and $S_2(x)$, containing respectively N_1 and N_2 events; or to compare $S_1(x)$ with a continuous known distribution of which the probability density function is f(x), and whose integral, the cumulative distribution function is F(x).

The Kolmogorov test statistic is a measure of the distance between the two distributions being compared. This measure is simply the largest distance (maximized with respect to x) between the two cumulative distributions:

$$D = \sqrt{N} \max|S(x)-F(x)| \qquad \text{where S has N events,}$$

or

$$D = \sqrt{(N_1N_2/(N_1+N_2))} \max|S_1(x)-S_2(x)|$$

It turns out that this test statistic is asymptotically distribution-free (that is, under the null hypothesis, the expected distribution of D is independent of S and F for large enough N.), and as written here it is correctly normalized to be also asymptotically independent of N, N_1, and N_2. The significance level α can be calculated from formulas or the table given in Eadie et al., page 270, or calculated by the CERN Program Library subroutine PROBKL.

Notice that this test involves no arbitrary binning, and is still very easy to calculate, although for very large data samples the ordering of the data may be longer than the histogramming required for the chisquare test.

6.3.3 The Smirnov-Cramer-Von Mises test

This test is very similar to the Kolmogorov test described above, except that the measure of the distance between two cumulative distributions is taken to be the integrated squared distance instead of the maximum:

$$W = N \int_{-\infty}^{+\infty} [S(x)-F(x)]^2 \, f(x) \, dx$$

$$= N \int_{0}^{1} [S(x)-F(x)]^2 \, dF(x)$$

The corresponding formula for comparing two experimental distributions is somewhat more complicated and is given in Eadie et al, page 269, as are formulas and tables for determining the significance level α.

The computational complexity of this test is clearly somewhat greater than for the Kolmogorov test, which probably explains why it is less popular. However it has the distinct advantage that the test is exactly distribution-free <u>for all values of N</u>, although the significance level is independent of N only for 'large' N (in fact $N>3$ is enough!). It is also more appealing (and probably more powerful against most alternatives) because it is really a function of all the data and not just the maximum distance.

Because it is free of binning, is sensitive to all the data, and is exactly distribution-free, the Smirnov-Cramer-Von Mises test is generally considered the most powerful goodness-of-fit test for one-dimensional data.

6.4 COMPARING MULTIDIMENSIONAL DISTRIBUTIONS

When the data are more than one-dimensional, goodness-of-fit testing becomes considerably more difficult.

6.4.1 Chisquare in d dimensions

In principle, the chisquare test for goodness-of-fit is dimension-free, in the sense that it can be defined independently of the dimensionality of the data. One merely compares the expected number of events in each data class (bin) with the actual number. The dimensionality of the bin does not enter into the theory.

In practice, however, multidimensional bins cause not only computational problems (when the boundary of the space is curved) but the number of events required to have a minimum number per bin becomes enormous, unless the number of bins is reduced to a very small number per dimension. The reason is that the number of bins increases exponentially with dimensionality. It is therefore clear that for multidimensional data we should prefer a test which does not require binning.

6.4.2 Kolmogorov-type tests in d dimensions

It is very appealing to try to extend tests based on order statistics to higher dimensionalities, since these tests (Kolmogorov, Smirnov-Cramer-Von Mises) use the data points as measured without binning. Such attemps however meet with several difficulties, both practical and theoretical, and such tests have not to my knowledge yet been used with success. The first difficulty is with the order statistics themselves, which lose some of their nice properties in higher dimensions, although they can still be defined in a straightforward way:

$$S(X_1,X_2,\ldots X_n) = \text{number of points } p \text{ such that } p_1<X_1,\ p_2<X_2,\ \ldots\ p_n<X_n$$

For example, they now depend on the orientation of the axes in the d-dimensional space. Another difficulty is the computational complexity, since the definition of F now requires multidimensional integration, and other multidimensional difficulties arise in defining the test statistic. Perhaps the most important barrier is that straightforward extension of both the tests given above are no longer distribution-free in many dimensions, so that one has no easy way to determine α.

If we are willing to give up distribution-free testing, the more recent permutation methods given in the next section are promising for both one- and many-dimensions, and for both large and small sample sizes.

6.5 PERMUTATION TESTS FOR COMPARING TWO POINT SETS

We wish to test the hypothesis that two sets of points are random samples from the same underlying distribution, where we do not know the underlying distribution. We wish to find a test valid for one or more-dimensional points, and the test should not involve binning. Two things are needed:

1. A test statistic Δ which will be a measure of the distance between the point sets.

2. A way to measure the significance of the value of Δ for the two point sets, giving the significance level α.

The permutation method allows us to measure the level of significance for any distance function Δ, without knowing the underlying distribution, assuming of course the null hypothesis. One first calculates the distance Δ_{12} between point sets one and two. Then the significance level is found as follows. Put both samples together to form a single sample of N_1+N_2 points. Under the null hypothesis, this will also be distributed like sample 1 or sample 2, and any (random) partitioning of this pooled sample into samples of N_1 and N_2 points should yield sets with the same distribution. We therefore make many different partitionings, each time choosing N_1 points from the total N_1+N_2 points, and calculate Δ for each of these partitionings. If N_1 and N_2 are small enough, we can actually enumerate all the partitionings possible.[7] Whether all partitionings are exhausted or only a random sample is used, the resulting distribution of the values of Δ can be used in an obvious way to determine the significance of the actual value Δ_{12} corresponding to the observed data points. The significance level α is simply the proportion of all Δ-values lying above the value Δ_{12}.

---

[7] The total number of partitions of N_1+N_2 points into two samples of N_1 and N_2 points is:

$$P_{12} = (N_1+N_2)!/N_1!N_2!$$

This increases very fast with N_1 and N_2; for example, the number of ways of dividing a sample of 10 into two samples of 5 is only 252, but the number of ways of dividing a sample of 20 into two samples of 10 is 184756.

Now that the significance level can be determined for any distance measure Δ, we return to the question of defining such a distance. Since the procedure outlined above is valid for any measure Δ, we are free to choose a measure with physical significance for the situation at hand. For example, suppose we want to test a theory which predicts that prompt muons produced in a beam dump experiment should on the average have higher momenta than prompt electrons, but there is no theory to predict the actual distributions. Then the distance measure could be the average muon momentum minus the average electron momentum. If the theory predicted only that a few muons should have exceptionally high momentum, we could use the difference between the average momentum of the fastest muons and the fastest electrons.

In spite of this extraordinary freedom in choosing a test statistic, it is still difficult to find good distance measures for multivariate data. A general procedure is given in Friedman (1974) using the k-nearest-neighbor concept. This is probably the best general method in use, although it suffers from at least two elements of arbitrariness:

1. The optimal number of nearest neighbors k is not generally known.

2. It is somewhat dependent on the metric of the space in order to find the nearest neighbors. However, it is hard to imagine any measure of the distance between two point sets which does not depend on the definition of the distance between two points.

If a good distance measure Δ is available for the physical problem at hand, the permutation technique gives a good way to evaluate its significance. The method is completely non-parametric, does not involve any attempt to actually estimate the point densities in either set, and is always exactly valid for the data sample at hand, no matter how large or small it is. On the other hand, the computation required is large by traditional standards, since one is in fact recalculating α each time instead of using a distribution-free test and a standard table. By modern computing standards however, this is a small price to pay for the advantages gained.

REFERENCES

Barrodale, I., and C.Phillips, ALGORITHM 495, Solution of an overdetermined System of Linear Equations in the Chebyshev Norm, ACM TOMS, 1 (1975),264-270.

Clopper, C.J. and Pearson, E.S., Biometrika 2 (1934) 404

Drijard,D., T.Ekelof, and H.Grote, On the reduction in space resolution of track detectors caused by correlations in the coordinate quantization, Nucl. Instr. and Meth., 176 (1980) 389.

Eadie,W.T., Drijard,D., James,F.E., Roos,M., and Sadoulet,B. Statistical Methods in Experimental Physics. North-Holland, Amsterdam (1971)

Friedman, J.H., Data Analysis Techniques for High Energy Particle Physics, Lectures presented at the CERN School of Computing, Godoysund, Norway, August 11-24, 1974, in CERN Yellow Report 74-23.

James,F., Function Minimization, in the Proceedings of the 1972 CERN Computing School (CERN 72-21), also available separately as a supplement to the long-write-up of CERN Program Library program MINUIT (D506).

James, F., and Roos, M., MINUIT - A system for function minimization and analysis of the parameter errors and correlations. Comp. Phys. Comm., 10 (1975) 343-367.

James, F. and Roos, M., Errors on ratios of small numbers of events, Nuclear Physics B172 (1980) 475-480

James, F., Interpretation of the shape of the likelihood function around its minimum, Comp. Phys. Comm., 20 (1980) 29-35

Rice,J.R. and J.S.White, Norms for smoothing and estimation, SIAM Review 6 (1964), 243

A THOUSAND TeV IN THE CENTER OF MASS:

INTRODUCTION TO HIGH ENERGY STORAGE RINGS[+]

J.D. Bjorken

Fermi National Accelerator Laboratory
P.O. Box 500
Batavia, Illinois 60510

I. INTRODUCTION

These lectures must begin with an apology. Normally at schools such as this, one expects the lecturer to be an acknowledged expert on the subject matter he is discussing. Here this is not the case. Design of high energy proton storage rings is not exactly my <u>forte.</u> Why am I doing this? There are several reasons, short of mental illness.*

1. I want to learn this subject myself and there is no better way than trying to teach it. And Ferbel didn't stop me.

2. There needs to be a broader knowledge of accelerator physics in the elementary-particle community. Experimentalists at the storage rings find themselves especially closely coupled to their machine and its operation. And theorists can find interesting and challenging questions which lie at the frontier of the very active field of nonlinear mechanics.

3. Straightforward extrapolation of existing acceleration techniques would seem to lead to very large, expensive machines. While we may envision one, perhaps two generations of future accelerators using essentially existing techniques, the question of how to go beyond that is a difficult one. There seems to be a growing feeling that it is not too soon to start to face up to the problem. A look at the alternative--as we do here--can only provide stimulation.

---

*See Appendix II.

[+]Lectures given at the 1982 NATO Advanced Study Institute, Lake George, N. Y., June 1982.

There is a famous plot ("Livingston Plot") of cms energy attained as a function of calendar year (Fig. 1). One sees, remarkably, a doubling-time of ~2-3 years. Can this be maintained? Extrapolating into the future, there exists UNK, the Soviet project, talk of a VBA (very big accelerator) with $E_{cms} \lesssim 50$ TeV (for colliding beams), and even the beginnings of discussion of a possible US machine on that energy scale.[1] Beyond that lies unknown territory, and a fundamental challenge to the natural push to even higher energy. This unknown territory of the distant future is where we shall reside during these lectures. We shall project ourselves into the years 2010-2020, and look[2] at an utterly unimaginative scaled up $p\bar{p}$ collider of 500 TeV + 500 TeV. Needless to say, such a machine is clumsy (circumference $> 10^3$km) and expensive (cost > 1000 Tevatrons) and must not be taken seriously. Nevertheless, the choice has the following advantages:

1. By stretching present ideas to (beyond?) the breaking point, we learn the scaling laws for more practical machines, i.e. how machine parameters scale with energy.

2. Once having grappled with such staggering energy scales, it is easier to interpolate back to "reasonable" (??) machines like the VBA.

3. It is an interesting exercise to see whether such a machine, even were it economically feasible, could work, or whether there are intrinsic technical limits to the energy-scale available to the present technology.

4. A 500 TeV proton ring is a nice pedagogical machine. In particular, synchrotron-radiation becomes quite important, and thus this proton machine shares features--and problems--characteristic of contemporary e^+e^- storage rings.

5. The machine is so big, so remote in time, and so unlikely to be built that no one could be misled into thinking that I take any of this seriously. To repeat, this machine is not to be taken seriously. <u>This machine is not to be taken seriously. THIS MACHINE IS NOT TO BE TAKEN SERIOUSLY.</u>

Our main purpose, after all, is pedagogy. In the next section, we shall try to outline in a rough semiquantitative way the big picture, i.e., we try to provide an overview of the material to follow. Section III is devoted to a more detailed discussion of linear optics and betatron motions. In Section IV we briefly survey questions of errors, tolerances and nonlinear resonances. In Section V we provide a very sketchy parameter list for the 500 TeV collider, and discuss some of the uncertainities. Section VI discusses some of the various demands upon the detection apparatus--especially the apparent inevitability of multiple interactions per bunch crossing. Section VII is devoted to concluding comments. An appendix provides a bibliography from which these lectures where prepared.

II. OVERVIEW

This section is divided into the following subsections:

1. Closed orbit
2. Vertical motion (betatron oscillations)
3. Horizontal motion (betatron oscillations and momentum dispersion)
4. Synchrotron radiation
5. Longitudinal phase space
6. Synchrotron damping of the phase space
7. Quantum excitation of the phase space.
8. Single beam instabilities
9. Luminosity
10. Beam-beam effect

1. Closed Orbit

Get a map and draw a circle. (This is a common pastime of laboratory directors.[3]) The bending radius ρ is proportional to momentum p and inversely proportional to magnetic field B:

$$p = eB\rho \tag{2.1}$$

The conversion factor is

$$0.3 \text{ GeV/c} = 1 \text{ T-m} = 10 \text{ kG-m} \tag{2.2}$$

(We shall often set $\hbar = c = 1$.) We shall choose 10T magnets, inasmuch as that has already been projected for the 20 TeV VBA. This gives for nominal radius ρ and circumference C:

$$\begin{aligned} \rho &= 170 \text{ km.} \\ C &= 1100 \text{ km.} \end{aligned} \tag{2.3}$$

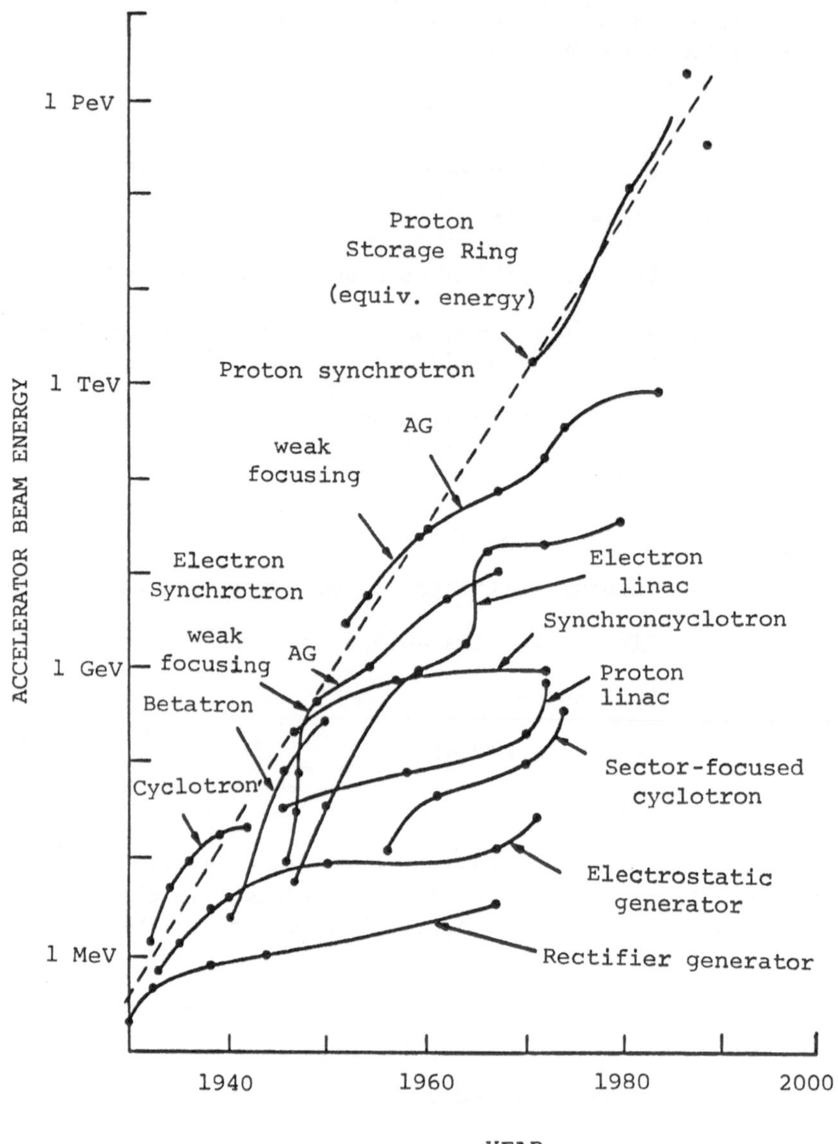

Figure 1. The "Livingston plot" showing growth of attained cms energy of hadron machines vesus time.

The actual values will be somewhat larger in order to account for the quadrupole magnets and straight sections.

2. Vertical Motion

No particle exactly follows the design orbit. The typical particle undergoes small oscillations about the design orbit. To a good approximation the vertical, horizontal, and longitudinal motions may be treated independently. Focussing is provided by quadrupole magnets; we write (with z the vertical coordinate and x the horizontal coordinate transverse to the direction of motion)

$$B_x = Gz \tag{2.4}$$

(implying, via Maxwell equations, B_z=+Gx, i.e. defocussing in the horizontal plane.)

The state-of-the-art maximum gradient G is ~1 T/cm; we take

$$G \simeq 2T/cm. \tag{2.5}$$

The transverse kick a particle gets in going through a quadrupole magnet of length ℓ is roughly

$$\Delta p_x \cong eG\ell z \tag{2.6}$$

For focussing in both planes one alternates focussing and defocussing quadrupoles which are spaced in such a way that, on average, the $<|z|>$ of particles at defocussing quads is smaller that the $<|z|>$ at focussing quads. Provided this can be arranged, then there will be net focussing. The condition for this is, in order of magnitude, that the focal length f of the quads be comparable to their spacing L. A typical particle orbit then is shown in Fig. 2.

From the above equation we see this implies

$$\Theta_z \sim \frac{z}{f} \sim \frac{\Delta p_z}{p} \sim \frac{eG\ell z}{p} \tag{2.7}$$

Thus the condition for stable, strong focussing is

$$eG\ell L \sim p = eB\rho \tag{2.8}$$

Normal economics implies that the investment in quads not be a large perturbation on the investment in dipole magnets

$$\frac{\ell}{L} \lesssim (0.1\text{–}0.2) \tag{2.9}$$

and hence

$$0.1eG\ell^2 \sim p = eB\rho \tag{2.10}$$

The spacing between F and D quads is then

$$L \sim 3\sqrt{\frac{B\rho}{G}} \tag{2.11}$$

and (for fixed magnet parameters) scales as $E^{1/2}$. We estimate

	Estimated	Actual		
$L \sim$	20 m	30 m	Fermilab TeVatron	(2.12)
	500 m	(400 m)	2020 Machine	

The wavelength of the betatron oscillations evidently also scales with L; actually the typical wave-number is $\sim L^{-1}$. A very fundamental machine parameter is the tune. It is defined as the number of betatron oscillations per revolution

$$\nu_z = \frac{\text{Circumference}}{\text{Betatron wavelength}} \sim \frac{2\pi\rho}{2\pi L} \sim \frac{\rho}{L} \tag{2.13}$$

We see that the tune also scales as $\sim E^{1/2}$. We find from the rough estimate

	Estimated	Actual		
$\nu_z \sim$	50	19.3	TeVatron	(2.14)
	350	(400)	2020 Machine	

(The discrepancy in the case of the TeVatron is accounted for by a smaller value of ℓ/L and a larger choice of betatron wavelength $\simeq 10L$.)

It is useful to consider the beam as a population in phase space. Vertical phase-space* is just z-p_z space. Provided the

---

*Canonical phase space here: the definition typically used by accelerator physicists contains a factor γ; cf Chapter III.

dynamics is derivable from a Hamiltonian (as is the case for particles moving in external electromagnetic fields--including time-dependent fields), the area in phase-space must be conserved, according to the Liouville theorem. Typically this area, as determined essentially by the low energy source, is

$$\Delta z\ \Delta p_z = \text{phase-space area} \simeq 1\ \text{MeV-cm.} \tag{2.15}$$

(In natural units this is $\sim 10^{10}\hbar$.)

We may now estimate the nominal beam size in a storage ring. Our previous estimate in Eqn. (2.6) relates Δp_z to Δz through the focal structure of the lattice (by <u>lattice</u> we mean the array of magnets comprising the ring)

$$\Delta p_z \sim eG\ell\ \Delta z \tag{2.16}$$

Then

$$\Delta z\ \Delta p_z \sim \frac{eG\ell}{L} \cdot [L(\Delta z)^2] \tag{2.17}$$

Only the right-hand factor is not a fixed quantity. Hence the nominal beam size* scales as

$$\Delta z \sim L^{-1/2} \sim E^{-1/4} \tag{2.18}$$

The nominal transverse momentum in the beam scales as

$$\Delta p_z \sim E^{1/4} \tag{2.19}$$

Putting in the numbers gives

		TeVatron	2020 Machine	
Δz	~	0.3 mm	70μ	(2.20)
Δp_z	~	30 MeV	140 MeV	

The shape of the population in phase space changes as one proceeds around the ring. It is typically elliptical, but the axes and orientation (but <u>not</u> area!) vary as one proceeds around the ring, as shown in Fig. 3.

---

*This $E^{-1/4}$ scaling of beam size in different storage rings should not be confused with the change in beam-size during acceleration. There the gradient G scales with B and hence with energy in order to keep a constant tune ν during the acceleration cycle. Therefore during acceleration, $\Delta z \sim E^{-1/2}$; $\Delta p_z \sim E^{+1/2}$.

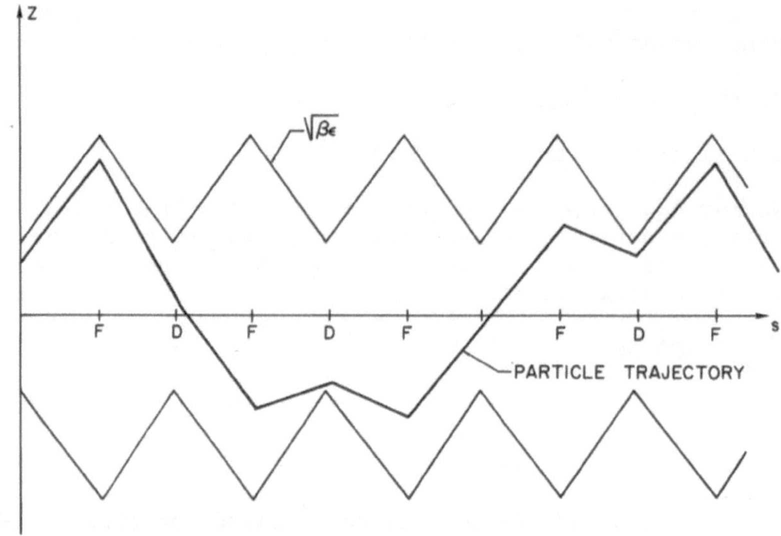

Figure 2. Typical betatron amplitude of a particle in a FODO lattice. Also shown is the amplitude function $\sqrt{\beta\epsilon}$ (cf. Section III).

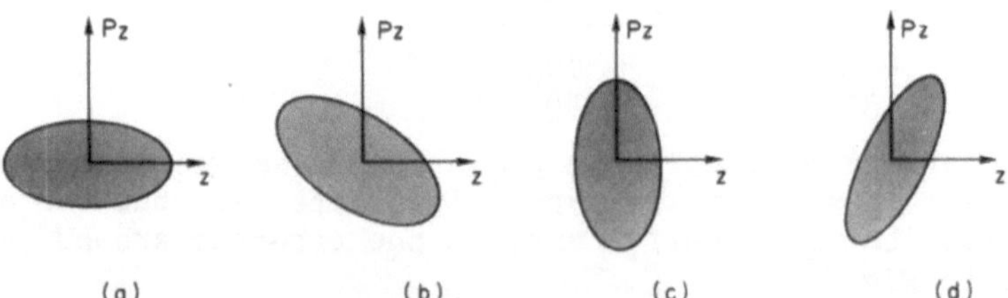

Figure 3. Typical z-p_z phase-space population at various points in the FODO lattice (moving downstream): (a) At the center of an F (focussing) quadrupole, (b) Between F and D (defocussing) quadrupoles, (c) At a D quadrupole, and (d) Between D and F quadrupoles.

3. Horizontal Motion

The description of horizontal betatron oscillations is very similar to vertical oscillations. Inclusion of the effect of curvature of the design orbit changes the tune ν_x by a negligible amount. A more important effect has to do with dispersion: an off-momentum particle has a different closed orbit (cf Fig. 4). If the momentum exceeds the design momentum by an amount Δp (in the high energy limit), the new closed orbit will lie at larger radius. Write, for a typical particle

$$x(s) = x_\varepsilon(s)+x_\beta(s) \tag{2.21}$$

where x_β is the betatron amplitude at coordinate s along the design orbit, and x_ε is the correction to the radius of the closed orbit. In linear approximation

$$x_\varepsilon = \eta(s)\frac{\Delta p}{p} \tag{2.22}$$

Let us estimate the nominal value of $\eta(s)$. In going through a cell (a cell is the basic element of the lattice consisting of F quad, D quad and the two sets of intermediate bending magnets), the off-momentum particle must be bent through the same angle; hence it must get an extra p_T kick from the quadrupoles in proportion to its momentum deviation $\Delta p/p$. We must have

$$\frac{\Delta p_{||}}{p_{||}} \sim \frac{\text{Extra } p_T \text{ kick}}{\text{Main } p_T \text{ kick}} = \frac{eG\ell x_\varepsilon}{2eBL} \tag{2.23}$$

This implies

$$\langle\eta(s)\rangle \sim \frac{2B}{G}\,\frac{L}{\ell} \tag{2.24}$$

a value independent of machine energy. Putting in the numbers gives

$$\langle\eta(s)\rangle \sim \begin{array}{lll} \underline{\text{Estimated}} & \underline{\text{Actual}} & \\ \text{m} & 2\text{ m} & \text{TeVatron} \\ 1\text{ m} & 0.7\text{ m} & \text{2020 Machine} \end{array} \tag{2.25}$$

For typical machine apertures of a few centimeters, a momentum spread $\Delta p/p \lesssim 10^{-2}$-$10^{-3}$ may be accepted independent of energy. The actual momentum spread in the beam must be found by considering longitudinal phase-space. We do this in part 5.

4. Synchrotron Radiation

In proton storage rings the beam may be bunched as it is during acceleration or as in $p\bar{p}$ colliders such as TeV I, where $\bar{p}$'s are in short supply. In other cases the beam may be unbunched ("coasting") as in the CERN ISR (or as in the ISABELLE design). In e^+e^- machines, the beams are necessarily bunched, because the energy loss from synchrotron radiation must be compensated by an RF accelerating system. In our 500 TeV machine, protons also emit a significant amount of synchrotron radiation. The handbook formula[4] for the energy loss per turn is (with $\hbar = c = 1$)

$$\text{Energy loss/revolution} \equiv U_0 = \frac{4\pi\alpha}{3\rho}\left(\frac{E}{m}\right)^4 \qquad (2.26)$$

With our parameters, this implies

$$U_0 \sim 3 \text{ GeV/turn} \qquad (2.27)$$

This radiation is emitted in a broad spectrum of photon energies, but the typical photon energy or critical energy E_c is given by

$$E_c \sim \frac{3}{\rho}\left(\frac{E}{m}\right)^3 \sim \frac{9U_0}{4\pi\alpha}\cdot\frac{m}{E} \qquad (2.28)$$

For the 2020 machine, this is

$$E_c \sim 300 \text{ keV} \qquad (2.29)$$

This gives the typical number of photons emitted per turn as

$$\text{Photons/revolution} \sim \frac{U_0}{E_c} \sim \frac{\gamma}{137} \qquad (2.30)$$

At 500 TeV, we get

$$\text{Emitted photons/revolution} \sim 4\times10^3 \qquad (2.31)$$

Thus a coasting proton would lose ≃1% of its energy in 2000 revolutions. Inasmuch as the revolution frequency is 250 Hz, 2000 turns is only 8 sec. A coasting beam is not possible. An RF system must be provided.

5. Longitudinal Phase Space

To define longitudinal coordinates, we may use the energy deviation ε (or momentum deviation $\Delta p \simeq \varepsilon$) and the distance Δs from a reference particle at the center of the bunch (or equivalently the arrival-time delay $(-\tau)$ of the particle at a given point of observation). As one might expect, ε and τ (or Δp and Δs) are canonically conjugate variables. Somewhere around the ring RF cavities must be placed to provide the acceleration. A particle with energy E entering the cavity at time τ leaves the cavity with energy $E+V(\tau)$, where by definition this defines the RF voltage $V(\tau)$. The RF voltage must be synchronous with the particle motion around the machine. Thus $V(\tau)$ must be periodic

$$V(\tau) = V\left(\tau + \frac{2\pi}{h\omega_0}\right) \tag{2.32}$$

where

$$f_0 = \frac{\omega_0}{2\pi} = \text{revolution frequency} \tag{2.33}$$

and

$$h = \text{integer} \equiv \text{harmonic number} \tag{2.34}$$

Typically (but not always) $V(\tau)$ is sinusoidal, and we shall assume it to be true here:

$$V(\tau) = V_0 \sin\,(h\omega_0\tau + \phi_0) \tag{2.35}$$

The choice of RF frequency is determined in large part by practical considerations beyond the scope of these lectures. At Fermilab, the RF frequency is ~50 Mhz; at CERN it is ~200 Mhz. The higher the frequency, the less bulky the cavities, and we shall provisionally pick (rather arbitrarily*) a frequency of 500 Mhz. This gives the harmonic number

$$h \sim 2\times 10^6 \tag{2.36}$$

The synchronous particle $(\tau=0)$ will gain energy $V_0 \sin \phi_0$ per turn. This must match the energy loss U_0:

---

*In retrospect, I think this is a mistake. Lower frequency seems preferable.

$$U_0 = V_0 \sin \phi_0 \tag{2.37}$$

We may now write down equations of motion for ε and τ, which may be functions of time. The time variation is assumed to be slow compared to the revolution frequency. Then we have

$$\frac{d\varepsilon}{dt} = \frac{\omega_0}{2\pi} \left[V_0 \sin(h\omega_0\tau+\phi_0)-U_0- \frac{\partial U_0}{\partial E} \varepsilon\right] \tag{2.38}$$

(The last term ultimately must be included. It provides damping; here we temporarily ignore it.)

Even in the absence of RF, the variable τ will change with time if the particle has a momentum error. This occurs because the revolution frequency depends upon momentum. We write

$$\frac{\Delta\omega_0}{\omega_0} = - \frac{\Delta T}{T} \equiv \eta \frac{\Delta p}{p} \tag{2.39}$$

where the "dilation factor" or "momentum-compaction" η is not the same $\eta(s)$ as introduced in connection with the dispersion. However this η (sometimes called α_p) is related to $\eta(s)$. At high energy,* evidently

*At lower energy, there is another contribution to η (of opposite sign) coming from the change in velocity with momentum. Since

$$\frac{\Delta T}{T} = \frac{\Delta \rho}{\rho} - \frac{\Delta v}{v} \tag{F.1}$$

and $\Delta v/v \simeq \gamma^{-2}(\Delta p/p)$, we get

$$\eta = \frac{\langle\eta(s)\rangle}{\rho} - \frac{1}{\gamma^2} \equiv \frac{1}{\gamma_T^2} - \frac{1}{\gamma^2} \tag{F.2}$$

At the value $\gamma=\gamma_T$, or transition energy, η changes sign and longitudinal motion becomes more nontrivial. This creates some complication during acceleration in lower energy machines. From Eqns (2.8), (2.13), and (2.24), we may see that $\gamma_T \simeq \nu$. For the 2020 machine, injection energy will be well above transition.

$$\eta \frac{\Delta p}{p} = \frac{\langle\Delta\rho\rangle}{\rho} = \frac{\langle\eta(s)\rangle}{\rho} \frac{\Delta p}{p} \tag{2.40}$$

or

$$\eta = \frac{\langle\eta(s)\rangle}{\rho} \tag{2.41}$$

For the 2020 machine, our rough estimate gives,

$$\eta \sim 6\times10^{-6} \tag{2.42}$$

Because of the dispersion, an off-energy particle will, in absence of other effects, change its position Δs, (or time τ) relative to the reference particle. We have

$$\frac{d\tau}{dt} = -\eta\frac{\Delta p}{p} \cong -\frac{\eta}{E}\,\varepsilon \tag{2.43}$$

This, together with Eqns. (2.37) and (2.38), leads to the equation of motion

$$\frac{d^2\tau}{dt^2} = -\frac{\eta}{E}\frac{\omega_0}{2\pi} V_0[\sin(h\omega_0\tau+\phi_0)-\sin\phi_0] \tag{2.44}$$

For ϕ_0=0 (no synchrotron radiation) this is the equation of a pendulum, with τ playing the role of an angle ("synchrotron phase"). For small amplitudes there is stability; for large amplitudes there is <u>not</u>, and τ on average increases linearly with time.

For nonvanishing ϕ_0, there is again phase-stability for small amplitudes. For large amplitudes τ increases (on average) quadratically with time, and consequently ε increases linearly with time, implying eventual loss of the particle.

In the limit of small τ, Eqn (2.44) is just an oscillator equation, and the angular frequency Ω_s (<u>synchrotron frequency</u>) is

$$\frac{\Omega_s}{\Omega_0} = \left(\frac{\eta h V_0\cos\phi_0}{2\pi E}\right)^{1/2} \tag{2.45}$$

For our 500 TeV machine, we get (choosing V_0~5 GeV) for the frequency of synchrotron oscillations,

$$\frac{\Omega_s}{2\pi} \sim 1 \text{ Hz} \tag{2.46}$$

justifying a posteriori our assumption of slow variation of τ with time.

It is again important to view all this in the longitudinal phase space. Considering first the case $\phi_0=0$, we see that in ε-τ space the small amplitude orbits are approximately circular (or elliptical), centered at $\tau=0$, $\pm 2\pi/h\omega_0$, $\pm 4\pi/h\omega_0$, (Fig. 5). Very large amplitude orbits are straight lines $\varepsilon \simeq$ const. In between is a special orbit, the separatrix, which comprises the boundary between oscillatory and non-oscillatory motion (it corresponds to the pendulum oscillation with ±180° excursion in angle). The equation for the separatrix is easily worked out, especially if one remembers the facts of life about pendula. The region in phase space of oscillatory motion which is enclosed by the separatrix is called, for obscure historical reasons, an RF bucket. The dimensions of the bucket are

$$\Delta t = \frac{2\pi}{h\omega_0} \text{ (full width)} \tag{2.47}$$

$$\Delta\varepsilon = \pm\left(\frac{2EV_0}{\pi h\eta}\right)^{1/2} \tag{2.48}$$

The typical phase-space area occupied by high energy proton beams is determined by the injectors

$$\Delta\varepsilon\,\Delta\tau \sim 1\text{-}2 \text{ eV-sec} \tag{2.49}$$

For the 2020 RF system, the nominal bucket area (still neglecting the synchrotron radiation) is much larger. We have

$$\Delta\varepsilon\,\Delta\tau \sim \frac{6\sqrt{EV_0}}{h^{3/2}\omega_0\eta^{1/2}} \tag{2.50}$$

Putting in the numbers with $\Delta\tau=2\times10^{-9}$sec gives,

$$\Delta\varepsilon_{max} \sim \pm 150 \text{ GeV} \tag{2.51}$$

and

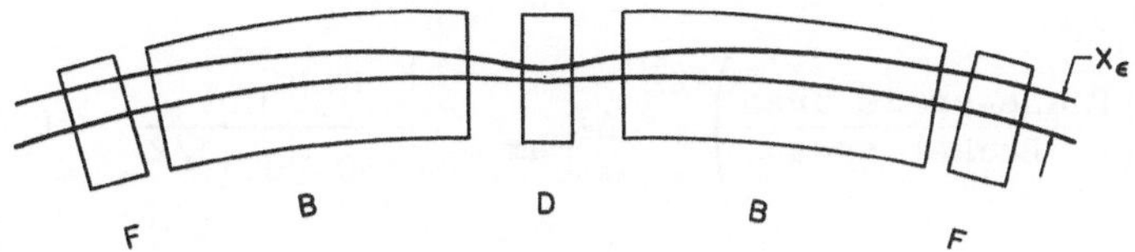

Figure 4. Closed orbit of an off-momentum particle in the FODO lattice.

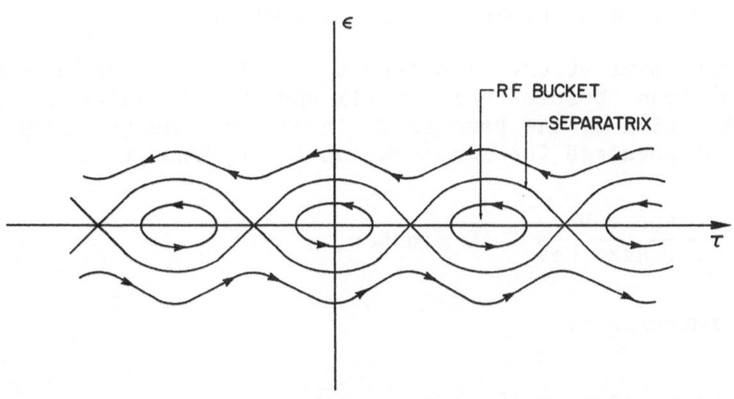

Figure 5. Structure of longitudinal phase space for a bunched beam (in absence of acceleration and/or synchrotron radiation).

$$(\Delta\varepsilon\ \Delta\tau)_{\text{RF bucket}} \lesssim 300 \text{ eV-sec.} \tag{2.52}$$

If one puts in a bunch with the nominal, small longitudinal phase space, it will evidently not occupy the full bucket area. Its dimensions will typically be in proportion to the bucket dimensions; hence (in the absence of synchrotron radiation effects)

$$\Delta\varepsilon \sim \left(\frac{\text{Phase-space area}}{\text{Bucket area}}\right)^{1/2} \Delta\varepsilon_{max} \sim \frac{150 \text{ GeV}}{(150\text{-}300)^{1/2}} \sim 10 \text{ GeV}$$

$$\Delta\tau \sim 10^{-10} \text{ sec.} \tag{2.53}$$

This would imply a bunch length

$$\Delta s \sim c\Delta\tau \sim 3 \text{ cm.} \tag{2.54}$$

If RF acceleration is present and $\phi_0 \neq 0$, the bucket size decreases and the bucket shape becomes similar to a fish, as shown in Fig. 6. However, the nominal order of magnitude estimates which we made will not be changed. An accurate discussion of this can be found in many places in the literature.[5]

We note that at energies less than 500 TeV, the synchrotron radiation loss becomes quite small and the RF system is suitable for acceleration of the beam from injection energy (~70 TeV??). The time Δt required for acceleration to top energy is

$$\Delta t \sim \frac{500 \text{ TeV}}{250 \text{ Hz} \times 3 \text{ GeV}} \sim 10 \text{ minutes} \tag{2.55}$$

which is reasonable.

To get an idea of the scale of this RF system, we note that in the LEP design at maximum energy the synchrotron loss per turn is ~2.4 GeV. The RF system (350 Mhz) is 1.6 km long and consumes ~100 MW. For the 2020 machine we therefore need ~3 km of RF with nominal power consumption (without considering future improvements such as superconducting RF) of ~200 MW. On the scale of this machine, these requirements are quite modest.

6. Radiation Damping

Emission of synchrotron radiation leads to damping of the phase space population of the beam. Consider first the vertical degrees of freedom. Synchrotron radiation is emitted along the direction of motion of the particle (to accuracy of order $\gamma^{-1}=(m/E)$). Thus when a photon is emitted, the transverse momentum is diminished by the same percentage as the longitudinal momentum. But only the loss in longitudinal momentum is compensated by the RF system. Hence the transverse momentum (better, phase-space area) diminishes in accordance with the rate of energy loss from synchrotron radiation.

$$\frac{1}{(\Delta p_z \Delta z)} \frac{d}{dt} (\Delta p_z \Delta z) \simeq \left(\frac{U_0}{E}\right)\left(\frac{\omega_0}{2\pi}\right) \tag{2.56}$$

or

$$(\Delta p_z \Delta z)_t = (\Delta p_z \Delta z)_0 \, e^{-t/\tau_v} \tag{2.57}$$

with the damping time τ_v given by

$$\tau_v = \frac{E}{U_0} \frac{2\pi}{\omega_0} \tag{2.58}$$

It is the time required for a particle to emit an amount of synchrotron-radiation energy equal to its own energy. For the 2020 machine, we get

$$\tau_v \sim 20 \text{ min.} \tag{2.59}$$

Notice that the scaling law is

$$\tau_v \sim \frac{R^2}{E^3} \sim \frac{1}{EB^2} \tag{2.60}$$

with B the magnetic field in the ring.

Horizontal motion is in principle more complicated because momentum dispersion and betatron oscillations are both damped, but[6] are in fact coupled. But the complications are inconsequential, and to good approximation

$$\tau_h \simeq \tau_v \tag{2.61}$$

The momentum spread is also damped. This can be readily seen by returning to Eqn. (2.38) and keeping the damping-term (obtained by expanding the energy loss due to synchrotron radiation through first order in ε.) Since

$$U_0 \sim \frac{(E_0+\varepsilon)^4}{R} \tag{2.62}$$

we have, to good approximation*

$$\frac{\partial U_0}{\partial \varepsilon} \cong 4\,\frac{U_0}{E_0} \tag{2.63}$$

Reconstruction of the oscillator equation, keeping track of the damping term, leads for small amplitudes to

$$\frac{d^2\tau}{dt^2} = -\,\Omega_s^2\tau + \frac{4U_0}{E_0}\,\frac{\omega_0}{2\pi}\,\frac{d\tau}{dt} \tag{2.64}$$

and thus

$$\tau(t) = \tau_0\, e^{i\Omega_s t}\, e^{-t/\tau_s} \tag{2.65}$$

with

$$\tau_s = \frac{1}{2}\tau_h = \frac{2\pi}{\omega_0}\,\frac{E_0}{2U_0} \tag{2.66}$$

7. Quantum Excitation

At this stage we would infer that with time the beam phase-space would shrink to a point, with a characteristic time equal to the time required for a particle to radiate its energy into synchrotron photons. That is too good to be true and other effects must intervene. The dominant effect (for beams of sufficiently low intensity) arises from the same source, namely the quantum nature of the emitted radiation. Quantum fluctuations produce noise; the particle energy random-walks away from the mean

---

*Strictly speaking, R is a function of ε. But this contribution is easily shown for our big machine to be small because of the smallness of $\langle\eta(s)\rangle$.

energy, being limited only by the damping time itself. The energy fluctuation is therefore

$$\Delta E \sim \sqrt{n_\gamma}\, E_c \tag{2.67}$$

where n_γ is the number of photons emitted during a damping period and E_c (eqn. 2.28) is the typical photon energy.

$$n_\gamma \sim \sqrt{\frac{E}{E_c}} \tag{2.68}$$

or*

$$\frac{\Delta E}{E} \simeq \sqrt{\frac{E_c}{E}} \tag{2.69}$$

For the 2020 machine this means

$$\frac{\Delta E}{E} \sim \sqrt{\frac{300\text{keV}}{500\text{TeV}}} \sim 2.5\times10^{-5} \tag{2.70}$$

and a horizontal spread

$$x_\varepsilon \sim \eta\, \frac{\Delta E}{E} \sim 20\mu \tag{2.71}$$

These energy fluctuations drive horizontal betatron oscillations; x_β can neither dominate x_ε nor be negligible with respect to x_ε:

$$x_\beta \sim x_\varepsilon \tag{2.72}$$

A detailed discussion can be found in Sands.[6]

So far the <u>vertical</u> motion is not affected by quantum excitation. While there is some vertical spread caused by the angular distribution of emitted photons, this is negligible in comparison to vertical spread induced by higher order effects (skew quadrupoles, nonlinear elements such as sextupole and octupole fields; noise, etc.) These are not easy (certainly not here) to quantify. Empirically one finds in e^+e^- machines beam heights ~1-10% of the beam width

$$\Delta z \sim (.01-.1)\Delta x \tag{2.73}$$

---

*This implies that $\Delta E/E \sim E/\sqrt{\rho}$. For <u>electron</u> machines, design considerations force $\rho \sim E^2$ in order to minimize the sum of RF costs ($\alpha\ E^4/\rho$) and ring costs ($\alpha\ \rho$). Hence $\Delta E/E$ is roughly machine-independent.

If true for the 2020 machine, this would imply

$$\langle \Delta z \rangle \lesssim 2\mu \qquad (2.74)$$

More important than the actual size is the distribution in z or x. The noisy nature of the quantum-fluctuation mechanism suggests a gaussian distribution. However, this driving mechanism, when coupled with the effects of nonlinear forces (beam-beam interaction, resonances, etc.), turns out to produce much longer (roughly exponential) tails. Formulation of diffusion-equations for the phase-space population is necessary to treat this question. Of course, the magnitude of the tail of the distribution at the machine aperture determines the rate of beam loss.

In proton machines at present energies, where synchrotron radiation is negligible, the tail of the transverse density distribution is much sharper. The dynamics which determines the nature of this distribution is obscure and has to do with nonlinearities in the optical properties of the lattice, the beam-beam interaction, the number and strength of nearby resonances, and sources of noise (power-supply ripple, RF noise, and gas scattering).

8. Collective (Single-Beam) Instabilities

There exist a large class of instabilities which occur because of the interaction of a single beam with its environment. The electromagnetic field of the beam induces currents in the walls of the vacuum chamber, which in turn create fields which drive the beam. If the beam intensity is high enough and the phase relation of the response to the source is "correct," there may be positive feedback and creation of instability.

There are both transverse and longitudinal instabilities. Some instabilities depend on coupling to resonant structures of high Q (eg RF cavities). Others are present even in a non-resonant environment. We shall not recite here the catalogue of instabilities. But the most serious ones are those with high frequency, where a single bunch executes complex internal motion of nontrivial "multipolarity," rather than rigid motion.

Rigid motion of an entire bunch may be monitored, and there is the opportunity to cure the instability via feedback. The high-frequency "microwave" instability must be cured by other means. The most serious such instability for the 2020 machine appears to be longitudinal and we briefly describe this one, in order to give some idea of how these are handled.

Consider first of all a coasting beam which has a small periodic density modulation as function of longitudinal coordinate s. (We ignore the transverse degrees of freedom.) The image charge and wall current will likewise be modulated, and exert a force back on the beam which is periodic with the same frequency. We consider* this frequency ω to be a multiple n of the revolution frequency ω_0. Then the beam finds itself in a RF potential $V(\omega)$ which may, if the phase relations are appropriate, tend to bunch the beam, thereby increasing the image currents, which in turn leads to increased bunching and ultimate instability. The induced RF voltage $V(\omega)$ will be linearly related to the perturbing current $i(\omega)$ (due to the linearity of Maxwell's equations)

$$V(\omega) = i(\omega) Z_{\parallel}(\omega) \tag{2.75}$$

This relation defines the (complex) impedance, a quantity which can be calculated by solving Maxwell's equations for the electromagnetic fields produced[8] in the vacuum chamber by the perturbed circulating beam.

For instability to ensue, $V(\omega)$ must be large enough to capture the beam, i.e. to bunch the beam and contain the beam phase-space within the (self-consistently induced) RF buckets. The bucket height is given by Eqn (2.48), with the harmonic number given by n:

$$(\Delta E)^2 \sim \frac{EV}{\eta n} = \frac{E}{\eta} \frac{Z_{\parallel}}{n} i \tag{2.76}$$

Thus large momentum spread and/or low currents are necessary to avoid the instability. (This is the Keil-Schnell criterion[9]):

$$\frac{Z_{\parallel}}{n} \cdot i \lesssim \eta E \left(\frac{\Delta E}{E}\right)^2 \tag{2.77}$$

For the "microwave" instability, the important wavelengths are smaller than the bunch length (Δs). The impedance on the other hand decreases as the frequency exceeds cd^{-1}, where d is the beam-pipe diameter. Then in evaluating Eqn (2.77), in this case, we must use the peak, or instantaneous current in the bunch, inasmuch as the instability is locally generated and would be

*If this is not the case the "induced" RF system can be thought of as rotating around the machine at the difference frequency.

equally important were the ring filled with bunches. That is, even though the Keil-Schnell criterion was derived[10] for coasting beams, it can be applied to the bunched beam case.

We shall not estimate the impedance $Z_{||}/n$ at all. It gets many contributions from miscellaneous elements around the ring (vacuum bellows, position monitors, RF cavities, etc.), especially those with discontinuities in radius or shape, and is reputed to be difficult to compute reliably. In any case $Z_{||}$ is roughly proportional to circumference. But for a given frequency ω, the harmonic number n is also proportional to circumference. Thus $Z_{||}/n$ is an intensive quantity, roughly independent of energy. Empirically, for present machines

$$\frac{Z_{||}}{n} \gtrsim 1\text{-}10 \text{ ohms} \tag{2.78}$$

at the relevant wavelength of order a few cm.

We may now attempt some numbers. We choose $Z_{||}/n \sim 3$ ohms, and take two cases. One is that of "short bunches," where the synchrotron damping has reduced the momentum spread and bunch length. This will, as we shall see, severely limit the number of particles per bucket. The opposite extreme is to fill the bucket (more or less; we take ~25%) thereby maximizing ΔE and Δs, and minimizing the peak current. The parameters we choose are

	Short Bunch	Long Bunch	
Bunch length Δs	3 cm.	30 cm.	
Energy spread ΔE	±10 GeV	±75 GeV	(2.79)

Then the number N of particles per bunch is limited by the Keil-Schnell criterion as follows.

$$N \lesssim \begin{cases} 3\times 10^{8} & \text{short bunch} \\ 2\times 10^{11} & \text{long bunch} \end{cases} \tag{2.80}$$

For the "natural" short bunch, this is quite a severe limitation; intensities of 3×10^{10} to 10^{11}/bunch are the norm in present machines. To see what is optimal, however, one must look with care at the questions of luminosity and of beam-beam tune shift.

9. Luminosity

Let us put one bunch of p's and one of $\bar{p}$'s into the machine. The luminosity $\mathscr{L}$ per crossing (for head-on collisions and ignoring density variations across the bunch) is

$$\mathcal{L} \simeq \frac{N^2}{A} \tag{2.81}$$

where A is the area of the bunches and N the number of particles per bunch (Luminosity x cross-section = no. of events). We get, for the limiting cases of short and long bunch we considered in the previous section, and using just the nominal area of the beams in the machine,

	Short Bunch	Long Bunch	
N	3×10^{8}	2×10^{11}	
Area	$\sim2\mu\times20\mu$	$70\mu\times70\mu$	
$\mathcal{L}$ (per crossing)	2×10^{23} cm^{-2}	8×10^{26} cm^{-2}	(2.82)

In the case of the long bunch, we have assumed a size as determined by the betatron motion. Typically the dispersion function $\eta(s)$ is designed to vanish at the collision point, so that the horizontal size at the collision region is controlled only by the betatron phase space.

The case of the "long bunch" would lead to a quite respectable luminosity already, when one considers that the revolution frequency is 250 Hz, that the beam can be focussed more strongly at the collision point than at typical points around the ring, and that we can contemplate having a large number of bunches stored in the ring. However, there is yet another limitation to consider. This is imposed by the beam-beam interaction. Before any serious optimization of luminosity can be attempted, the beam-beam limit must be taken into account.

10. Beam-beam limit

The proton beam, through which the antiprotons necessarily pass, is a (nonlinear) focussing element for the antiprotons (and vice versa). Let us estimate the focussing strength. To simplify the discussion, approximate the proton beam by a uniform slab of charge of half-width Δx and half-height $\Delta z \ll \Delta x$. Consider only the vertical force, which for this "slab" geometry is dominant. Then the vertical impulse received by an antiproton with impact parameter z is easily worked out (most easily in the rest frame of the proton bunch).

$$\Delta p_z = e\int ds\ E_T\left(\frac{z}{\Delta z}\right) = \frac{Ne^2 z}{2\Delta x \Delta z} \qquad (z \leq \Delta z) \tag{2.83}$$

We may compare this kick to that given by a lattice quadrupole. This is (cf. Eqn. (2.8)

$$(\Delta p_z)_{quad} \sim eG\ell z \sim \frac{p}{L} z \tag{2.84}$$

The (Δp_z) from this quad changes the betatron tune ν_z by an amount $\Delta\nu$ which is O(1). Thus we get*

$$(\Delta\nu)_{Beam-beam} \simeq \frac{(\Delta p_z)_{Beam-beam}}{(\Delta\ p_z)_{quad}} \sim \frac{e^2 N}{2\Delta x \Delta z} \cdot \frac{L}{p} \tag{2.85}$$

The (linear) tune shift can be compensated by retuning quadrupoles. However, some of the beam (and after many turns, almost all of the particles) have impact parameters $z>\Delta z$ and suffer a smaller $\Delta\nu$ than the core. Thus $\Delta\nu$ is better regarded as a <u>tune spread.</u> As we discuss later, in order to maintain stability the tune of the machine cannot be an integer or integer plus a vulgar fraction p/q (p,q small integers). In practice the tune must be controlled to a few percent of an integer. For e^+e^- machines, the empirical (and to some extent theoretical)[11] limit on $\Delta\nu$ is $\lesssim .03$ for PEP/PETRA/CESR and $\lesssim .06$ for SPEAR/DORIS.

For proton rings at present-day energies it is believed that the maximum allowed $\Delta\nu$ probably is smaller than that, owing to the lack of synchrotron-radiation damping as a stabilizing** influence. Recent experience at the CERN $p\bar{p}$ collider indicate stable operation at $\Delta\nu \sim 2\text{-}3\times 10^{-3}$, in accordance with theoretical estimates.

---

*We have assumed the focussing strength at the collision point is typical of that around the ring. This is typically not the case. Very strong local focussing is used to increase the luminosity. If the "focal length" is smaller, then L should be replaced by the local focal length (more precisely the β-function at the collision point, to be discussed in the next section). However, then the area $A=\Delta z\Delta x$ should be replaced by the (smaller) local value as well. It turns out that A is also proportional to L, so that the beam-beam tune shift is independent of the local focussing strength.

**After a damping time the electron beam forgets its past. Protons at present energies, like elephants, never forget. But for electrons the radiation excitation introduces more noise into the phase-space, a destabilizing influence.

Our 2020 machine is an intermediate case; we assume $\Delta\nu \lesssim .01$ is the limit (good to a factor 3). Putting in some rough numbers then gives, for single bunches and our previous parameters,

$$\Delta\nu_{\text{beam-beam}} \overset{?}{\sim} \begin{array}{ll} .06 & \text{"short" beam} \\ 0.3 & \text{"long" beam} \end{array} \tag{2.86}$$

This is unacceptable, and we must optimize the luminosity taking simultaneously into account the synchrotron damping, the limit on single-beam current from the microwave instability, and the limit from beam-beam tune shift.

We note here that the beam-beam limit, Eqn. (2.83), implies a maximum value for the transverse current density. To increase luminosity it is advantageous to make the beam bigger. In particular, dividing Eqn. (2.80) by (2.83), we get

$$\mathcal{L} \sim \frac{N\Delta\nu}{2\pi\alpha} \cdot \frac{p}{L} \qquad \text{(per crossing)} \tag{2.87}$$

Thus for fixed $\Delta\nu$ and p, our options are

1) Increase N, at the same time somehow increasing the area A.

2) Decrease L, i.e. make the focussing at the collision point as strong as possible.

We shall not pursue these issues further here, but will wait until we have built up more formalism. Suffice it to say that we do not yet have any reliable luminosity estimate.

III. OPTICS

1. Vertical Motion; Hill's Equation

In this section we shall discuss in more detail how one describes the optics of the machine. We begin as before with vertical motion, and write down the basic equations:

$$z' \equiv \frac{dz}{ds} \tag{3.1}$$

$$\frac{dz'}{ds} = -\,k(s)z \tag{3.2}$$

Here

$$z' \equiv \frac{p_z}{p} = \Theta_z \tag{3.3}$$

the derivative of the vertical coordinate is used as a momentum variable rather than p_z. Thus phase space is conventionally taken to be z-z′ space.

The focussing function k(s) is (to first approximation), for an ideal alternating gradient lattice, nonvanishing only within quadrupole magnets, where

$$k(s) \equiv \pm \frac{eG}{p} = \pm \frac{G}{B\rho} \tag{3.4}$$

The two first-order equations combine to produce an oscillator-like equation.

$$z'' + k(s)z = 0 \tag{3.5}$$

This is known as Hill's equation. The focussing function is periodic, k(s) = k(s+C), with C the orbit circumference. But general solutions, of course, need <u>not</u> be periodic.

Were k a constant, we would have oscillatory motion.

$$z \sim z_0 \sin\left(\frac{s}{\beta} + \phi_0\right) \tag{3.6}$$

$$z' \sim \frac{z_0}{\beta} \cos\left(\frac{s}{\beta} + \phi_0\right) \tag{3.7}$$

with

$$\beta = \frac{1}{\sqrt{k}} \tag{3.8}$$

a constant. The set of all orbits of constant amplitude would form in z,z′,s phase space a tube (or torus if z is closed back upon itself: z+C=z), with individual orbits following helical paths around the tube, as shown in Fig. 7.

<u>Provided stable solutions exist</u>, the general solution to Hill's equation has a similar form. It is conventionally written

$$z = \sqrt{\beta(s)\varepsilon}\,\sin\phi(s) \tag{3.9}$$

where the amplitude function β(s) is periodic and depends only upon the lattice, i.e. the focussing function k(s). The constant ε, called the Courant invariant, determines the normalization of the amplitude. The phase function ϕ(s) is determined by β(s) as follows

$$\phi(s) = \int_{s_0}^{s} \frac{ds}{\beta(s)} \tag{3.10}$$

This latter relation can be obtained by substituting the "solution", Eqn (3.9) into Hill's equation, and demanding the coefficient of cos ϕ vanish.* The vanishing of the coefficient of sin ϕ then produces a nonlinear second order differential equation for β that we shall not bother to write down. There are other convenient ways of obtaining the β-function, which we shall describe later on.

The amplitude function $\beta(s)$ is a most important function; it determines the (linear) optical properties of the lattice, and essential properties of the vertical motion. In particular the tune ν (the number of betatron oscillations per revolution) now has a precise definition

$$\nu = \frac{1}{2\pi} \oint \frac{ds}{\beta(s)} \tag{3.11}$$

2. Linear Maps

We may obtain more insight into the motion and determine $\beta(s)$ as well by going back to the Hamiltonian form of two first-order equations for z and z′. In this form the content of the Liouville theorem is more directly seen.

First order linear equations can (like the Schrodinger equation) be formally integrated. Define

$$\xi(s) = \begin{pmatrix} z(s) \\ z'(s) \end{pmatrix} \tag{3.12}$$

and relate $\xi(s+ds)$ to $\xi(s)$. From Eqns (3.1) and (3.2)

$$\xi(s+ds) = \left[\begin{pmatrix} 1 & 0 \\ 0 & 1 \end{pmatrix} + \begin{pmatrix} 0 & 1 \\ k(s) & 0 \end{pmatrix} ds\right] \xi(s) \equiv (1+Tds)\xi(s) \tag{3.13}$$

*The constant of integration one gets is set to zero. This cleans up the equation for β.

and hence $\xi(s')$ can be obtained from $\xi(s)$ by multiplying by a product of matrices each of which depend only on the lattice. For sufficiently small Δs we have, schematically

$$\xi(s') = M(s',s)\xi(s) = [\prod_{\Delta s} (1+T\Delta s)]\xi(s) \tag{3.14}$$

[A veteran particle theorist will write

$$M(s',s) = P \exp \int_s^{s'} ds''T(s'') \tag{3.15}$$

where P is the path ordering operator.]

The matrix $M(s',s)$ is known as the <u>transport matrix.</u> Because a lattice is composed of a sequence of basic elements and because transport matrices satisfy the group property

$$M(s'',s')M(s',s) = M(s'',s) \tag{3.16}$$

we need only know the matrices for the basic elements. For example

$$M = \begin{pmatrix} 1 & 0 \\ -\frac{1}{f} & 1 \end{pmatrix}$$ Thin focussing quadrupole* (For defocussing, $f \to -f$)

$$M = \begin{pmatrix} 1 & L \\ 0 & 1 \end{pmatrix} \tag{3.17}$$ Bending magnet or straight section of length L.

Here the focal length is

$$\frac{1}{f} = \frac{eG\ell}{p} = \frac{G\ell}{B\rho} \tag{3.18}$$

Then ξ can be propagated around the ring by multiplying these 2×2 matrices together. Note that the determinant of the matrix for a basic element is unity; hence so also is a product of them

$$\det M(s',s) = 1 \tag{3.19}$$

[This is directly related to the Liouville theorem.]

*For accuracy, one must go beyond the thin lens approximation. For our purposes this is hardly necessary. The reader is invited to work out the correction using the basic elements given here.

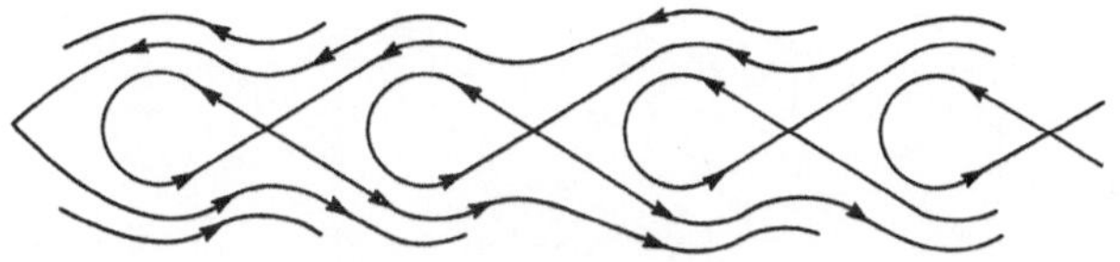

Figure 6. Structure of longitudinal phase space during acceleration or in presence of synchrotron radiation.

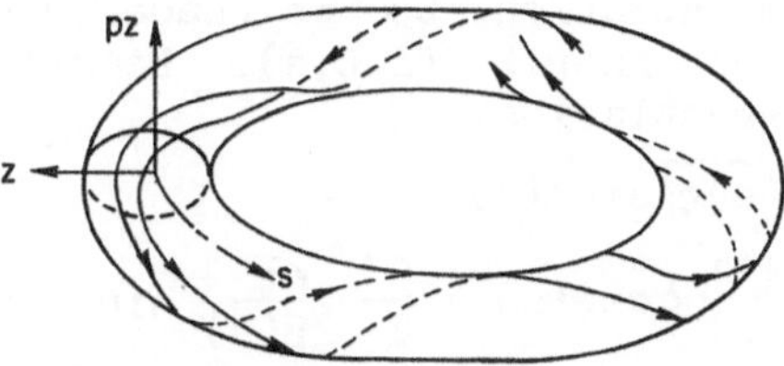

Figure 7. "Invariant torus" for particles of a given emittance traversing a constant focussing structure.

Now consider the matrix for a complete circuit around the ring. After N revolutions the coordinate is

$$\xi^{(N)} = M^N \xi^{(0)} \tag{3.20}$$

If we diagonalize M

$$SMS^{-1} = \begin{pmatrix} \lambda_1 & 0 \\ 0 & \lambda_2 \end{pmatrix} \tag{3.21}$$

we must have

$$\lambda_1 \lambda_2 = 1 \tag{3.22}$$

and for stability

$$\begin{aligned} \lambda_1 &= e^{i\phi} \\ \lambda_2 &= e^{-i\phi} \end{aligned} \tag{3.23}$$

(Otherwise $\xi^N \sim \lambda^N{}_{max}$ grows exponentially with N)

$$\mathrm{Tr}\ M(s+C,s) = \mathrm{Tr}\ SMS^{-1} = 2\cos\phi < 2 \tag{3.24}$$

is sufficient to ensure stability. It takes little imagination to guess that ϕ is essentially the tune

$$\phi = 2\pi\nu \tag{3.25}$$

[We leave the demonstration to the reader.] We may also obtain $\beta(s)$ from the transport matrix M(s+C,s). Recall, from Eqn. (3.9), along with one differentiation,

$$\xi = \begin{pmatrix} z(s) \\ z'(s) \end{pmatrix} = \begin{pmatrix} \sqrt{\beta(s)\varepsilon}\ \sin\phi(s) \\ \sqrt{\frac{\varepsilon}{\beta(s)}}\ \cos\phi(s) + \frac{\beta'}{2}\sqrt{\frac{\varepsilon}{\beta(s)}}\ \sin\phi\ (s) \end{pmatrix} \tag{3.26}$$

Now choose the initial phase such that $\phi(s)=0$ and initial amplitude $\varepsilon=\beta$. Then

$$\xi(s) = \begin{pmatrix} 0 \\ 1 \end{pmatrix} \tag{3.27}$$

After one revolution

$$\xi(s+C) = \begin{pmatrix} \beta(s)\sin 2\pi\nu \\ \cos 2\pi\nu + \frac{\beta'}{2}\sin 2\pi\nu \end{pmatrix} \tag{3.28}$$

But from the definition of the transport matrix and Eqn. (3.27)

$$\xi(s+C) = \begin{pmatrix} M_{12} \\ M_{22} \end{pmatrix} \tag{3.29}$$

and thus the upper right-hand element of M(s+C,s) determines directly β(s). This clearly provides a convenient numerical procedure. Using the constraints on Tr M and det M, the general form of M(s+C,s) is

$$M = \cos 2\pi\nu + J\sin 2\pi\nu = e^{2\pi\nu J} \tag{3.30}$$

with

$$J = \begin{pmatrix} -\frac{\beta'}{2} & \beta \\ -\frac{1}{\beta}\left(1 + \frac{\beta'^2}{4}\right) & \frac{\beta'}{2} \end{pmatrix} \tag{3.31}$$

Note that $J^2=-1$.

We repeat:

To obtain the tune ν and the β-function β(s):

Compute the transport matrix for a complete revolution. Then

$$\text{Tr } M(s+C,s) = 2\cos 2\pi\nu$$

$$M_{12} = \beta(s)\sin 2\pi\nu \tag{3.32}$$

The reader should not delude himself into thinking that this sketch provides a strict derivation of the results we have presented. However, the raw material for providing the missing links has been given.

3. Nature of the Motion

We have already described the nature of the motion in the case of a constant focussing function k(s). For the general case of stable motion there is in z,z′, s space again an "invariant torus" on which all orbits with a given value of Courant-invariant ε lie; these orbits again spiral around the torus. At any given s, the cross-section of the torus is an ellipse, as can be seen from Eqn (3.26) and $\cos^2+\sin^2=1$

$$\varepsilon = \frac{z^2}{\beta(s)} + \beta(s)\ (z' - \frac{\beta'}{2\beta}\ z)^2 \tag{3.33}$$

The area of the ellipse is independent of s (Liouville again!) and is πε. At extrema of β(s) we have an erect ellipse; in between extrema it is skew (cf. Fig. 8). Note the β-function does not describe the orbit of a given particle but rather the envelope of many particle orbits.

For a beam, i.e. a phase space population, the typical value of the phase space area of the particles in the beam is called the emittance of the beam (careful! sometimes what is quoted is 1σ; other times 2σ):

$$\text{emittance} = \langle\varepsilon\rangle = \pi\sqrt{\langle z^2\rangle\langle z'^2\rangle} \tag{3.34}$$

The maximum value of emittance which survives in the machine is called admittance or aperture.

4. FODO Lattice

We now compute β(s) for a FODO lattice. This is, as already described in the previous section, a regular sequence of focussing and defocussing quads separated by bending magnets. We first compute the transfer matrix for a cell (the basic element of the lattice consisting of one F quad, one D quad, and the intervening magnets). We again assume the quads are separated by distance L and start the transfer matrix from the center of an F quad, where by symmetry we expect an extremum of the β-function. Then first calculate [cf. Eqn. (3.17)]

$$\sqrt{F}\ \mathrm{O}\ \sqrt{D} \equiv \begin{pmatrix} 1 & 0 \\ -\frac{1}{2f} & 1 \end{pmatrix} \begin{pmatrix} 1 & L \\ 0 & 1 \end{pmatrix} \begin{pmatrix} 1 & 0 \\ \frac{1}{2f} & 1 \end{pmatrix}$$

$$= \begin{pmatrix} 1 + \frac{L}{2f} & L \\ -\frac{L}{4f^2} & 1 - \frac{L}{2f} \end{pmatrix} \tag{3.35}$$

Next calculate $\sqrt{D}O\sqrt{F}$ which is easy; just change $f \to -f$. Then multiply the two together to get

$$\sqrt{F}\,O\,D\,O\,\sqrt{F} = \begin{pmatrix} 1 - \frac{L^2}{2f^2} & 2L(1 + \frac{L}{2f}) \\ \text{etc.} & 1 - \frac{L^2}{2f^2} \end{pmatrix} \tag{3.36}$$

By symmetry the β-function is periodic over a cell as well as over the entire machine. Therefore we can define a phase advance per cell μ such that

$$2\pi\nu = \mu\, n_{cells} \tag{3.37}$$

and use the transport matrix for a cell rather than for the whole machine to obtain the β-function β_{max} at the focussing quad. According to our recipe

$$\cos\mu = 1 - \frac{L^2}{2f^2}$$

$$\beta_{max}\sin\mu = 2L(1 + \frac{L}{2f}) \tag{3.38}$$

or

$$\beta_{max} = 2f\sqrt{\frac{1 + \frac{L}{2f}}{1 - \frac{L}{2f}}} \tag{3.39}$$

Note that, as expected, $\beta' = 0$. Also by replacing $F \to D$ we get β_{min} and

$$\frac{\beta_{max}}{\beta_{min}} = \frac{1 + \frac{L}{2f}}{1 - \frac{L}{2f}} = \frac{1 + \sin\frac{\mu}{2}}{1 - \sin\frac{\mu}{2}} \tag{3.40}$$

We may observe that for stability we must have

$$\frac{L^2}{2f^2} < 2 \tag{3.41}$$

or

$$f > \frac{L}{2} \tag{3.42}$$

That is, we must not overfocus. In practice a phase advance per cell of $\sim 90^0$ (sometimes a little less)

$$\mu \lesssim \frac{\pi}{2}$$

$$f \lesssim \frac{L}{\sqrt{2}} \tag{3.43}$$

is chosen. We shall take $\mu=\pi/2$ for definiteness. Note also that, because

$$\frac{1}{f} = \frac{eG\ell}{p} \tag{3.44}$$

the above condition is

$$\frac{eG\ell L}{p} = \frac{G\ell L}{B\rho} = \sqrt{2} \tag{3.45}$$

which is essentially what we saw in the previous section (cf.Eqn.(2.8)). Note also that

$$\langle\beta\rangle \sim 2f \sim \sqrt{2}\,L \tag{3.46}$$

so that the fuzzy parameter L used in the previous section should be replaced by the basic optical parameter $\beta(s)$, which measures the local betatron wavelength according to the relation, Eqn.(3.10):

$$\Delta\phi = \frac{\Delta s}{\beta(s)} \tag{3.47}$$

The β-function in between F and D magnets can be obtained by the same technique but by employing a different starting point. This is left as an exercise.

5. Horizontal motion and dispersion

The betatron motion in the horizontal plane, as mentioned in the previous section, is almost the same as for the vertical plane. Inclusion of the curvature correction in Hill's equation gives

$$\frac{d^2x}{ds^2} + k(s)x + \frac{x}{\rho^2} = 0 \tag{3.48}$$

i.e. bending magnets produce a little focussing. (This should be put into the matrix O for a bending magnet.)

The dispersion function $\eta(s)$ describing the closed orbit deviation x_ε for off-momentum particles

$$x_\varepsilon = \eta(s)\frac{\Delta p}{p} \equiv \eta(s)\sigma_\eta \tag{3.49}$$

may now be calculated. To first order, contributions to the dispersion function, which evidently is periodic over a cell, come only from bending magnets (which are analogous to prisms). A magnet of length L contributes deviations δx and $\delta x'$ given by (cf. Fig.)

$$\begin{pmatrix} \delta x \\ \delta x' \end{pmatrix} = \begin{pmatrix} \frac{L^2}{2\rho^2} \\ \frac{L}{\rho^2} \end{pmatrix} \delta\rho = \begin{pmatrix} \frac{L^2}{2\rho} \\ \frac{L}{\rho} \end{pmatrix} \sigma_\eta \equiv \Delta \tag{3.50}$$

Let

$$\xi(s) = \begin{pmatrix} \eta(s) \\ \eta'(s) \end{pmatrix} \sigma_\eta \tag{3.51}$$

be the orbit deviation. The change in orbit deviation over a cell is then expressed by

$$\xi(s+2L) = M(s+2L,s)\xi(s) + \sqrt{F}\Delta + \sqrt{F}OD\Delta \tag{3.52}$$

where the last two inhomogeneous terms are contributed by the magnets in the cell, and the first term propagates the orbit error through the cell. By symmetry

$$\xi(s+2L) = \xi(s) \equiv \xi_{max} \tag{3.53}$$

and we get

$$\xi_{max} = \frac{1}{1-M} \sqrt{F}(1+OD)\Delta \tag{3.54}$$

This determines the maximum value of $\eta(s)$. Evaluation of the right hand side is done via the following steps:

1. Explicity work out the numerator.

2. Determine eigenvectors of M.

3. Expand the numerator in terms of these eigenvectors.

4. Evaluate ξ_{max}.

One finds

$$\eta_{max} = \frac{L^2}{\rho \sin^2 \frac{\mu}{2}} \left(1 + \frac{1}{2} \sin \frac{\mu}{2}\right) \qquad (3.55)$$

with η_{min} obtained, as usual by changing +1/2 to -1/2 in the numerator. Putting in numbers for $\mu=\pi/2$ gives

$$\eta_{max} = 2.7 \frac{L^2}{\rho} \qquad \eta_{min} = 1.3 \frac{L^2}{\rho} \qquad (3.56)$$

From Eqn. (3.45) (for $\mu=\pi/2$)

$$\sqrt{2} = \frac{eG\ell L}{p} = \frac{G\ell L}{B\rho} = \frac{G}{B} \cdot \frac{\ell}{L} \cdot \frac{L^2}{\rho} \qquad (3.57)$$

Hence, for 90° phase advance/cell

$$\eta_{max} \cong 4 \left(\frac{B}{G} \cdot \frac{L}{\ell}\right) \qquad (3.58)$$

which is, as advertised, independent of energy, and of order meters.

A more systematic way of treating dispersion[12] is to enlarge the transport matrix to a 3x3 matrix which acts upon x, x', and σ_η.

6. Chromaticity

Not only does the closed orbit change for off-momentum particles but also the focal properties of the lattice – in particular the tune ν. The optical analogue is chromatic aberration. Thus the chromaticity ξ is defined as the percent change in tune per percent change in momentum

$$\frac{\delta\nu}{\nu} = \xi \frac{\delta p}{p} = \xi \sigma_\eta \qquad (3.59)$$

The natural chromaticity, which is the contribution to ξ of the normal lattice, can be computed directly. From Eqn. (3.38).

$$\cos \mu = 1 - \frac{L^2}{2f^2} \qquad (3.60)$$

and

$$\frac{\delta f}{f} = -\frac{\delta p}{p} \tag{3.61}$$

one finds easily that

$$\frac{\delta \nu}{\nu} = \frac{\delta \mu}{\mu} = -\frac{2}{\mu} \tan \frac{\mu}{2} \frac{\delta p}{p} \tag{3.62}$$

For 90° phase advance, this gives

$$\xi = -1.3 \tag{3.63}$$

This is too large and, as it turns out, also of the wrong sign. Single-beam instabilities are sensitive to chromaticity, and stability requires a positive value. The chromaticity is adjusted by adding sextupole magnets around the ring. Evidently the sextupole strength (per cell) needed to do this scales with the quadrupole strength, independent of energy.

IV. ERRORS AND NONLINEAR RESONANCES

With such a big machine, we might expect it to be impossible to align. We briefly investigate here the effect of errors and nonlinearities.

1. Closed orbit error

Suppose one magnet at position s_0 provides the wrong bending field by an amount ΔB. The normal bend angle $\Delta\theta$ in the magnet is

$$\Delta\theta = \frac{eB\Delta s}{p} = \frac{\Delta s}{\rho} \tag{4.1}$$

Then the angle change $\delta x'$ caused by the error is

$$\delta x' = \frac{\Delta B}{B} \Delta\theta \sim \frac{\Delta B}{B} \frac{\Delta s}{\rho} \tag{4.2}$$

Therefore the perturbed closed orbit is just a betatron oscillation which has a kink in slope of magnitude $\delta x'$ located at s_0. We need therefore to simply locate a phase point $\xi(s_0)$ such that $x(s_0+C)=x(s_0)$ and $x'(s_0+C)=\delta x'+x'(s_0)$. Write

$$x(s_0)=\sqrt{\beta\varepsilon}\cos \phi_0 \tag{4.3}$$

Then

$$x(s_0+C) = \sqrt{\beta\varepsilon}\cos (\phi_0+2\pi\nu) \tag{4.4}$$

implying $\phi_0=-\pi\nu$. We now calculate $\delta x'$:

$$x'(s_0) = \sqrt{\frac{\epsilon}{\beta}} \sin \phi_0 + \frac{\beta'}{2\beta} x(s_0) \tag{4.5}$$

and

$$x'(s_0+C) = \sqrt{\frac{\epsilon}{\beta}} \sin(\phi_0+2\pi\nu) + \frac{\beta'}{2\beta} x(s_0+C) \tag{4.6}$$

Subtraction gives us the amplitude $\sqrt{\epsilon\beta}$

$$\delta x' = x'(s_0+C) - x'(s_0) = 2\sqrt{\frac{\epsilon}{\beta}} \sin\pi\nu = \frac{\Delta B}{B\rho} \Delta s \tag{4.7}$$

and thus the solution for x (Green's function!) is

$$x(s) = \frac{\sqrt{\beta(s)\beta(s_0)}}{2 \sin \pi\nu} \left(\frac{\Delta B}{B}\right)_{s_0} \frac{\Delta s}{\rho} \cos \left(\pi\nu - \int_{s_0}^{s} \frac{ds'}{\beta(s')}\right) \tag{4.8}$$

Note that (1) the amplitude is largest if the field error is located in a region of large β, and that (2) the amplitude blows up if ν is an integer. This is the simplest kind of resonance. If on each revolution the betatron phase at s_0 is the same, the error kick will always increase the amplitude in the same way; one has just a resonantly driven oscillator.

The total contribution to the closed orbit is obtained, via superposition, by adding the contributions of the individual magnet errors. If they are random, then the rms error is

$$\langle x^2\rangle_{rms} = \frac{\beta(s)\bar{\beta}}{8 \sin^2 \pi\nu} \left\langle \left(\frac{\Delta B}{B}\right)^2 \right\rangle_{rms} \frac{\ell^2_{magnet}}{\rho^2} N_{magnet} \tag{4.9}$$

The scaling law is

$$\Delta x \sim \frac{\beta}{\rho} \cdot \left(\frac{\Delta B}{B}\right) \cdot \ell_{magnet} \cdot \sqrt{N_{magnet}} \sim \sqrt{\frac{N_{magnet}}{\nu}} \tag{4.10}$$

For fixed magnet type and magnet quality, and for $\nu \sim \sqrt{E}$, we find Δx is independent of energy. Choosing $\ell_{mag} \sim 7$m, $N \sim 1.4\times10^5$, $\nu \sim 400$, and $\Delta B/B \sim 10^{-3}$, we get $\langle\Delta x\rangle \sim$ 5mm. Correction elements are a necessity, but the problem (on paper) does not worsen with energy.

Not all errors need be random. If fourier components (in longitudinal coordinates) of the error field peak near the betatron wavelength, one can get enhancements. This is in fact one way of correcting closed orbit errors; one analyzes the Fourier spectrum of the particle orbit deviations, and applies correction (dipole) fields having the dominant Fourier component.[13]

2. Tune Shifts

Suppose a quadrupole at s_0 has the wrong field. Then the tune will be modified. To estimate this, we look at the modification to the transport matrix

$$M(s_0+C,s_0)\rightarrow\begin{pmatrix}1 & 0\\ \ell_{quad}\,\delta k & 1\end{pmatrix}M(s_0+C,s_0) \tag{4.11}$$

We need only recalculate the trace

$$\begin{aligned}\mathrm{Tr}M &\rightarrow \mathrm{Tr}M + \ell_{quad}\,\delta k\cdot M_{12}\\ &= 2\cos 2\pi\nu + \beta(s_0)\,\ell_{quad}\,\delta k\cdot\sin 2\pi\nu\end{aligned} \tag{4.12}$$

and thus

$$\begin{aligned}\Delta\nu &= \frac{\ell_{quad}}{4\pi}\,\beta(s_0)\delta k\\ &= \frac{1}{4\pi}\,\beta(s_0)\left(\frac{\Delta G}{G}\right)\cdot\frac{1}{f}\cdot\frac{\ell_{quad}}{\ell}\end{aligned} \tag{4.13}$$

The last factor takes into account the fact that for the 2020 machine (unlike present machines) what we call a focussing element F consists of a <u>sequence</u> of quadrupole magnets, not an individual magnet. Again, assuming random errors and averaging around the ring gives

$$\langle\Delta\nu\rangle \sim \frac{1}{4\pi}\sqrt{\frac{\beta^2_{max}+\beta^2_{min}}{2}}\;\frac{\Delta G}{G}\;\frac{\sqrt{N_{quad}}}{f}\left(\frac{\ell_{quad}}{\ell}\right) \tag{4.14}$$

Since

$$\frac{1}{f\ell} = \frac{eG}{p} = \frac{G}{B\rho} \tag{4.15}$$

we have

$$\langle\Delta\nu\rangle \sim \frac{\beta\sqrt{N_{quad}}}{\rho}\cdot\left(\frac{\Delta G}{G}\right)\cdot\frac{G}{B}\cdot\ell_{quad} \tag{4.16}$$

Again, with

$$\beta \sim \sqrt{E}$$
$$\rho \sim E$$
$$\sqrt{N_{quad}} \sim \sqrt{E} \tag{4.17}$$

we find a tune-shift independent of energy. Putting in numbers for the 2020 machine gives

$$\langle\Delta\nu\rangle \sim 3\langle\frac{\Delta G}{G}\rangle \tag{4.18}$$

3. Miscellaneous Errors

Other such errors can be treated similarly. A short compendium can be found in a contribution by King[14] for ICFA studies of a 20 TeV machine. They include

1. Vertical plane misalignments of dipoles.
2. Quadrupole position errors.
3. Quadrupole tilts.
4. Stray magnetic fields at injection.
5. Gradient errors in dipole magnets.

Examination of the formulae again shows that none of them scale with energy in such a way that the problems worsen at higher energy. Typically the number of sources scale linearly with energy, giving a net deviation growing as $\sqrt{N}\sim\sqrt{E}$. The betatron wavelength also scales as $\sqrt{E}$, and magnetic rigidity (a good word!) as E. The net deviation then scales as $(\sqrt{E})^2/E \sim 1$.

4. Tune Shift from Machine Nonlinearities

The magnetic fields in the lattice are not ideal. Higher order nonlinear terms are present and must be kept under control. To begin, write, for corrections from dipole magnets,

$$\Delta B_x = B_0(1 + \sum_n b_n x^n) \tag{4.19}$$

There are also cross terms in x and z (which are typically more important!), but we come back to those later. The modification to Hill's equation is

$$\frac{d^2x}{ds^2} + k(s)x = \frac{\Delta B(s)}{B\rho} = \frac{1}{\rho}\sum_n b_n(s)x^n \tag{4.20}$$

We have a change in focussing strength

$$\Delta k \sim \frac{-1}{\rho}\sum_n b_n x^{n-1} \tag{4.21}$$

which is, as one could have guessed, most important for large amplitudes. To estimate the change in tune, we may borrow Eqn. (4.13):

$$\Delta\nu = \frac{ds}{4\pi}\,\beta(s)\Delta k \tag{4.22}$$

and obtain

$$\Delta\nu = -\frac{1}{4\pi\rho}\oint ds\ \beta(s)\sum_n b_n(s)x(s)^{n-1} \tag{4.23}$$

The analysis then proceeds as before. Assuming uncorrelated contributions from the different magnets (even more dangerous here??), and taking a single term in the sum,

$$<(\Delta\nu_n)^2> = \frac{1}{16\pi^2\rho^2}\,N_{mag}\,L_{mag}^2 <b_n^2><\beta^2x^{2n-2}> \tag{4.24}$$

Recalling that $<x^2>=\frac{1}{2}<\beta\varepsilon>^2$, we get, roughly

$$\sqrt{<(\Delta\nu_n)^2>}_{rms} \sim \frac{\beta}{4\sqrt{2}\,\pi\rho}\sqrt{N_{mag}}\,L_{mag}\sqrt{<b_n^2>}_{rms}<\beta\varepsilon>^{\frac{n-1}{2}} \tag{4.25}$$

The scaling with energy follows, not surprisingly, the pattern we have seen before, and for higher moments (on paper) even improves, inasmuch as $\beta\varepsilon\sim E^{-1/2}$. The lower values of n, say 3 and 4, are contributed by sextupole and octopole fields, and are relatively controllable, inasmuch as such fields are deliberately included in the lattice to control chromaticity and help tame single-beam collective instabilities.

Higher orders in n are in less control.

The tune shifts (spreads) are very sensitive to large values of β, putting an especially high premium on field quality in the neighborhood of focussing quadrupoles.

5. Resonances

We have already seen that it is unwise to choose an integer for the machine tune inasmuch as dipole field errors drive resonances. Field errors of higher multipolarity drive higher order resonances at tune values, as we shall see, equal to an integer plus a vulgar fraction p/q, with p and q small integers. After the integer resonances, the simplest case is that of linear coupling resonances. These are contributed by tilted (skew) quadrupoles, for example.

$$\frac{d^2x}{ds^2} + k_x(s)x = \delta k(s)z$$

$$\frac{d^2z}{ds^2} + k_z(s)z = \delta k(s)x \tag{4.26}$$

Extension of the transfer-matrix method (to 4×4 matrixes) handles this case quite easily and exactly. The analysis is pretty.[15] It turns out that if

$$\nu_x + \nu_z = n \tag{4.27}$$

with n integer, there exists instability and emittance growth. For

$$\nu_x - \nu_z = n \tag{4.28}$$

there also exists resonance. Energy is transferred back and forth between horizontal and vertical motion in a manner similar to coupled degenerate pendula. But the betatron amplitudes remain bounded.

Nonlinear resonance phenomena are very rich and interesting. The dynamics of a single isolated nonlinear resonance can be worked through in a reasonably systematic way. To go beyond that point is to enter the active research field of 20th-century nonlinear mechanics. Here we shall only partially treat the case of a single nonlinear resonance. One begins with the perturbed Hill's equation

$$\frac{d^2z}{ds^2} + k(s)z = -\frac{1}{\rho}\sum_n b_n(s)z^n \tag{4.29}$$

and change variables ("Floquet transformation")

$$z = \sqrt{\beta(s)}\, v$$

$$d\phi = \frac{1}{\nu}\frac{ds}{\beta(s)} \quad (0\leq\phi\leq 2\pi \text{ around the ring}) \tag{4.30}$$

This smooths out the betatron motion into that of a harmonic oscillator:

$$\frac{d^2v}{d\phi^2} + \nu^2 v = -\frac{\nu^2}{\rho}\sum_n \beta(s)^{\frac{3+n}{2}}\, b_n(s)v^n$$

$$= -\sum_n \left(\sum_p A_p^{(n)} \cos p\phi\right)v^n$$

$$= -\sum_n A^{(n)}(\phi)v^n \tag{4.31}$$

The nonlinear term on the right-hand side is periodic in ϕ and drives the resonance. The above equation can be obtained from the Hamiltonian (think of ϕ as time!)

$$H = \frac{p^2}{2} + \frac{\nu^2 v^2}{2} + \sum_n \frac{A^{(n)}(\phi)}{n+1} v^{n+1} \tag{4.32}$$

The conventional method of solution starts with this Hamiltonian and performs successive canonical transformations until the dominant component of the nonlinearity (for given tune choice) is isolated. The remaining piece of the nonlinear Hamiltonian is then thrown away and the solution obtained by additional canonical transformations.

Nowadays the techniques of Hamiltonian mechanics are more familiar to many of us in the context of the quantum theory. We shall use that language here to motivate the procedure. The nonlinear term contains, in terms of creation and destruction operators a and a^+, terms of order $(a^+)^{n+1}$, inducing transitions with $\Delta E=(n+1)\nu$. The driving term $A_n(\phi)$ contains only integer fourier components exp $\pm ip\phi$. If the energy p of a quantum delivered by the driving term equals the excitation energy ΔE of the oscillator, there will be resonance

$$p = (n+1)\nu \tag{4.33}$$

or

$$\nu = \frac{p}{n+1} = \text{vulgar fraction} \tag{4.34}$$

Note that n is determined by the multipolarity of the magnetic field:

n = 1 quadrupole
 2 sextupole
 3 octupole
 . .
 . .
 . .

while the resonance is driven (essentially)* by the p'th (circumferential) harmonic of the nonlinear force around the ring.

For coupled resonances, the Hamiltonian contains terms which have the form

$$H' \sim z^{n+1} x^{m+1} \tag{4.35}$$

These contain terms $\sim(a_z^+)^{n+1}(a_x^+)^{m+1}$, which by the previous argument gives the resonance condition

$$p = (n+1)\nu_z + (m+1)\nu_x \tag{4.36}$$

There can also be difference resonances (n,m,p integer) as well as sum resonances, driven by terms $\sim(a_z^+)^{n+1}(a_x)^{m+1}$. In this case, when the z oscillator is excited, the x oscillator is simultaneously de-excited. This is a relatively inefficient way to pump energy into the betatron-oscillators, and we infer that, in general, <u>sum resonances will be more dangerous than difference resonances.</u>

To get some more detailed insight into the nature of these resonances, we return to the uncoupled case of a pure sextupole term (n=2) in the oscillator calculation Eqn. (4.31). We expect

*The irregularity of the focussing structure (i.e. β- function) also contributes to the driving term.

that the following steps will set up the simplified calculation.

1. The resonance will be important when $\nu \equiv N+1/3$ or $N+2/3$, N integer, implying $p=3N+1$ or $p=3N+2$.

2. We shall find the equation of motion for the creation operator a^*, which we define as usual

$$v = \frac{1}{\sqrt{2\nu}}(ae^{-i\omega\phi} + a^*e^{i\omega\phi})$$

$$p = i\sqrt{\frac{\nu}{2}}(ae^{-i\omega\phi} - a^*e^{i\omega\phi}) \tag{4.37}$$

where we anticipate the need to remove some "time"-dependence (Remember! "time" is ϕ) from a^* by putting in the $e^{i\omega\phi}$ factor. The frequency ω will be chosen later; it will be approximately — but not precisely — equal to the natural frequency ν.

3. On the right-hand side, we keep only the term proportional to $e^{-ip\phi}a^2$, because that is the only one which will approximately match the frequencies and produce the resonance condition, Eqn. (4.33).

Then the equation of motion for a^* is easily found to be

$$\frac{da^*}{d\phi} = -i(\nu-\omega)a^* - A_2e^{-ip\phi}e^{i3\omega\phi}a^2 \tag{4.38}$$

We choose

$$\omega = \frac{p}{3} \equiv \nu_{res} \tag{4.39}$$

to rid the equation of oscillatory factors and finally obtain

$$\frac{da^*}{d\phi} = -i(\nu-\nu_{res})a^* - A_2a^2 \tag{4.40}$$

Now complex a-space is essentially rescaled phase-space. We see the following features:

1. The phase-space trajectories have a 3-fold symmetry; if $a(\phi)$ is a solution, so also is $a(\phi)\exp(2\pi iM/3)$ with M integer.

(For the generic n^{th} nonlinear term, this clearly generalizes to an (n+1) -fold symmetry.

2. There exist fixed points, i.e. "time-independent" solutions of $da/d\phi=0$. They satisfy

$$a^3 = -i \frac{\Delta\nu}{A_2} \qquad (4.41)$$

and are at a distance from the origin given by

$$|a| = \left(\frac{\Delta\nu}{A_2}\right)^{1/3} \qquad (4.42)$$

That is, they are far away unless

$$\Delta\nu \lesssim A_2 \qquad (4.43)$$

which is, roughly, the condition on tune shift obtained earlier (Eqn.4.23).

3. On resonance, when $\Delta\nu=0$, the fixed points converge to the origin. One may then find simple radial solutions of the equation of motion. Writing

$$a(\phi) = \rho(\phi)e^{i\alpha} \qquad (4.44)$$

with the phase α kept constant, we get

$$\frac{d\rho}{d\phi} = A_2 e^{3i\alpha}\rho^2 \qquad (4.45)$$

For self consistency, we need $d\rho/d\phi$ to be real so that we must have

$$e^{3i\alpha} = \pm 1 \qquad (4.46)$$

which gives six solutions for α, corresponding to rays emanating from the origin at $\theta=0^0$, 60^0,... Motion on three of the rays will be outward from the origin, and inward on the other three. In between these rays it is easy to guess the phase-space trajectories (Fig. 9a). The rays are clearly separatrices. [For nth-order resonance there are 2(n+1) such separatrices.]

4. Moving back off-resonance, the phase-space trajectories near the origin will be circular, oscillator like. Far from the

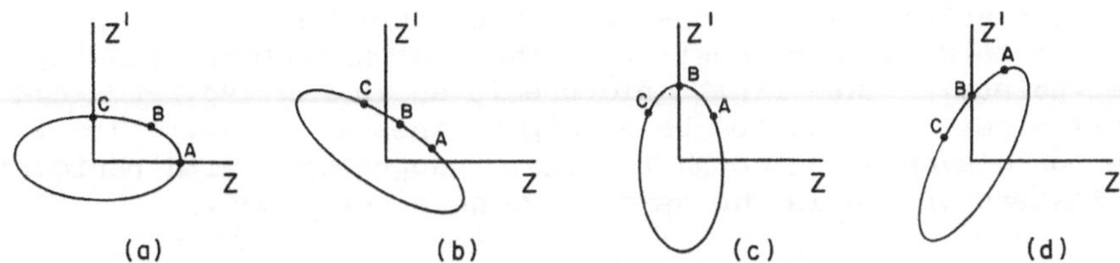

Figure 8. Phase-space positions of particles of a given emittance at various locations in a FODO lattice. The curves are loci of constant emittance. Compare Figs. 3 and 8. (a) F quad, (b) between F and D, (c) D quad, and (d) between D and F.

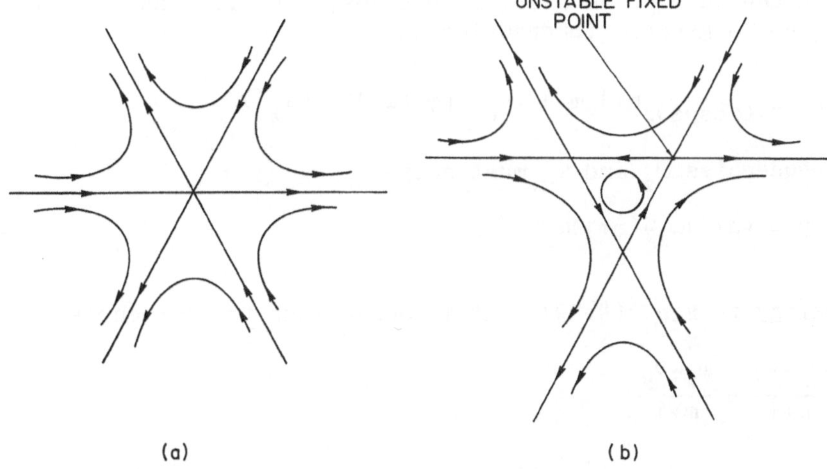

Figure 9. Regions of z-p_z phase space near a 1/3-integer resonance: (a) on resonance, and (b) just off resonance.

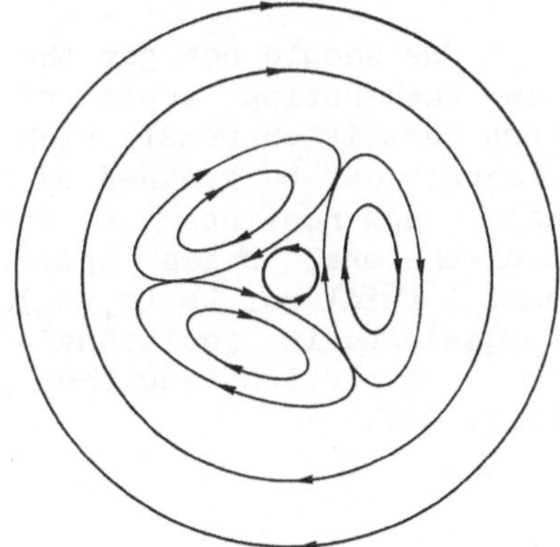

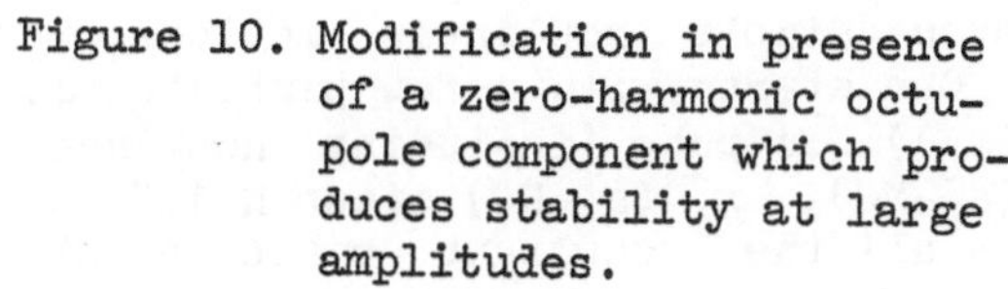

Figure 10. Modification in presence of a zero-harmonic octupole component which produces stability at large amplitudes.

origin, well beyond the fixed points, the small change in tune will not affect the motions we deduced for the on-resonance case. Thus, in between there must be other separatrices linking the fixed-points. The final picture is as in Fig. 9b. For n=2 the separatrices turn out to be straight lines as shown. This is not true for higher n, although the basic geography – with periodicity n+1 instead of 3 – is the same as we have described.

Often there will be other focussing forces of higher multipolarity present. For example if there is a zero-harmonic octupole component which dominates the sextupole component at large amplitudes the outer separatrices will become closed, creating "islands of stability" (Fig. 14). This is a common pattern for the phase-space structure.

In the case of coupling resonances, a similar analysis can be made. For a generic interaction term

$$H' \sim (\cos p\phi) z^{n+1} x^{m+1} \sim e^{-ip\phi} (a_z^+)^{n+1} (a_x^+)^{m+1} \tag{4.47}$$

the frequencies ω_z and ω_x must satisfy

$$p = (n+1)\omega_z + (m+1)\omega_x \tag{4.48}$$

in analogy to Eqn. (4.39). It is convenient to also choose

$$\frac{\omega_z - \nu_z}{n+1} = \frac{\omega_x - \nu_x}{m+1} \tag{4.49}$$

in order to obtain fixed-points in the 4-dimensional a_z-a_x phase space. (These "fixed"-points actually are one-dimensional paths.) We shall, however, not go further into describing the structure of this phase space.

One should not get the idea that we have even begun to cover the fascinating topic of nonlinear resonances. The phase space structure is extremely rich – in fact fractal. Some idea of this richness can be gleaned by looking at computer- generated iterated maps – now nonlinear – for even simple nonlinear motions in 2-dimensional phase space.[16] One starts with a few particles at some initial points $(x_i(s_0), x_i'(s_0))$, compute (including nonlinear forces) their positions $(x_i(s_0+NC), x_i'(s_0+NC))$ after $N=1,2,\ldots 10^{5\pm1}$ revolutions and then plots all the resultant phase points (Fig. 11).

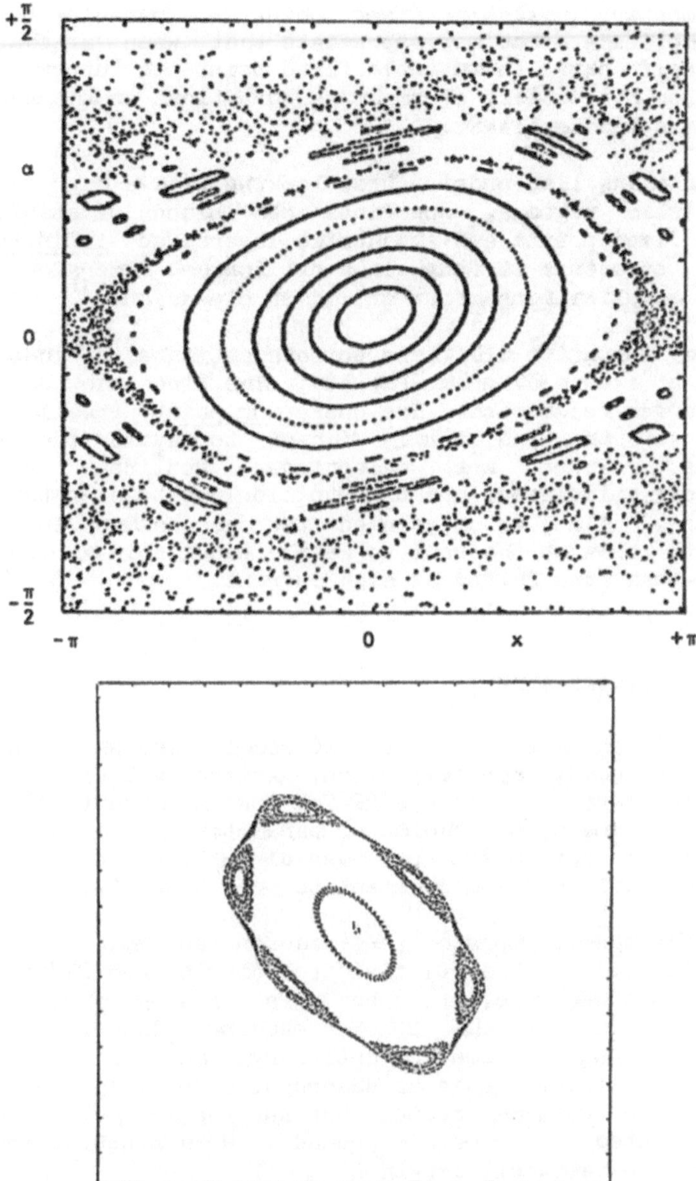

Figure 11. Iterated maps in presence of large nonlinearity. From Reference A6, pp. 232 and 285.

In regions of stability one sees reasonably deterministic orbits. Near separatrices one sees "stochastic layers" emanating from the unstable fixed points. If one recalls the nature of the most elementary unstable fixed point — a pendulum balanced at maximum amplitude — one may appreciate that when a phase point finds itself near an unstable fixed point its future is vitally dependent on the details of small perturbations in the present and tends toward indeterminacy.

But within chaos lies order. Upon looking with a magnifying glass at stochastic regions, one finds more islands of stability with their own fixed points and stochastic layers, ad infinitum.* And all this structure is found in a one (space) dimensional system. Essential complications occur in higher dimensions.[16]

We may summarize all these anxious estimates of instabilities in terms of a tune diagram (Fig.12). The lines show danger zones from low order resonances. The operating point (really a small area) must be judiciously chosen to avoid the important resonances. There are exceptions: Nonlinear resonances, deliberately stimulated, are used to slowly extract a beam from an accelerator.[17] Also, we may need the resonances in the 2020 machine to blow up the beam to keep a manageable current density (low beam-beam tune shift) at high current.

V. MACHINE PARAMETERS

The main parameters of the 2020 machine are shown in Table I. We have included, for comparison, various numbers from the CERN PS, Fermilab Tevatron, and the 20 TeV ICFA machine (VBA). We now comment in turn on the choice of parameters in those cases where it is not obvious from what has been directly presented. Beware! None of the numbers are consistent to more than 20%.

1. The magnet lengths are chosen as 7m, which is the state-of-the-art. The 10T field for dipoles and 2T/cm for quads is not state-of-the-art, but is assumed by ICFA as state-of-the-art for the 20 TeV machine. Therefore we use it. The choice of tune is somewhat arbitrary, and is based on the fraction of circumference in quadrupoles, ℓ/L, chosen to be 10%, somewhat larger than the custom, but not the maximum which might be contemplated on economic grounds. With $\ell/L=0.1$, Eqn. (3.45) yields for the half-cell length L

*Not infinitum! Quantum mechanics soon intervenes.

$$L^2 = \sqrt{2}\,\frac{B\rho}{G}\left(\frac{L}{\ell}\right) = (340\text{m.})^2 \tag{5.1}$$

Dividing into the circumference gives the number of cells, and thereby the tune, which is 1/4 the number of cells. (Remember, betatron phase advance/cell is 90^0.)

To obtain the maximum and minimum values of β in the normal lattice, we use Eqn. (3.38) directly. Likewise, the nominal value of the dispersion (or off-momentum) function $<\eta(s)>$, and "dilation factor" η is found from Eqns. (3.56) and (2.41).

2. Collision Region

Thus far, we have not inserted straight sections into the regular lattice of the machine. These are required for beam injection, beam abort, and RF, as well as for the collision regions.

a. β and β' should be matched at the input and output of the straight section in order that the optics of the rest of the machine not be disturbed. This typically implies an integer contribution of the insertion to the tune.

b. It is deemed desirable to design the insertion region so that the dispersion function $\eta(s)$ in the straight section vanishes. The most straightforward way to change η in some region is to add an opposed pair of dipole doublets as shown in Fig. (13). But this is unnecessarily elaborate; it suffices to decrease (by a factor ~2) the bending strength of the dipoles in the cells adjacent to the straight sections to accomplish the goal.

In the collision straight-sections, one naturally wants to focus the beams especially strongly, i.e. reduce the β-function at the collision point to a small value. One (or more) pairs (or triplets) of strong, large-aperture quadrupole magnets are utilized for this purpose. The rough behavior of the β-function in the straight section is shown in Fig. (14). The β-function in a drift space has a quadratic behavior:

$$\beta(s^*+\ell) = \beta^*+\frac{\ell^2}{\beta^*} \tag{5.2}$$

where ℓ is the distance from the collision point s^*. [To see this observe that the transport matrix for one revolution starting a distance ℓ from s^* is

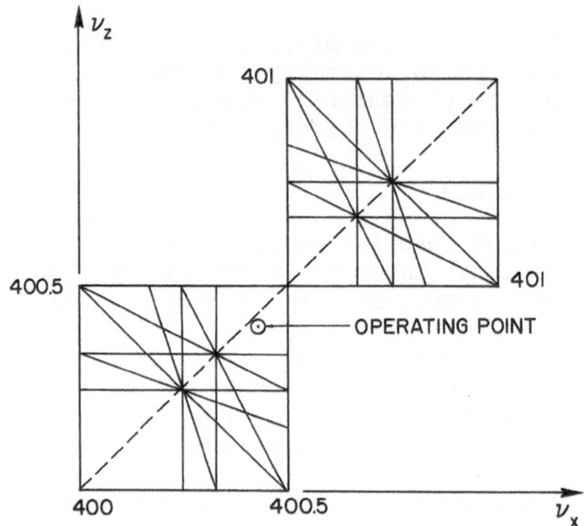

Figure 12. Tune diagram showing dangerous resonances.

Figure 13. Simple dispersion suppressor.

$$M(s^*+\ell+C,s^*+\ell) = \begin{pmatrix} 1 & \ell \\ 0 & 1 \end{pmatrix} \begin{pmatrix} \cos 2\pi\nu & \beta \sin 2\pi\nu \\ -\beta^{-1} \sin 2\pi\nu & \cos 2\pi\nu \end{pmatrix} \begin{pmatrix} 1 & -\ell \\ 0 & 1 \end{pmatrix} \tag{5.3}$$

Calculation of the upper right hand element produces Eqn. (5.2)).]

Thus the maximum value of β, which we call $\hat{\beta}$, occurs in the neighborhood of the final focussing quadrupoles, is roughly

$$\hat{\beta} = \frac{\ell^{*2}}{\beta^*} \cong \frac{(L_{straight} - \ell^*)^2}{\langle\beta\rangle} \tag{5.4}$$

where $\langle\beta\rangle$ is a typical (or slightly less-than-typical) value of β in the normal lattice. For $\beta^* \ll \langle\beta\rangle$, $L_{straight} \gg \ell$, and

$$L_{straight} \sim \sqrt{\langle\beta\rangle\hat{\beta}}$$

one has

$$\ell^* \sim \sqrt{\beta^*\hat{\beta}} \tag{5.5}$$

The maximum allowed β is determined by aperture limitations of the quad, focal power (the length of the doublet must be less than ℓ^*), and chromaticity. Inspection of Eqns. (3.62) and (3.40) show that, in the very-strong focussing limit of phase advance of ~180°/cell, a single doublet in the lattice contributes an amount $\delta\xi$ to chromaticity of

$$d\xi \sim \frac{1}{\nu} \sqrt{\frac{\beta_{max}}{\beta_{min}}} \tag{5.6}$$

With 4 collision regions (and two quad pairs per straight section) we should have

$$\frac{8}{\nu} \sqrt{\frac{\beta_{max}}{\beta_{min}}} = \frac{8}{\nu} \sqrt{\frac{\hat{\beta}}{\beta^*}} \ll 1 \tag{5.7}$$

in order that the natural chromaticity not be dominated by interaction-region quads. Thus we guess

$$\frac{\hat{\beta}}{\beta^*} = \frac{\beta_{max}}{\beta_{min}} \lesssim 500 \tag{5.8}$$

Guessing

$$\begin{aligned} \hat{\beta} &= 10 \text{ km} \\ \beta^* &= 20 \text{ m} \\ \langle\beta\rangle &= 500 \text{ m} \end{aligned} \tag{5.9}$$

gives

$$\begin{aligned} \ell^* &= 100 \text{ m} \\ L_{straight} &= 2.5 \text{ km} \end{aligned} \tag{5.10}$$

These estimates are very superficial, and may be quite wrong. Some clues on how to do better may be found in discussions by Keil.[18]

In some machines, one must also ensure, because of the behavior, Eqn. (5.2), of β near the collision point, that the bunch length not exceed β^*. This is evidently not a problem in the 2020 machine.

3. RF System and Synchrotron Radiation

These have already been discussed, and we have nothing much to add here. Because of the problems with the collective, single-beam instabilities one might want to increase bunch length by lowering the RF frequency. This would allow, according to the Keil-Schnell criterion, Eqn. (2.77), more protons per bunch. For a more-or-less uniformly filled RF bucket, the bunch length and hence the number of protons in the bucket is inversely proportional to the RF frequency.

4. Injection

Superconducting magnets exhibit a kind of hysteresis associated with persistent currents generated in the superconducting wire. This limits the injection field to >10-20% of the peak field. Thus a reasonable injection energy is ~70 TeV. This implies that the 2020 machine would probably be *third* generation: 1TeV → 10TeV → 70TeV → 500TeV. At injection energy the beam emittance will be larger by a factor ~7, and beam size by a factor ~2-3, than at peak energy. Thus, unlike what we have done, tolerances are typically most severe for injection conditions. This will not affect our estimates of how tolerances scale with energy, inasmuch as the ratio of injection to peak energy is, to first order, independent of machine.

We have neither addressed the question of how beam is injected nor how it is aborted, i.e. extracted in no more than one turn if trouble is sensed. This is done with pulsed magnets (kickers); I see no epic problems in doing that, although in the case of aborts the demands on beam dump design may be heavy.

5. Beam Parameters

The emittance of the FNAL machine at transition (~20GeV) is $\varepsilon \sim \pi$ mm-mrad. The quoted numbers for the 2020 machine are scaled up by the E^{-1} factor implied by Liouville's theorem. They should be considered as roughly 2σ estimates, unless otherwise stated. This incudes the longitudinal emittance.

We should not take seriously the beam parameters after synchrotron- radiation damping. They are only relevant for low intensity beams. We shall instead assume bunches of 5×10^{11} particles which are "artificially" maintained at an optimal transverse size by, say, occasionally tuning onto an appropriate resonance to blow up the beam. Likewise we assume the RF buckets can be uniformly filled either by manipulations at injection and/or by using the microwave instability itself.

6. Limits from Single-Beam Instability

We shall assume that the principal limitation on bunch intensity comes from the longitudinal "microwave" instability which we have already treated in Section II.7. We shall take the "long bunch" case, with rms bunch length σ_s=10 cm. This produces an rms momentum spread of order $\sigma_p \cong 30$ GeV or $\sigma_\eta = \sigma_p/p \cong 6\times10^{-5}$. The Keil-Schnell criterion, Eqn. (2.77), for these parameters gives, for $Z_{||}/n \sim 3$ ohms and $\Delta p = 2\sigma_p \sim 60$ GeV, a limiting peak current of $i_{peak} \lesssim 100$ ampere. With

$$i_{peak} \simeq \frac{N}{\sqrt{2\pi}\sigma_s} \tag{5.11}$$

this implies

$$N \leq 5 \times 10^{11} \tag{5.12}$$

As shown in the Supplement, we may hope to put one bunch per betatron wavelength into the machine; this would imply ~400 bunches, or $\sim2\times10^{14}$ p and $\bar{p}$ stored in the ring. This in turn requires an improvement in antiproton production by a factor 10^2-10^3 over existing sources. [If necessary, this problem could be overcome by building ~100 sources; this would not peturb greatly the total cost of the machine!!]

7. Luminosity

The formulae for luminosity and beam-beam tune-shift differ sightly from the rough estimates we have previously used, if one considers gaussian beams. We use (compare with Eqn. (2.81))

$$= \frac{N_p N_{\bar{p}} n_b}{4\pi\sigma_x^*\sigma_z^*}\left(\frac{\omega_0}{2\pi}\right) \qquad (\sigma_z \ll \sigma_x) \tag{5.13}$$

where

N_p = No. of protons per bunch

n_b = No. of bunches in ring

$\sigma_{x,z}$ = rms beam size at collision point (5.14)

The vertical beam-beam tune shift per crossing $\Delta\nu$ is, instead of Eqn. (2.85),

$$\Delta\nu \cong \frac{\alpha N \beta_z^*}{2\pi E_p \sigma_x^* \sigma_z^*} \qquad (\sigma_z \ll \sigma_x) \tag{5.15}$$

We shall assume that the beams are separated except in the four collision regions. This is discussed in the Supplement; electrostatic deflectors are used to separate the beams except at collision points. We shall furthermore assume that the beam-beam effects from different collision regions add incoherently, so that $\Delta\nu \sim \sqrt{N_{crossings}} = \sqrt{8}$. There is little theoretical justification for this, although it is not too bad empirically.

The strategy is therefore fixed:

a) $N_{p,\bar{p}}$ is determined by the limit on beam current from longitudinal microwave instability.

b) β_z^* is minimized via lattice-optics considerations.

c) $\sigma_x^*\sigma_z^*$ is determined by the condition (good to factor ~3!!)

$$\Delta\nu = .01 \tag{5.16}$$

It will be larger than the natural value from synchrotron damping. Putting numbers into Eqn. (5.15) gives

$$\sigma_x^*\sigma_z^* = 8 \times 10^{-6}\ \mathrm{cm}^2 \tag{5.17}$$

Assuming $\sigma_x^* = 2\sigma_z^*$ for definiteness gives for emittance

$$\frac{\varepsilon}{\pi} = \frac{(2\sigma_z^*)^2}{\beta^*} \sim 20\ \mu\text{m-}\mu\text{rad} \tag{5.18}$$

This is only slightly less than what exists immediately after injection and acceleration. We must assume that the beam can be artificially excited and the size controlled.

This problem exists already in e^+e^- storage rings. It has proved difficult to blow up the transverse phase space by external means; the synchrotron damping dominates. However, in those cases the damping time is of order milliseconds instead of a fraction of an hour, so the problem for the 2020 machine may not be as severe. Mechanisms for enlarging the phase space might be tuning onto weak betatron resonances or adding RF noise.

d) With n_b of order ν, as determined by the scheme for electrostatic separation of the beams, we may now directly estimate the luminosity. We obtain

$$\mathscr{L} = 10^{33}\ \mathrm{cm}^{-2}\ \mathrm{sec}^{-1} \tag{5.19}$$

and per crossing

$$\mathscr{L} = 10^{28}\ \mathrm{cm}^{-2}/\text{crossing} \tag{5.20}$$

There is a lot of uncertainly in this estimate. A tune shift per crossing of .005 may be too big by a factor 2-3. This influences linearly the luminosity. Putting 5×10^{11} particles in a short bunch may be overoptimistic. The diminution in luminosity goes linearly with N (provided the beam-size at the collision point can be re-optimized).

But perhaps the most severe constraint comes from detection problems. There are, with this luminosity, many interactions per crossing. These cannot be resolved by existing detection techniques, and thus the physics opportunities are constrained. These problems are discussed in the next section.

VI. DETECTION PROBLEMS

A traditional design requirement for colliders is that there be no more than one interaction per bunch crossing. This implies, for a total cross-section (within factor 2) $\sim 10^{-25}$ cm^2, a luminosity per crossing $\sim 10^{25}$cm^{-2}. Furthermore, we have seen that the bunch spacing for a $p\bar{p}$ collider is (optimistically!) ~ one bunch per betatron wavelength. Since the betatron wavelength increases with energy this implies (using these ground rules) that the net luminosity is ~

$$\mathcal{L} = 10^{25} \nu \frac{\omega_0}{2\pi} \tag{6.1}$$

For the 2020 machine with $\nu \sim 400$ and $\omega_0/2\pi = 250$ sec^{-1}, this gives

$$\mathcal{L} \sim 10^{30} \text{cm}^{-2} \text{sec}^{-1} \tag{6.2}$$

which is hardly adequate for investigation of hard collisions with subprocess cross-sections (at the energy scale appropriate to this machine (~10-300 TeV), which may be estimated (just from dimensional analysis) to be

$$\sigma \lesssim \frac{1}{s} \tag{6.3}$$

i.e.

$$\sigma \lesssim (4\times10^{-36} \text{ to } 4\times10^{-39} \text{cm}^2) \tag{6.4}$$

If one allows multiple collisions per crossing, one cannot expect charged particle tracking or-at these energies - even muon identification to be viable. One is left with electromagnetic and hadron calorimetry. In the case of calorimetry, a certain amount of pileup can be tolerated without losing too much information. The situation is described in more detail in the Supplement. Here we summarize a few of the salient points. First of all, we can only hope to see very high p_T jets. In principle calorimeter resolution for energies above 1 Tev is not much of a problem, although the clean isolation of a hadronic jet in the presence of QCD gluon bremsstrahlung may be a considerable nuisance. We now estimate the pileup underneath an observed high-p_T jet. First of all we assume the jet is contained within 0.1 steradian. The distribution of energy into 0.1 steradian (at 90^0 cms) is empirically bounded above by an exponential

$$\frac{dN}{dE_T} \lesssim \frac{c}{\langle E_T \rangle} e^{-\frac{E}{\langle E_T \rangle}} \tag{6.5}$$

with $\langle E_T \rangle$ an increasing function of total cms energy. The value of $d\langle E_T \rangle/d\Omega$ versus energy is shown in Fig. 15. The value at $\sqrt{s} \sim 10^3$ TeV might be as large as ~5 GeV/steradian. Then 1000 collisions per bunch crossing could put, <u>on average</u>, 500 GeV into each calorimeter element subtending <u>0.1</u> steradian. The

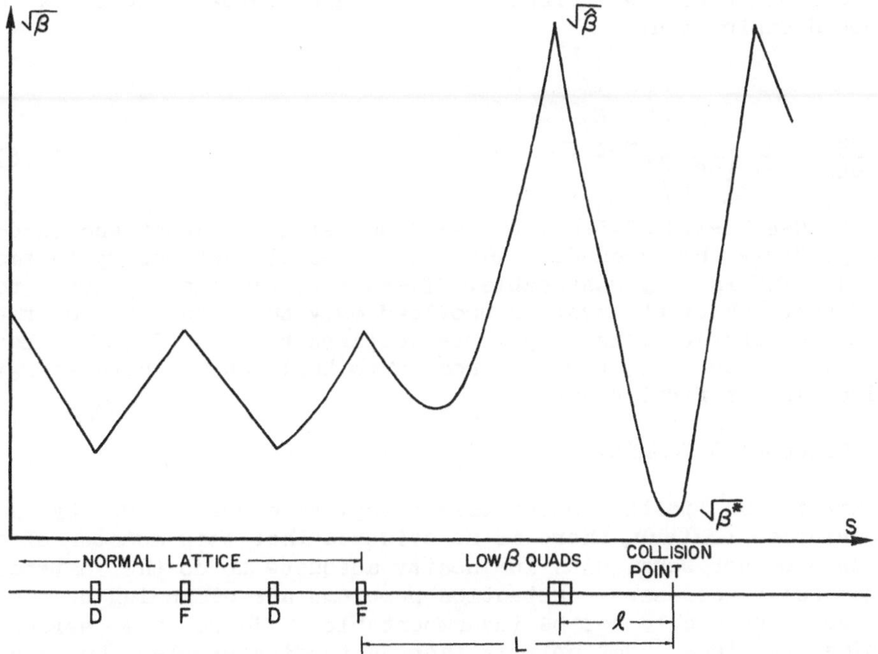

Figure 14. Schematic of optics in collision straight section.

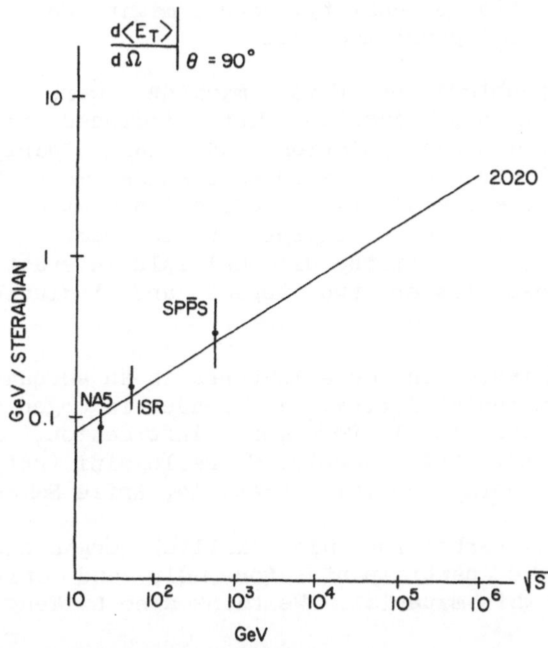

Figure 15. Mean transverse energy into one steradian at 90 versus collision energy.

fluctuations about the average, however, are bounded above by a Poisson distribution.

$$\frac{dN}{dE_T} \lesssim \frac{1}{N!}\left(\frac{cE_T}{<E_T>}\right)^N e^{-\frac{E_T}{<E_T>}} \tag{6.6}$$

For $E_T >> N<E_T> = 10^3 \times 0.5$ GeV the exponential is dominant and leads to negligible background. Thus a threshold jet energy in the range 1-3 TeV is very reasonable. There thus remains a hope to reconstruct multijet systems provided only the total mass of the system and subsystems is very large compared to, say, 3 TeV. But that is, after all, the main reason to build such a high energy collider in the first place.

VII. CONCLUDING COMMENTS

Within the limitations of this study, as carried out by an inexperienced amateur, there is no evidence that this monster of a machine does not work, with luminosity adequate to do the physics it naturally addresses. Detection problems are demanding because of pileup, but should not be insurmountable. There are evident practical problems. Not only is there a funding problem, but also a system problem. Not one of the 2×10^5 cantankerous superconducting magnets and their complex support systems can fail. This question of quality control might well be the most demanding technical problem of all.

Various problems of this machine were addressed by participants of the school. These included the question of disposal of synchrotron radiation (use warm "scrapers" located between magnet strings), beam abort systems (conventional methods seem to work), the electrostatic deflection system to keep p and $\bar{p}$ bunches from colliding except in interaction regions, and detection problems associated with multiple interactions per bunch crossing. These latter two topics are included here in two supplements.

A major omission in these lectures is an adequate discussion of single-beam instabilities. The reader is urged to consult the references in Appendix I for more information, in particular A. Chao (ref. A7, SLAC School), C. Pellegrini (ref. A7, Fermilab School), and A. Hofmann et.al. (ref. A4, Erice School).

We thank T. Ferbel for his skillful organization of this school, and the participants for help and criticism in the preparation of this material. We thank also L. Teng for helpful criticism.

APPENDICES

I. Reference Material

There is considerable material on high energy accelerators and storage rings, ranging from textbooks to monographs and lecture series, to published papers, and to preprints and internal laboratory memoranda. (The latter component often seems to be the dominant one!) I am not enough of a scholar to provide an authoritative bibliography, but shall simply list here a few which were used in preparation of these notes. This listing should provide the reader with an avenue into more detaled papers on specific topics.

REFERENCES

A1. H. Bruck, "Circular Particle Accelerators," PUF, Paris (1966). A good general reference. (Unfortunately the only copy I could locate was in French.) Translated by LASL LA-TR-72-10 Rev.

A2. M. Sands, SLAC report SLAC-121; also "Physics with Intersecting Storage Rings," ed. B. Touschek, Academic Press, N.Y. 1971. This is a splendid introduction to e^+e^- rings and especially problems of synchrotron radiation.

A3. E. J. N. Wilson, CERN 77-7, "Proton Synchrotron Accelerator Theory." CERN academic-training lectures; a very nice introduction.

A4. "Theoretical Aspects of the Behavior of Beams in Accelerators and Storage Rings," CERN 77-13. A quite comprehensive discussion of many topics at a fairly advanced level. (This is the proceedings of an Erice school on accelerators.)

A5. Proceedings of the ICFA Workshops on Possibilities Limitations of Accelerators and Detectors. The first (1978) is a Fermilab report; the second (1979) a CERN report. These consider design problems of very big (e.g. 20 TeV) rings and are very useful source material for considering the 2020 machine.

A6. "Nonlinear Dynamics and the Beam-Beam Interaction," AIP Conference Proceedings No. 57. This nicely explores the interface between accelerator design and 20th century nonlinear mechanics.

A7. "Physics of High Energy Particle Accelerators," AIP Conference Proceedings No. 87 and No. xxx, ed. R. Carrigan, R. Huson, and M. Month. This contains the proceedings of a 1981 Fermilab Summer School. A second school was held at SLAC in 1982; proceedings have not yet appeared.

A8. "Techniques and Concepts of High Energy Physics," ed. T. Ferbel (Plenum, New York, 1981). This is the 1980 NATO Summer Institute Proceedings, and contains lectures by M. Month on specialized topics including beam-beam interaction and single-beam instabilities.

APPENDIX II: Limerick to a Theoexperimentalist

A well known theorist, Bjorken,
Field theory his great claim to fame;
Did an experiment one day,
Couldn't re-normalise away,
And something just snapped in his brain.

The result of this now must be
We henceforth accept G-U-T;
That theorists may change,
To plumbers ain't strange,
Especially at quadrillion eV.

Robert J. Wilson

APPENDIX III. The Missing Supplement

Regrettably the supplement mentioned in the text was unable to be completed. It was meant to consist of two parts. The material for the first, on bunch avoidance using electrostatic separators, was worked out by Nikos Giokaris. The scheme is practical and appears to cause no great difficulty, although it imposes geometrical constraints on the lattice design.

The material for the second part was provided by Geoffrey Taylor. It considers the interesting problem of detector design in the presence of multiple interactions per beam crossing. For jet transverse momenta well in excess of 1 TeV, problems of pileup appear to be small.

It was my intention to assist in editing and assembling this material together with the authors. But other pressures and commitments crashed in as the publication deadine approached, and there was no time available to properly complete the work. I deeply regret this failure and apologize to Nikos and Geoffrey, and thank them for their hard work.

TABLE I: PARAMETER LIST††

	Scaling Rule	2020
Energy per beam	~E	500 TeV
Peak magnetic field	~B	10 T
Bending radius	~E/B	170km
Nominal circumference	~E/B	≳12600m
Dipole magnet Length		7m
Number of dipoles		170,000
Peak quadrupole Gradient	G	2T/cm.
Quadrupole Length		7m
Cell length	$\sim(EB/G)^{1/2}$	~700m
Betatron phase advance/cell		90°
Betatron tune (horizontal~vertical)	$(EG/B)^{1/2}$	400
Transition energy	$(EG/B)^{1/2}$	~400 GeV
Length of quads per half-cell		~35m
Number of quads per cell		10
Total number of quads		16,000
Maximum β	$(E/BG)^{1/2}$	~1200m
Minimum		~200m
Nominal dispersion $\langle\eta(s)\rangle$	$\sim G^{-1}$	0.7m
Momentum compaction η	(B/EG)	4×10^{-5}
No. of long straight collision regions		8

Full length of long straight section	$(EB/G)^{1/2}$	~3km
β* at collision point		~20m
Free space around collision point		±100m
Maximum β at quads		10km
Revolution frequency	B/E	250Hz
RF frequency		500Mhz
Harmonic number		2×10^6
RF voltage		5GeV
Synchrotron frequency		~1Hz
Maximum Δp		±150GeV
Maximum Δp/p		3×10^{-4}
Bucket area		~600 eV-sec
Synchrotron radiation energy loss/turn	$\sim E^3B$	3 GeV
Photon critical energy	$\sim E^2B$	300 keV
Longitudinal damping time	$\sim E^{-1}B^{-2}$	~10 min
Transverse damping time	$\sim E^{-1}B^{-2}$	~20 min
Injection energy		70 TeV
Transverse emittance		
at injection	E^{-1}	~300 μm-μrad
at peak energy*		~40 μm-μrad
Nominal rms beam size:		
at injection		~0.2 mm
at peak energy*	$(EBG)^{-1/4}$	~70μ

(Continued)

TABLE 1 (continued)

Nominal rms p_T-spread:		
at injection		±30 MeV
at peak energy*	$(EBG)^{1/4}$	±70 MeV
Longitudinal emittance at injection	~1	2eV-sec
Rms momentum spread Δp*	~1	±15 GeV
Rms Δp/p at 500 GeV*	E^{-1}	$\pm 3\times 10^{-5}$
Rms bunch length*	~1	±3 cm
Nominal rms beam size^{+}		
Horizontal	$(B^3/EG^2)^{1/2}$	20μ?
Vertical		2μ?
Emittance‡		
Horizontal	$(B^7/E^3G^3)^{1/2}$	1μm-μrad?
Vertical		10^{-2}μm-μrad?
Rms momentum spread‡		
Rms Δp/p	$(EB)^{1/2}$	$\pm 3\times 10^{-5}$
Rms bunch length‡		±3m
Rms beam size at collision‡		
Horizontal		4μ
Vertical		0.5μ
Optional rms beam size (500TeV)		
Horizontal		90μ
Vertical		90μ
Optimal rms emittance		
Horizontal		20πμm-μmrad

Vertical	20πμm-μmrad
Optimal rms momentum spread Δp/p	$\pm 6\times 10^{-5}$
Optimal rms bunch length	±5cm
Optimal rms beam size at collision	
Horizontal	40μ
Vertical	20μ
No. of $p(\bar{p})$ per bunch	5×10^{11}
No. of bunches	400
Beam-beam tune shift/crossing	.01
Luminosity	$10^{33}cm^{-2}sec^{-1}$
Energy stored in each beam	20GJ
RF power	200MW
Refrigeration power	3GW?

*Prior to synchrotron damping.
‡At 500 TeV, dilute beam, after synchrotron damping.
††The parameters for ICFA, TeV I, and PS were hard for me to assemble and verify. They have been dropped for the table.

REFERENCES

Note: Reference numbers preceded by an A are to be found in Appendix I.

1. See the discussions by L. Lederman and others at the 1982 Snowmass Summer Study.

2. For background see J. Bjorken, reference A7 (AIP. 92), p. 25.

3. For example, R. R. Wilson, Revs. Mod. Phys. 51, 259 (1979).

4. Particle Properties Data Booklet, Particle Data Group [Physics Letters 111B, (1982).

5. For example, A1, A3, A4.

6. See reference A2.

7. For example, see A. Piwinski, DESY Report M-81/03.

8. For example, se A. Chao, 1982 SLAC School, reference A7.

9. E. Keil and W. Schnell, CERN-ISR-TH-RF/69-48 (1969).

10. H. Hereward, Proc. of 1975 Isabelle Summer Study, p. 555.

11. A comprehensive review is given by J. Schonfeld, 1982 SLAC School, reference A7.

12. See E. Keil, Ref. A4.

13. See Ref. A3.

14. N. King, 2nd ICFA Workshop (Reference A5, Appendix I), p. 155.

15. See E. Courant, Fermilab Summer School (Reference A7, p. 1 AIP No. 87.

16. See contributions by R. Helleman and J. Moser, for example, in reference A6.

17. See for example L. Teng, reference A7 (AIP No. 87), p. 62.

18. E. Keil, reference A7 (AIP No. 87,) p. 405.

NEW DEVELOPMENTS IN GASEOUS DETECTORS

F. Sauli

CERN

Geneva, Switzerland

1. PRELUDE

The multiwire proportional chamber (MWPC) has been introduced more than a decade ago as a powerful, fast detector suited for high-energy physics experimentation (Charpak et al., 1968). Since then, an impressive amount of work has been realized both on the improvement and the better understanding of the original detector, and on the development of various derived structures. For a review of detectors' performances and a general discussion of their use in experimental physics, see, for example, Charpak (1970), Rice-Evans (1974), Sauli (1977), Fabjan and Fisher (1980).

It is clear that a detailed knowledge of all aspects of charge drift and multiplication is necessary for the understanding of the behaviour of a given detector configuration. In what follows, I will describe in some detail only the processes directly connected with the main subject of interest; the reader is referred to the quoted papers and to the references there given to more detailed studies.

2. POSITION DETECTORS: PROPERTIES, LIMITATIONS AND POSSIBLE DEVELOPMENTS

Localization accuracy in the original MWPC was restricted to the anode wire spacing, typically 2 mm, although it was soon realized that a complementary measurement of electron collection time could largely improve localization (Charpak, Rahm and Steiner, 1970; Walenta, Heintze and Schürlein 1971). Many geometries have been developed to better profit from this property, leading to various kinds of thin layer multiwire drift chambers with localization accuracies around 100 μm or better. A considerable advance in the multiple track recognition capability has been realized by the development of large volume drift chambers such as ISIS (Allison et al., 1974), Time Projection Chamber (Clark et al., 1976; Nygren and Marx, 1978), JADE (Drumm et al., 1980), UA1 Imaging Chambers (Barranco Luque et al., 1980) and similar devices. In this kind of detector, ionized trails produced in a large gas volume are drifted to a plane of proportional wires where recording of position, drift time, pulse height can provide a very detailed description of the event pattern (see Figs. 1 and 2).

In this section I will analyse some of the factors that contribute to the localization accuracy obtainable in a gaseous device, and that can indeed spoil the performance of the detector if not properly taken into account. They are, in order of relevance: correct knowledge of the drift velocity, including the distorting effects of magnetic fields; lateral and transverse diffusion of the electrons in the ionization trail while drifting; positioning error in the induced cathode readout due to angular and track density effects; primary energy deposition statistics; positive ions interference.

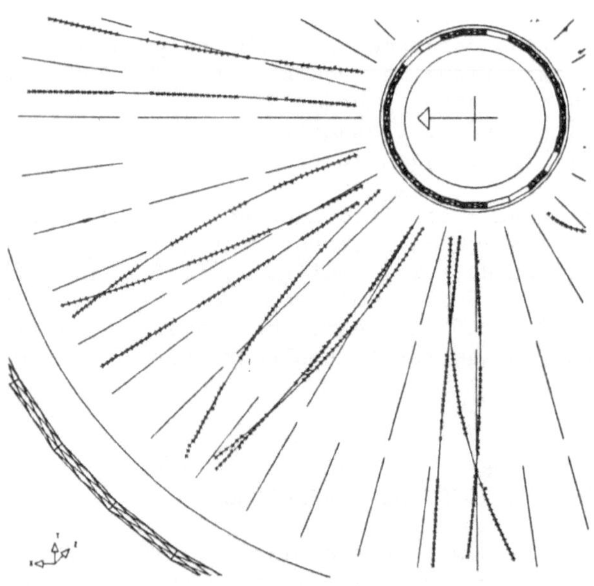

Figure 1

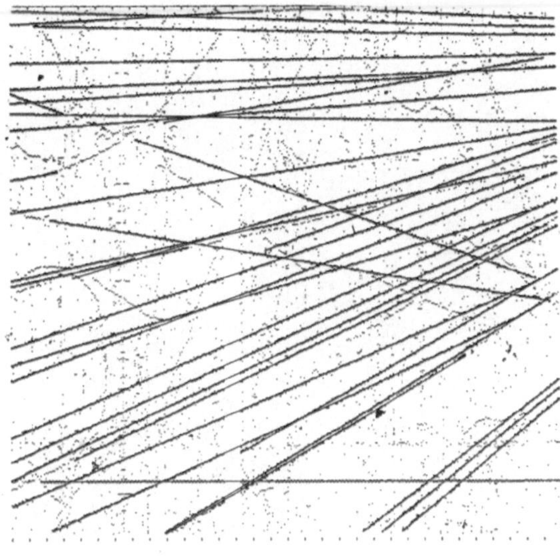

Figure 2

The drift velocity has been measured and computed in a large number of gases and gas mixtures; for general reviews see, for example, Palladino and Sadoulet (1975), Shultz and Gresser (1978), Jean-Marie et al. (1979). As an example, Fig. 3 shows the comparison between computed and measured drift velocity in various mixtures of argon and isobutane, from Shultz and Gresser. The effect of the magnetic field can also be taken into account (Ramanantsizehena, 1979). A large software effort is however required to extract from the measured drift time the actual coordinate, especially when the electric field is not uniform (as always the case around the anode wires) and in presence of magnetic fields; Figure 4 shows the computed electron trajectories and equal time distances from a wire in the JADE jet-chamber (Drumm et al., 1980). The space-time correlation is obviously asymmetric and all but linear. Moreover, it is not always obvious that somebody else's measurements can be used, due to the different calibration methods used for the gas mixture; temperature and pressure can also influence the calibration. In general, the programmer is reduced to adjusting painfully many parameters to obtain a satisfactory fit to the data.

As compared to the systematic shift of coordinates due to a poor knowledge of drift velocity and trajectories, electron diffusion represents an intrinsic limit to localization accuracy. Figure 5 from Palladino and Sadoulet (1975) shows the dependence

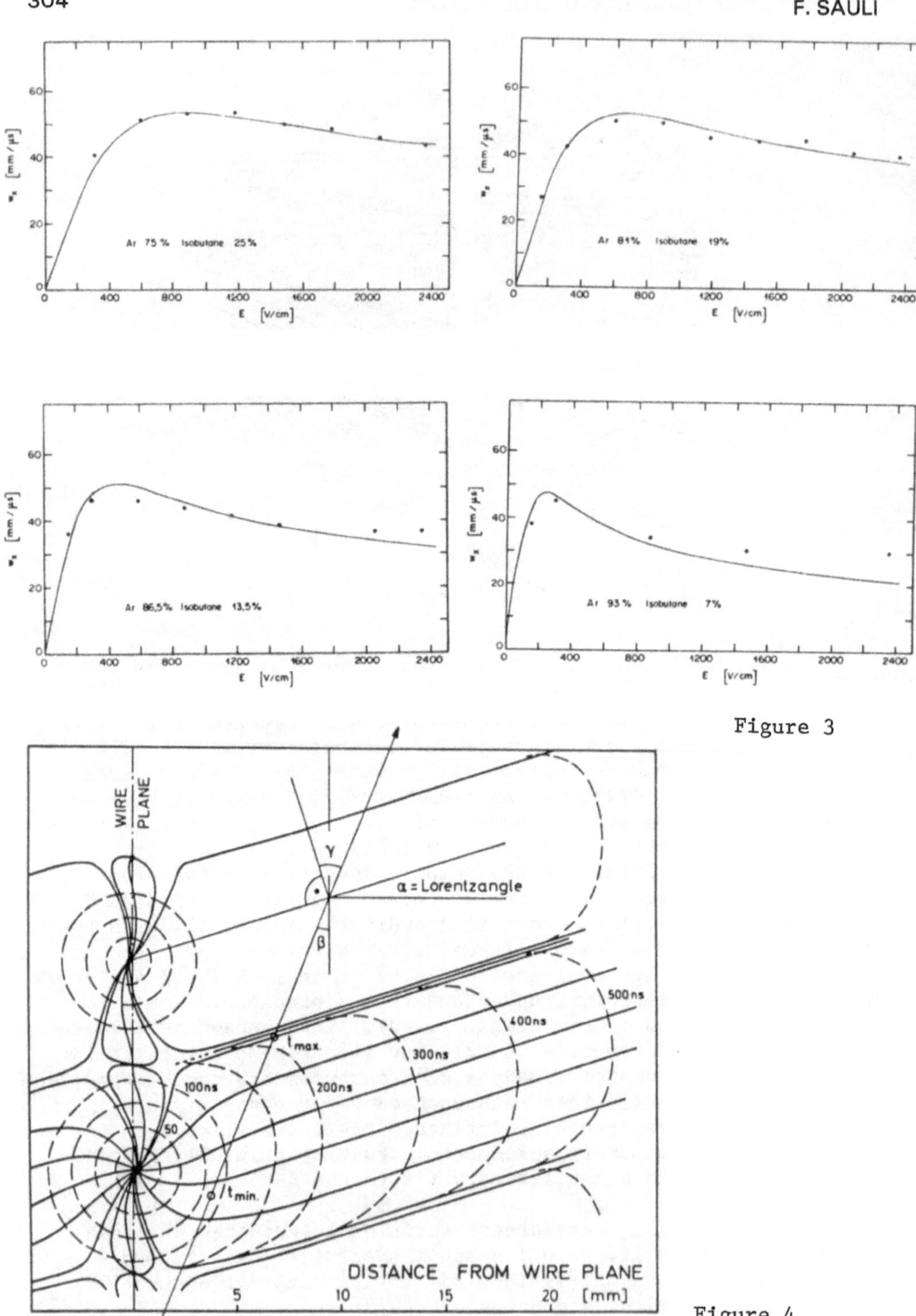

Figure 3

Figure 4

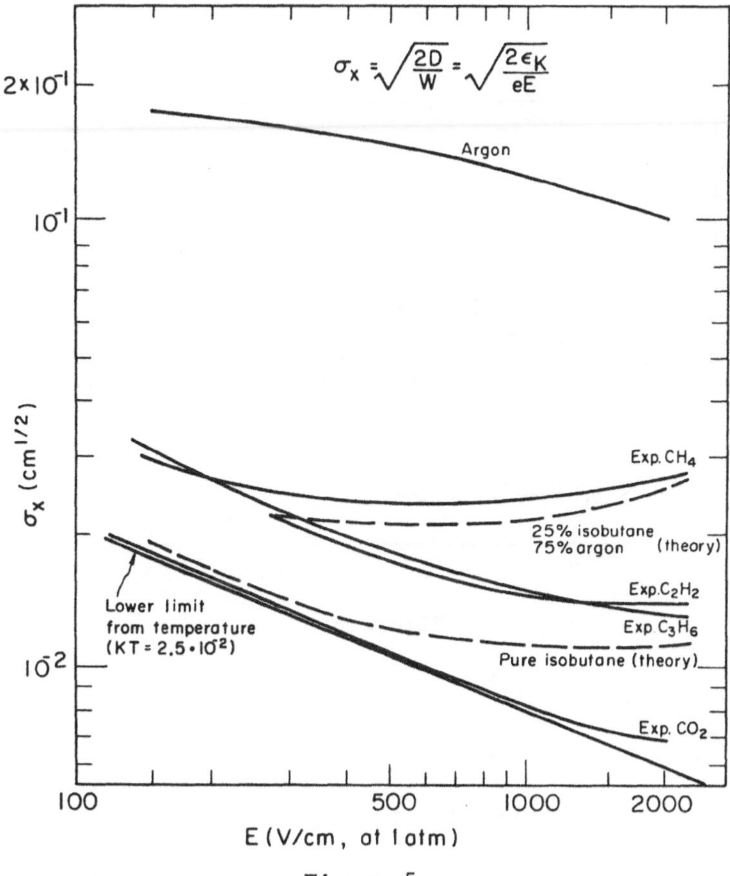

Figure 5

of the space diffusion on the electric field for several gases and mixtures and 1 cm of drift; it equals about 100 μm for common mixtures. Since the curves refer to a single electron, while segments of tracks contain several of them, the accuracy obtained in practice can be slightly better, around 60 μm for 1 cm of drift in ordinary conditions, see Fig. 6 (Filatova et al., 1977).

An increase in the operating gas pressure, for equal field conditions, directly reduces the diffusion in proportion to the square root of the pressure; Figure 7 shows the effect of going from one to four atmospheres (Farr et al., 1978). Both TPC and JADE operate at pressures higher than atmospheric, 10 and 4 atm, respectively, to increase the energy resolution, and this is beneficial to the resulting accuracy. It should be noted also that generally the drift time dispersion is determined by the longitudinal and not the transverse diffusion, whose value is given by the classic theory of drift (as in Fig. 5). In some gases, and at high values of the field, the longitudinal diffusion coefficient is considerably

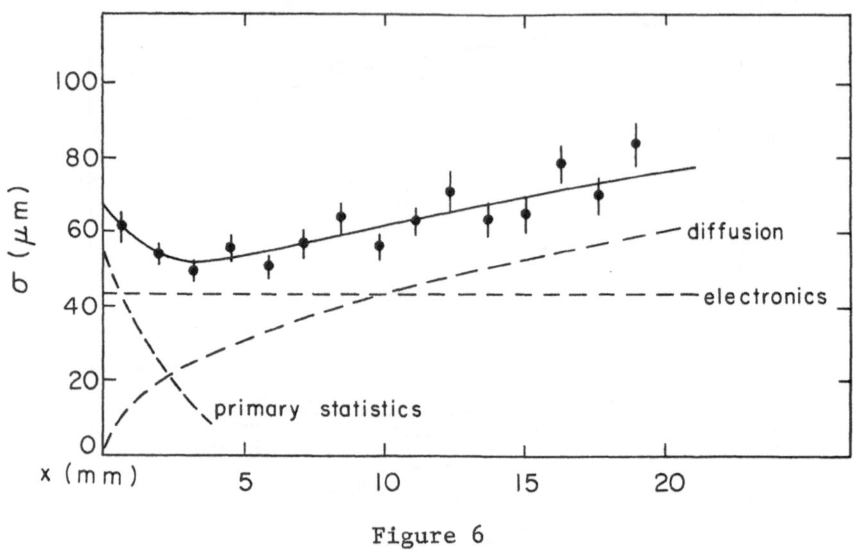

Figure 6

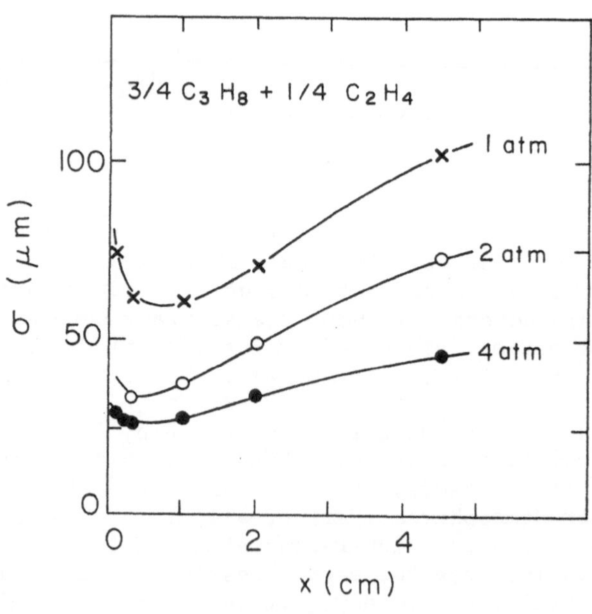

Figure 7

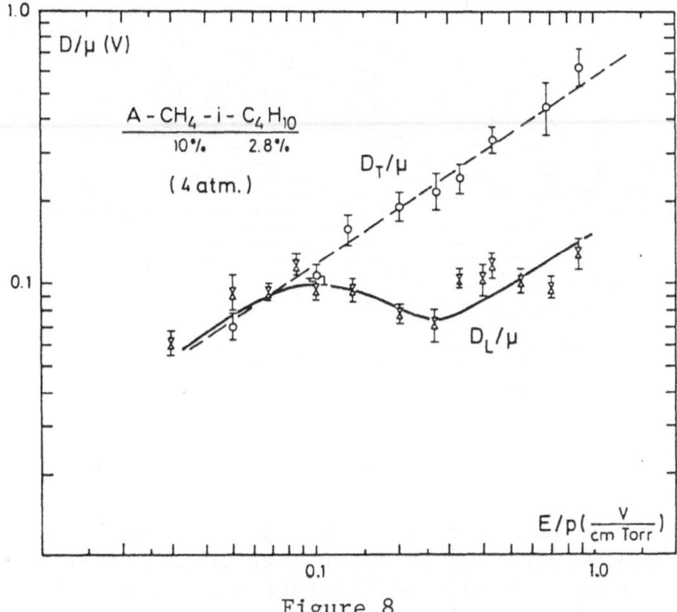

Figure 8

smaller than the transverse; Figure 8 shows that the measured values for D_L in the JADE chamber are six times smaller than D_T at the operational fields. Since the space accuracy is proportional to $\sqrt{D_L}$, a considerable gain can be expected: a more systematic investigation on the matter is desirable.

Presence of a strong magnetic field also reduces the diffusion in some gases, but only in the transverse direction: this is exploited in the TPC to keep a small lateral track size but does not influence the space accuracy determined by the drift time measurement.

A localization method used often in conjunction with the drift time measurement is the determination of the centre of gravity of the induced signal on cathode strips or pads, a method which is known to allow intrinsic accuracies of 30-40 μm for soft X-rays (Charpak et al., 1978).

It consists of recording the charge profile of the pulses induced by an avalanche on suitably stripped cathode planes, and computing the centre of the distributions (Fig. 9). In imaging chambers, the cathode wire strips are replaced by individual pads facing some wires which allows the determination of the coordinate along the wires themselves. Preserving a good accuracy in running conditions and for charged particles is, however, all but easy, for technical as well as physical reasons. On the one hand, obtaining an accuracy corresponding to a fraction of the pad width from ratios of pulse heights requires calibrations and overall stabilities

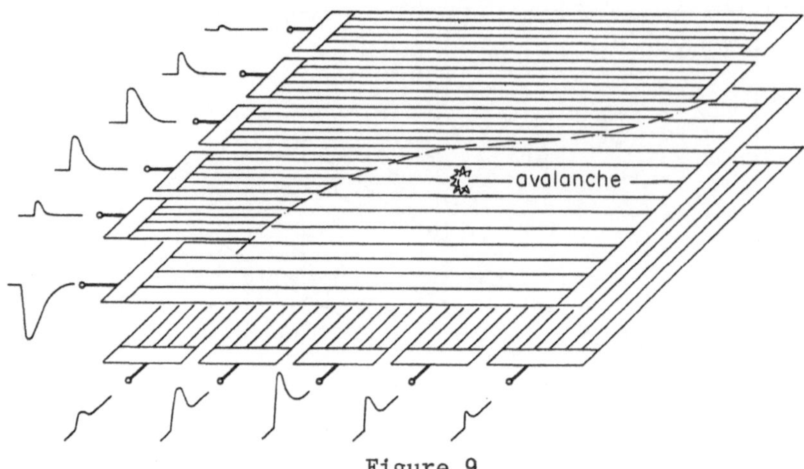

Figure 9

of a few percent or so. Moreover, the discrete structure of the pads and the cross-talks due to capacitive couplings introduce various kinds of modulations and interactions. The scatter plot of Fig. 10, representing the difference between real and measured coordinates in a cathode readout chamber, shows the modulation effect due to a 1.4% cross-talk between adjacent strips, before and after correction (Piuz, Roosen and Timmermans, 1982). After correction, the space resolution has about 60 μm r.m.s. (this measurement refers to tracks perpendicular to the wire planes). For inclined tracks, a dispersive effect of physical nature appears however. It has to do with the large fluctuations in the rate of energy deposition by fast particles in thin layers of gases (the Landau distribution). Consider indeed the case of a slanted ionizing track traversing a chamber and interesting the same anode wire. The induced pulse distribution on cathodes at any given instant reflects the asymmetry in the ionization density; let one side contain two times as many charges than the other (a likely happening with typical energy loss resolutions of 100%!), the actual charge centre of gravity will be displaced by almost $1/4$ of the gap width for 45° tracks. The effect on localization accuracy is visible in Fig. 11 for increasing incidence angles ($\theta = 0°$ meaning tracks perpendicular to the wire planes). At 30° the average accuracy has moved to almost 300 μm, and is obviously the worse for events in the tail of the energy loss distribution (Charpak et al., 1979).

Let me analyse now in some detail the process which is at the origin of the detection process, i.e. the energy loss by ionization of charged particles in matter. The average rate of energy loss reaches a minimum at high energies (around 1 GeV) and has then a small increase, the relativistic rise, at even higher energies; for more details on this region, see the next section, devoted to particle identification.

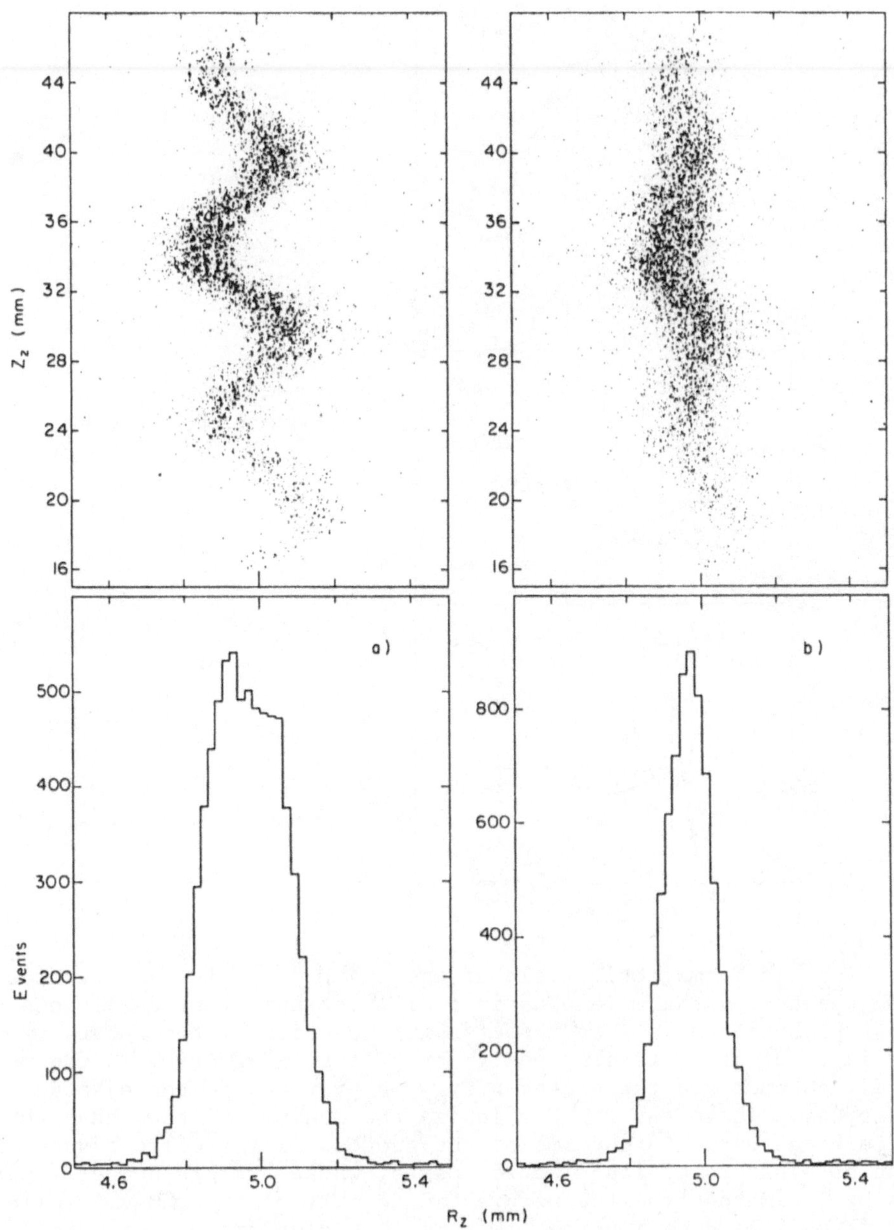

Z_z (mm)
44
40
36
32
28
24
20
16
Events
500
400
300
200
100
0
800
600
400
200
0
a)
b)
4.6
5.0
5.4
R_z (mm)

Figure 10

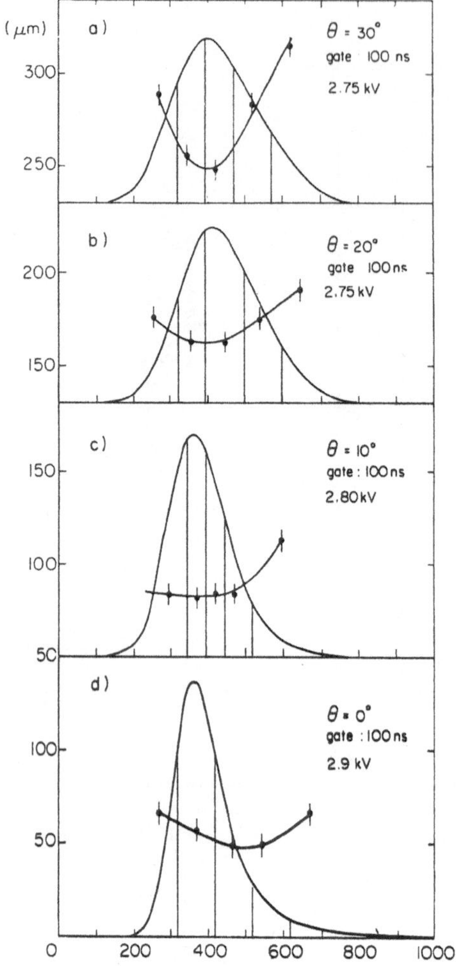

Figure 11

The electromagnetic field of the charged particles interacts with the one of the molecules in a discrete number of collisions, whose probability is a fast decreasing function of the energy transfer implied; at increasing energy transfers, ionization is the most likely outcome and the excess energy is provided to the ejected electrons. These can further ionize the medium, if they have sufficient energy; the outcome of the process is therefore a succession of clusters of ion-electron pairs, whose average number (that can be estimated from the nature and density of the medium) follows a Poisson-like distribution. Within each cluster, one or more electrons are produced depending on the energy transfer in the primary collision: the average is between two and three for most gases. In some cases the ejected electron has enough energy to travel considerably in the medium (the so-called δ-electrons);

for such events, the energy loss on a given layer of gas can exceed several times the average leading to the typical Landau distribution of energy in thin samples of gas (Fig. 11). For a modern formulation of the theory of energy loss for fast particles, see, for example, Talman (1979), Allison and Cobb (1980), Lapique and Piuz (1980).

Presence around the primary trail of secondary ionization clusters due to long-range electrons spoils the localization accuracy of most gaseous detectors, since the two components cannot in general be separated. It is relatively straightforward to obtain an approximate expression for the number of δ electrons having an energy equal or larger than E_0 in a layer of material of thickness x and density ρ (see, for example, Ritson, 1961):

$$N(E \geq E_0) \simeq \frac{K}{\beta^2} \frac{Z}{A} \frac{1}{E_0} \rho x ,$$

where Z and A are the medium atomic number and mass, and k = = 0.15 MeV g^{-1} cm^2. Figure 12 shows the computed value of N for 1 cm of argon at normal conditions, where the average number of primary ionizing collisions is $n_p \simeq 29$. One can see, for example, that about 5% of the tracks contain one electron with an energy exceeding 2 keV, and whose range in the gas (Fig. 13) is about 150 μm.

Consider now a typical drift chamber structure, as in Fig. 4. The coordinate of the track is deduced from a measurement of drift time, i.e. of the lap between the ionizing encounter and detection of the electron trail at the wire. A long-range δ-ray on one side of the trail will obviously result in a shorter measured time. An asymmetric localization accuracy distribution is indeed observed in high accuracy drift chambers, see Fig. 14 (Charpak et al., 1973), with a tail extending as far as 600 μm from the centre! About 5% of the events are contained in the non-Gaussian tail. The question may arise as to whether an increase of pressure in the detector would help: it does so by reducing the range of electrons, but unfortunately the number of δ electrons produced in a given thickness of gas increases as well. Obviously in a large volume drift chamber where a given track is measured many times in small segments, a fit procedure eliminates such tail events, at the expense of a reduced number of coordinates effectively used.

The last source of troubles I would like to discuss has to do with the presence in the detector of positive ions, produced copiously in the avalanches; they interfere with the localization process at two levels. To begin with, ions are responsible for the detected signal in the chamber, as a result of their motion away from the anode wire. For typical values of the time constant of the amplifiers mounted on cathode strips, the signal induced by a given avalanche lasts for about 100 ns. During this time,

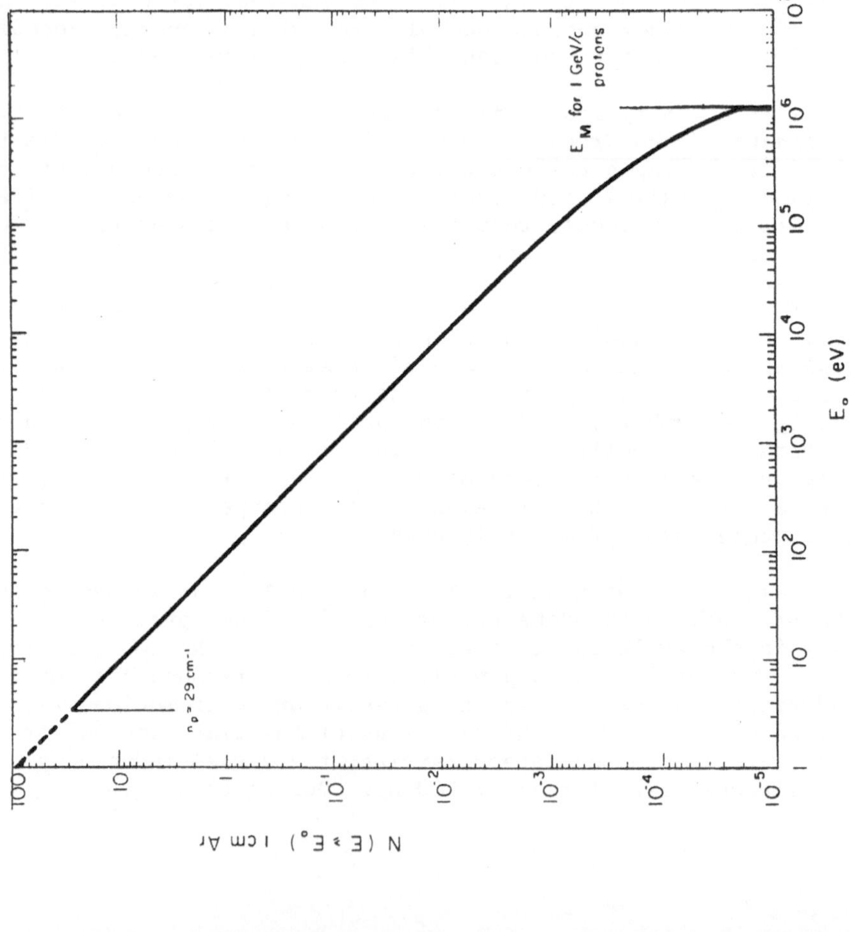

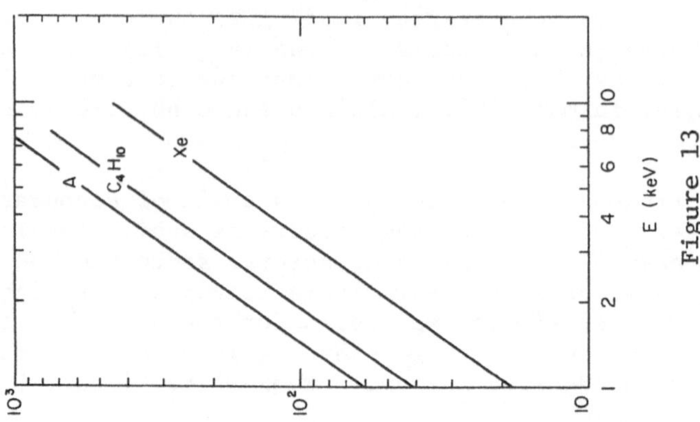

Figure 13

Figure 12

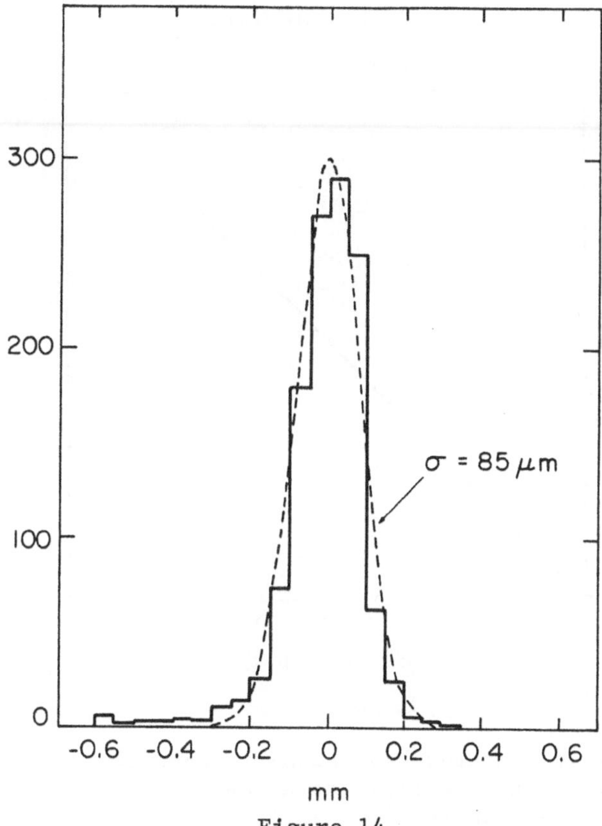

Figure 14

electrons from a section of the same track around 5 mm away keep reaching the anode and producing avalanches. For an inclined track drifting into the proportional chamber detector, this means a systematic displacement, at any given time, of the recorded charge profile as compared to the actual position of the track segment (identified by the drift time measurement). This effect has received so far little attention. Ions drifting into the large volume of detection of imaging chambers take a long time to clear the region (several ms). At large enough counting rates, this can create a permanent charge density, localized around the hot spots of the detector, and resulting in substantial modifications of the electric field strength and direction. Electrons drifting close or into these hot spots follow distorted trajectories. Figure 15 is a measurement of the lateral displacement on the recorded position for a cluster of electrons drifting 1 cm aside a column of positive ions with the indicated density in a 10 cm thick drift volume. The average particle rate producing, at gains around 10^5, the indicated density is also shown (Friedrich et al., 1979).

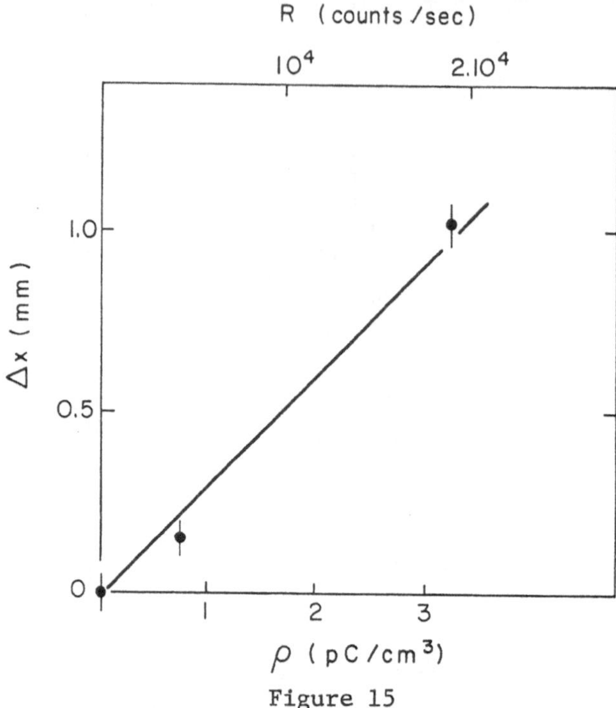

Figure 15

Various electrode structures have been proposed to gate on and off the charges entering the multiplying element, so as to reduce the build up of positive ions (Allison et al., 1973; Breskin et al., 1980).

The fact that charge detection in proportional counters relies on the copious production of positive ions has another important consequence, apart from the ones mentioned. The charge signal detected on the anodes is mainly due to the motion of ions away from the wires, towards regions of lower and lower fields; on infinite impedance, the signal would have the shape shown in Fig. 16a, lasting for many μs. With a short differentiating RC constant on the amplifier, at best a signal decay time of 20-30 ns can be obtained (Fig. 16b), and this limits the two-track resolution, with the usual values of drift velocity, to 1-2 mm. With the increasingly complex topology of events at high energies, this may seriously limit the detector performances. Moreover, the localized high density positive ion cloud produced in the first avalanche can decrease, if not suppress, any further amplification in a given spot of the wire. A possible solution to the second problem was found several years ago in the so-called scintillating drift chambers (Fig. 17) (Charpak, Majewski and Sauli, 1975). Tracks were detected through the light flash produced, in suitable conditions, on a photomultiplier, in a condition of moderate charge gain; no

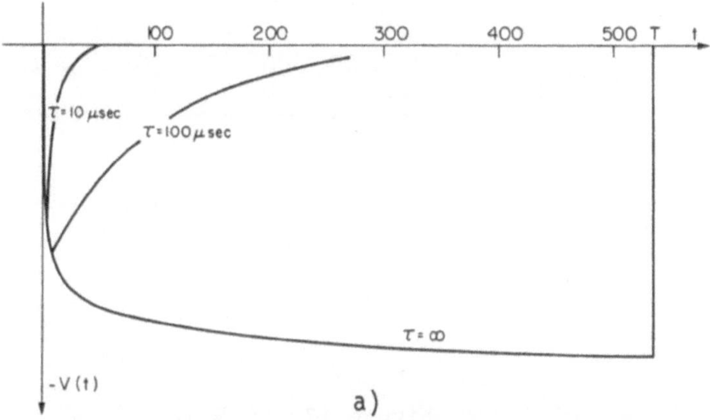

a)

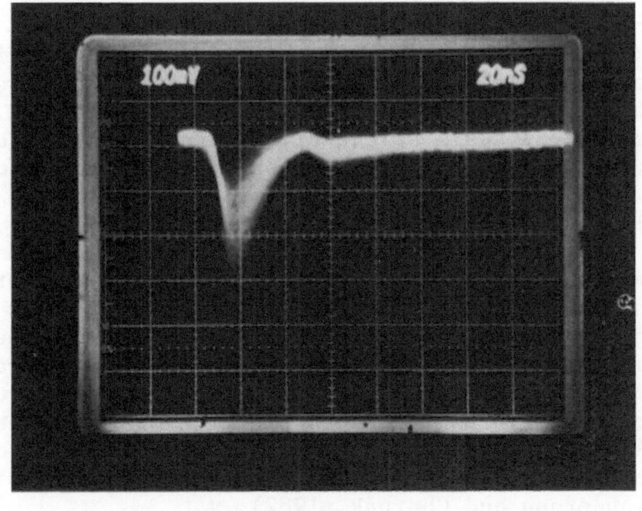

b)

Figure 16

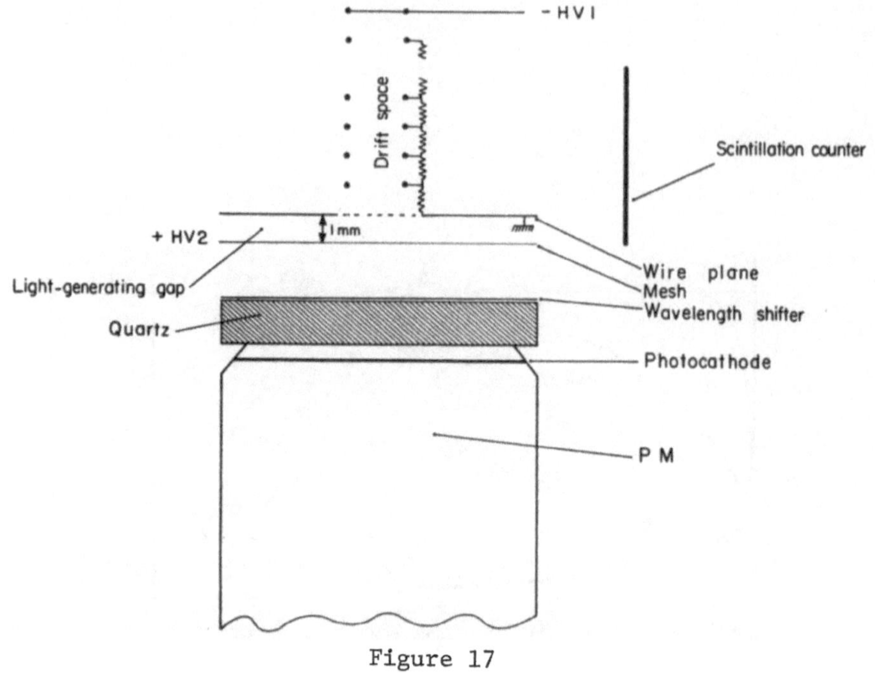

Figure 17

space charge effects were noticed up to very high fluxes (above 10^6 counts $s^{-1} \cdot mm^{-2}$). In the geometry and for the gas filling used, however, the average width of the detected light flash was around 50-100 ns, thus offering no improvement on the two-track resolution. The limit seemed intrinsic at the time, since light emission goes through many long-lived molecular de-excitation modes (for a review on secondary gas scintillation, see, for example, Salete Leite, 1980 and Policarpo, 1981). A recent work has completely changed the picture (Siegmund et al., 1982). By a proper choice of the gas mixture, these authors could obtain light flashes as short as 2-3 ns at the base in the detection of single electrons. Developed to allow single electron counting in low energy X-ray spectroscopy, the device may allow considerable progress in the two-track separation. One can indeed imagine a TPC-like drift chamber, operated at very high pressures so as to reduce electron diffusion, and where detection of tracks through the light flash produced in a parallel plate chamber at the end would allow space resolutions of perhaps 200-300 μm (Anderson and Charpak, 1982).

3. PARTICLE IDENTIFICATION THROUGH ENERGY LOSS

Fast charged particles traversing a medium lose energy through ionizing collisions at a rate which is a function of their velocity. After a fast decrease, the differential energy loss reaches a minimum around $\beta \simeq 0.9$ and then slightly increases until saturation is

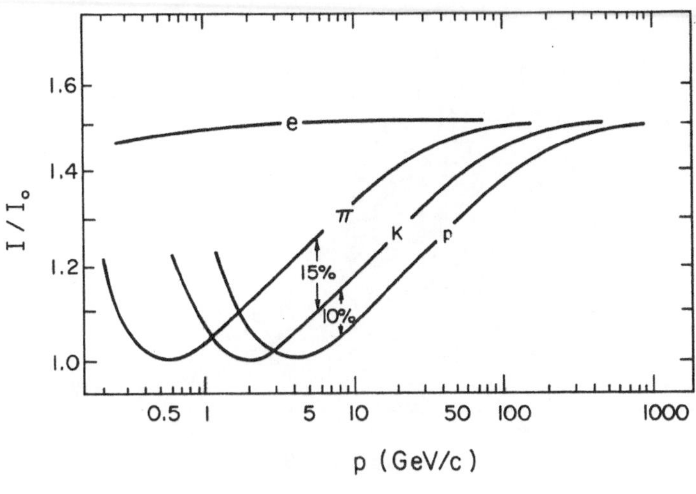

Figure 18

reached; the increase is relevant (40-50%) in gases, see Fig. 18 (in argon at normal conditions). A good measurement of dE/dx in the relativistic rise region allows, in principle, the determination of the particle's velocity and the identification of its mass if the momentum is known (through magnetic analysis). The practical problem resides in the large fluctuations of the energy loss in thin samples of gas caused by the ejection of energetic δ electrons in the primary interactions; with an average loss of 2 keV cm^{-1}, as in argon at NTP, a fast particle may eject a 2 keV electron per cm with 10% probability, thus doubling the loss in that sample (see Fig. 12). In Fig. 19 the two energy loss distributions measured in 1.5 cm of argon correspond to 25 GeV/c protons and π, whose averages differ by about 20%, see Fig. 18 (Allison et al., 1976). The distributions largely overlap and obviously only a multiple sampling followed by adequate fitting procedures may allow identification; a truncated mean is generally used, i.e. the average on a given fraction of the smallest energy losses (40 to 60% of the total).

The analysis of the expected detector performance begins with a good knowledge of the energy loss process for fast particles. This has been discussed recently both analytically (Talman, 1979) and computed for some noble gases using phenomenological model calculations (Allison and Cobb, 1980; Lapique and Piuz, 1980). In Fig. 20 the experimental histogram of energy loss measured for

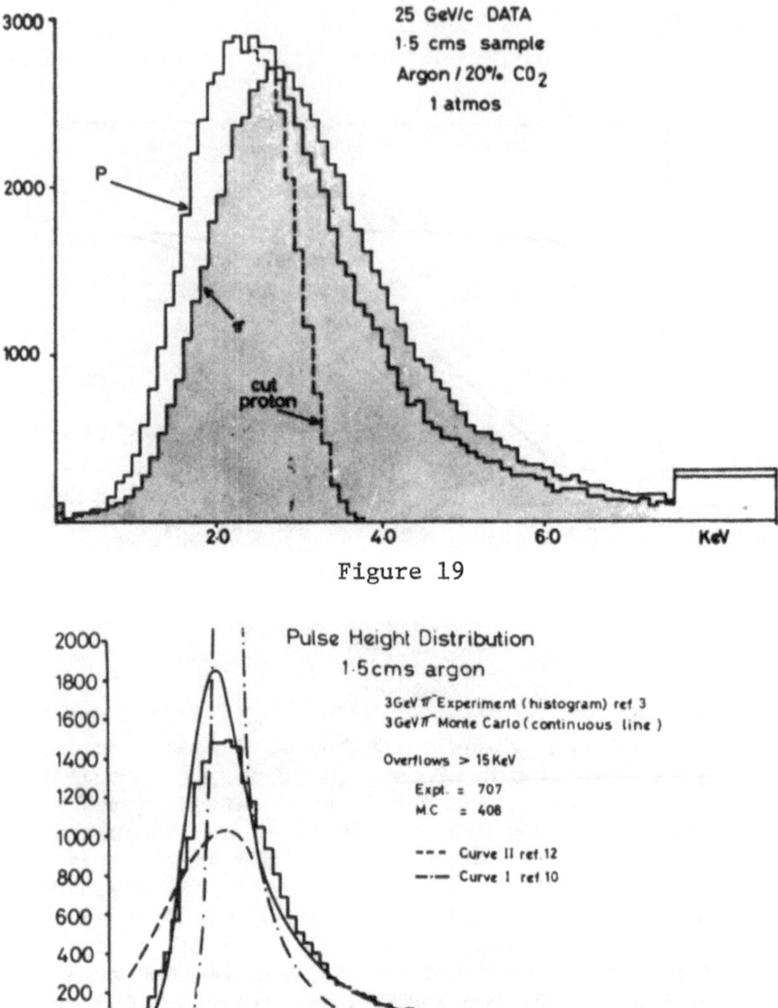

Figure 19

Figure 20

3 GeV/c π^- in 1.5 cm of argon is compared with the Landau (dotted curve), Blunk and Leisegang (dashed) and phenomenological photo-absorption model (full curve, from Cobb, Allison and Bunch, 1976). Clearly the last model provides a distribution rather closely matching the experimental one, and has been used for the optimization of the ISIS detector; we will see some outcomes of the model later.

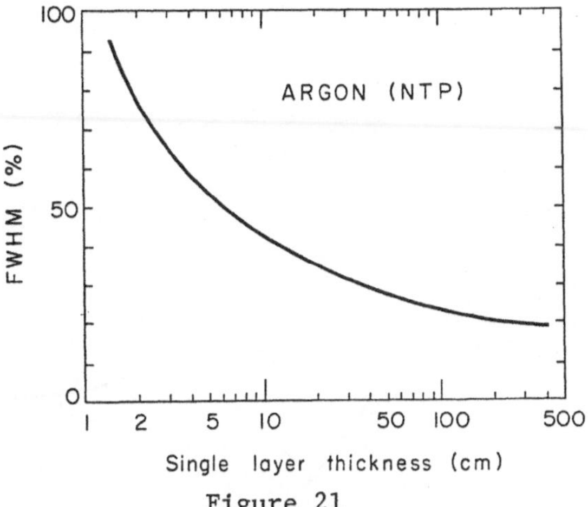

Figure 21

From the knowledge of the single layer resolution for a given gas, one can compute the expected separation power for a multilayer device. For argon, the experimental resolution (the FWHM divided by the average) is shown in Fig. 21 (Lazeyras, Lehraus, Mattewson and Tejessy, 1979); using these values, the curves of Fig. 22 have been computed, giving the expected dE/dx resolution as a function of sample size and number and of total detector thickness (Walenta, 1981). Letters in the graph refer to existing detectors, in project or already in operation: point (b) corresponds to the Time Projection Chamber, (c) to ISIS, (f) to JADE; for a complete list of references, see the quoted work. As an example, the Time Projection Chamber (TPC), where 180 ionization samples are measured in a little less than 1 m long detector operated at 8 atm, has a computed resolution of 5.5% FWHM, sufficient to identify π and k within 3 standard deviations between 1.5 and 20 GeV/c (compare with Fig. 18). The first data from the TPC group show indeed that the goal is reached, at least for a restricted class of events.

Note that resolution is very quickly lost if the number and thickness of samples is far from optimum, or if the detector suffers from systematic dispersions. In the case of JADE the expected 10% resolution, computed for 48 ionization samples 1.5 cm thick each, turned out in practice to be around 14% FWHM at the beginning of the operation; the result is shown in the scatter plot of Fig. 23, where obviously particles of different masses largley overlap in dE/dx in the relativistic rise region.

In what follows, I would like to discuss the various components that add up to define the dE/dx resolution obtainable in an ionization sampling device, and to describe some proposed improvements on the technique.

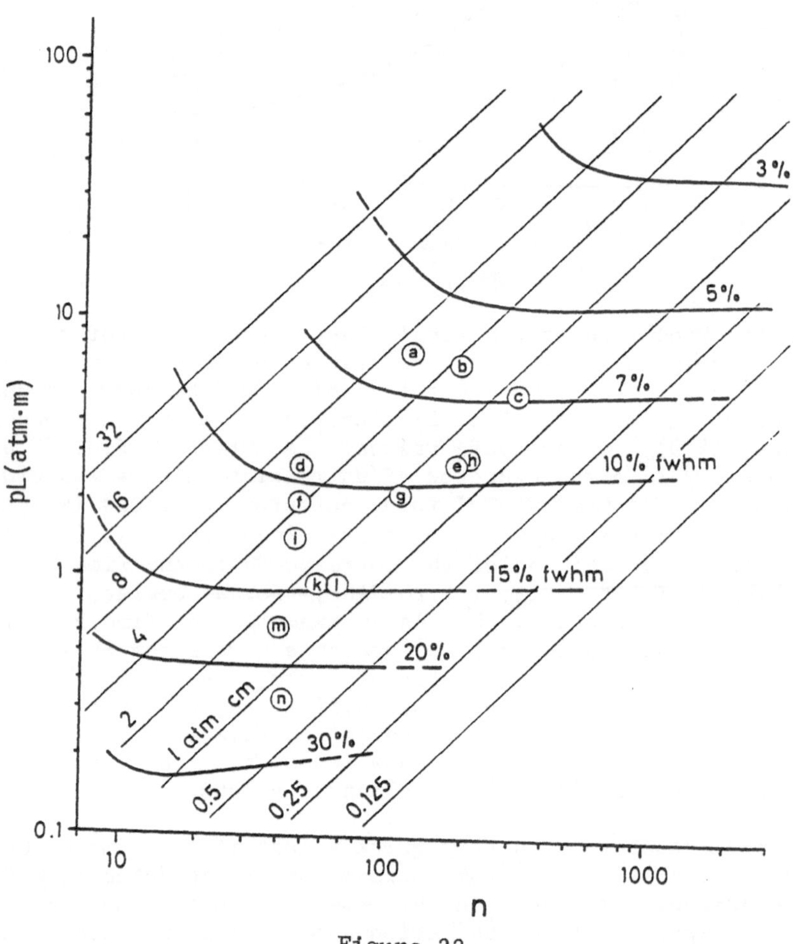
100
10
1
0.1
pL(atm·m)
3%
5%
7%
10% fwhm
15% fwhm
20%
30%
32
16
8
4
2
1 atm cm
0.5
0.25
0.125
a
b
c
d
e
h
f
g
i
k
l
m
n
10
100
1000
n

Figure 22

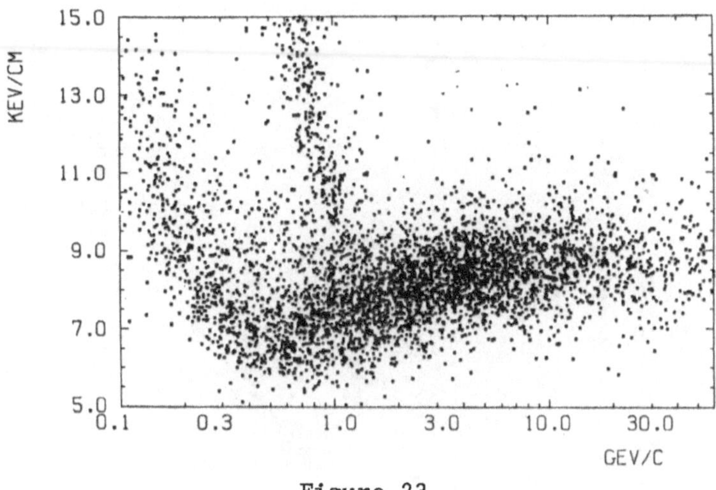

Figure 23

First, let us consider the influence of the choice of the gas filling and pressure on the expected resolution. From Fig. 22 it is obvious that, given the detector length and sample size, an increase in gas pressure improves the resolution. The amount of relativistic rise before saturation, however, decreases due to the increased density of the medium, as shown in Fig. 24 (from Lehraus, Matthewson and Tejessy, 1981). The choice therefore depends on the range of momenta where a given level of separation is demanded.

Another aspect of the choice concerns the intrinsic energy resolution in thin samples, that was given for argon in Fig. 21. It has been realized by several authors that in hydrocarbons and in CO_2 the single layer resolution is better, for a given thickness, than for the noble gases (Walenta, Fisher, Okuno and Wang, 1979; Lehraus, Matthewson and Tejessy, 1982). As shown in Fig. 25, the relative resolution measured for identical sample thickness can be a factor of two better in some hydrocarbons as compared to argon. It has been found (Walenta, 1981) that the relative resolution in all measured gases is a simple function of the reduced absorber length $\alpha t/I$, where I is the mean ionization potential of the medium (Fig. 26). Is it then obvious that in all existing devices one should replace noble gas filling with hydrocarbons? This is not so for two reasons. The first is purely technical: the operating voltage for a proportional chamber would have to be almost doubled, and generally this is impossible if not foreseen in the original design (both because of localized breakdowns and of electrostatic instabilities). The second reason is however more fundamental: what happens to the relativistic rise in hydrocarbons? The

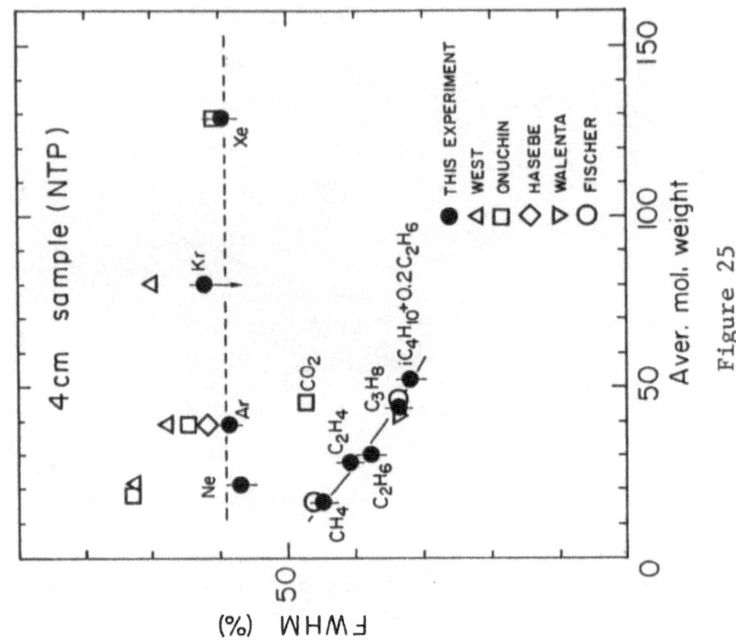

Figure 25

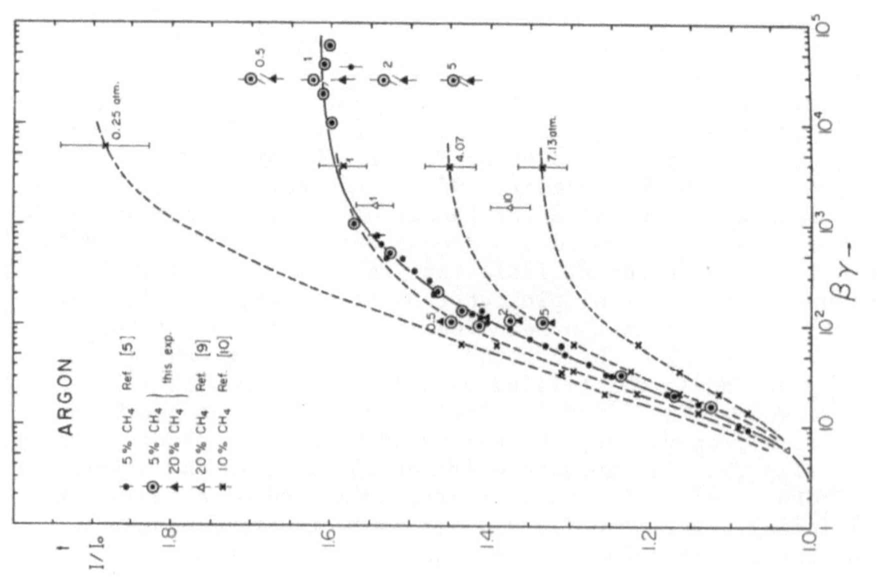

Figure 24

unfortunate answer is that the rise is smaller and that saturation is reached faster, see Fig. 27 (Allison, Bunch and Cobb, 1978). Again there seems to be a simple dependence of the amount of relativistic rise on the molecular properties of the absorbing medium which does not leave much space for "new" gases, as shown in Fig. 28 (Walenta, 1981; for the definition of the gas constant k that depends on the average energy loss and ionization potential, see the quoted reference). Figure 29 shows examples of measured πp and $e\pi$ identification power, given as a ratio between the difference in average energy loss and resolution, for different gases and mixtures at 15 GeV/c (Lehraus, Matthewson and Tejessy, 1982). The detector used for the measurement is the External Particle Identifier (EPI) where 64 samples, 4 cm thick each are measured for each track.

There appears then to be no best general choice and the operating conditions depend on the technological constraints (size and pressure of the detector) and on the momentum range in which the particle separation has to be optimized. For an extended discussion on the subject, see Allison (1982). Special mention deserves the region of decreasing energy loss, before the minimum; it seems obvious that the use of pure hydrocarbons as detecting media in this case is bound to provide better particle identification in existing detectors, or else allow the design of detectors with a smaller number of samples for a given resolution.

Having described some limits in resolution intrinsic to the energy loss process, I would like now to mention several non-trivial sources of dispersion connected with the detecting instrument, a multiwire gas proportional counter. Because of the fluctuations in the avalanche process, the gas multiplication adds dispersion to the energy loss sample. In a simple formulation, one can write for the variance of the recorded pulse height:

$$\left(\frac{\sigma_P}{P}\right)^2 = \left(\frac{\sigma_N}{N}\right)^2 + \frac{1}{N}\left(\frac{\sigma_A}{A}\right)^2 ,$$

where σ_N/N is the relative dispersion of the original charges, and σ_A/A the variance of the avalanche produced by a single electron. The second term is gain-dependent, and has been computed and measured for several gases; Figure 30 shows a typical behaviour in argon with $f = (\sigma_A/A)^2$; $\bar{K}$ is the average amplification and pr_a the product of pressure times wire diameter (Alkhazov, 1970). For minimum ionizing particles, where $N \simeq 100\ cm^{-1}$ in typical samples and $\sigma_N/N \sim 50\%$, the dispersion due to gas gain is negligible unless the sample size is very small (this is not so in X-ray detection). When one looks to experimental data, however, the outcome is very

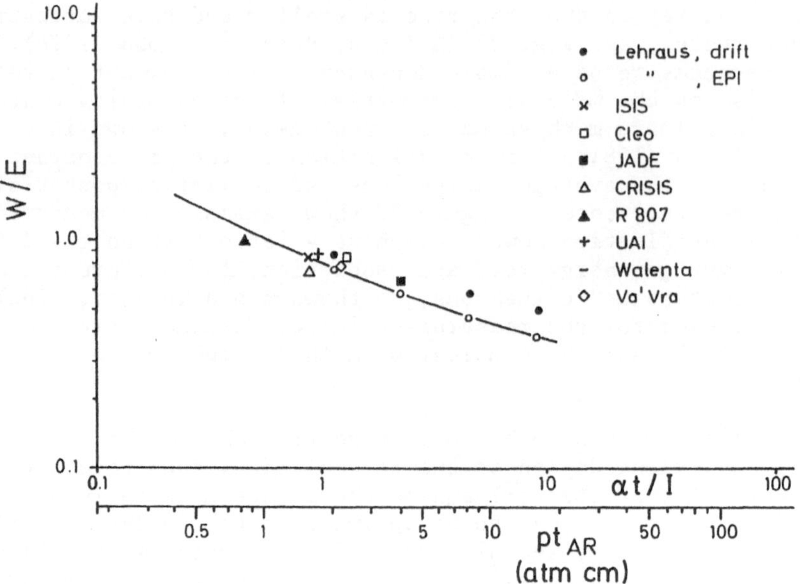

Figure 26

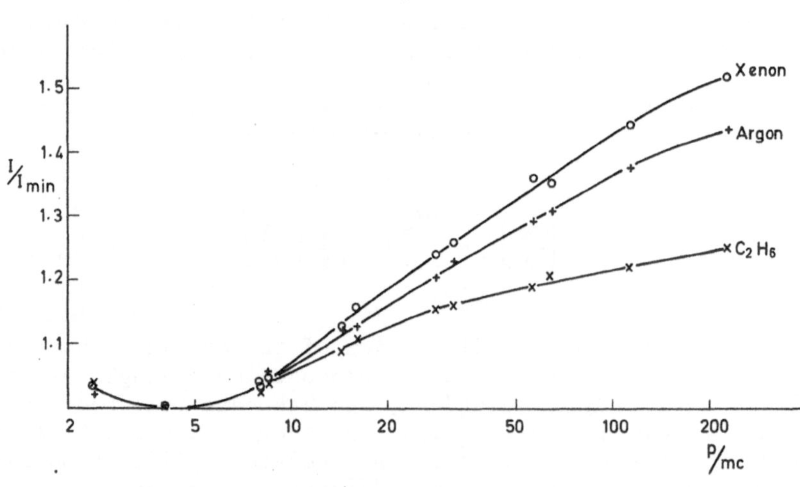

Figure 27

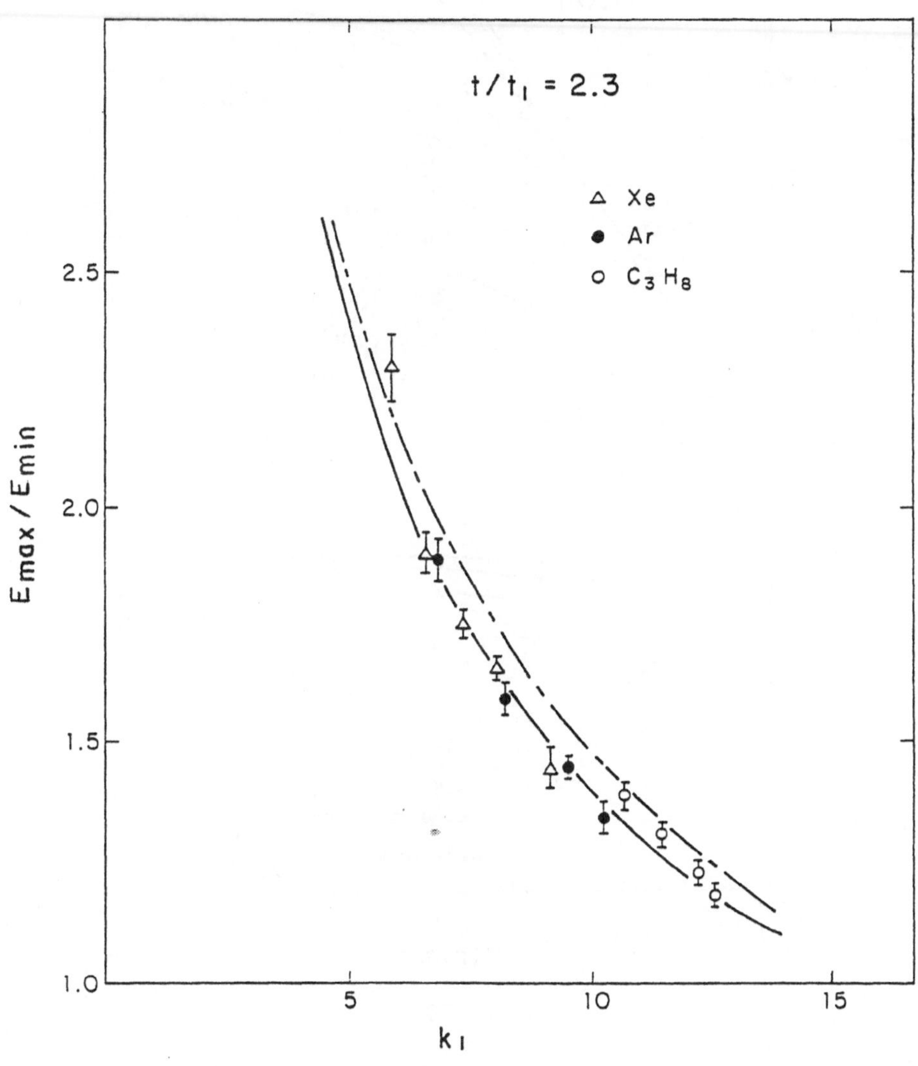

Figure 28

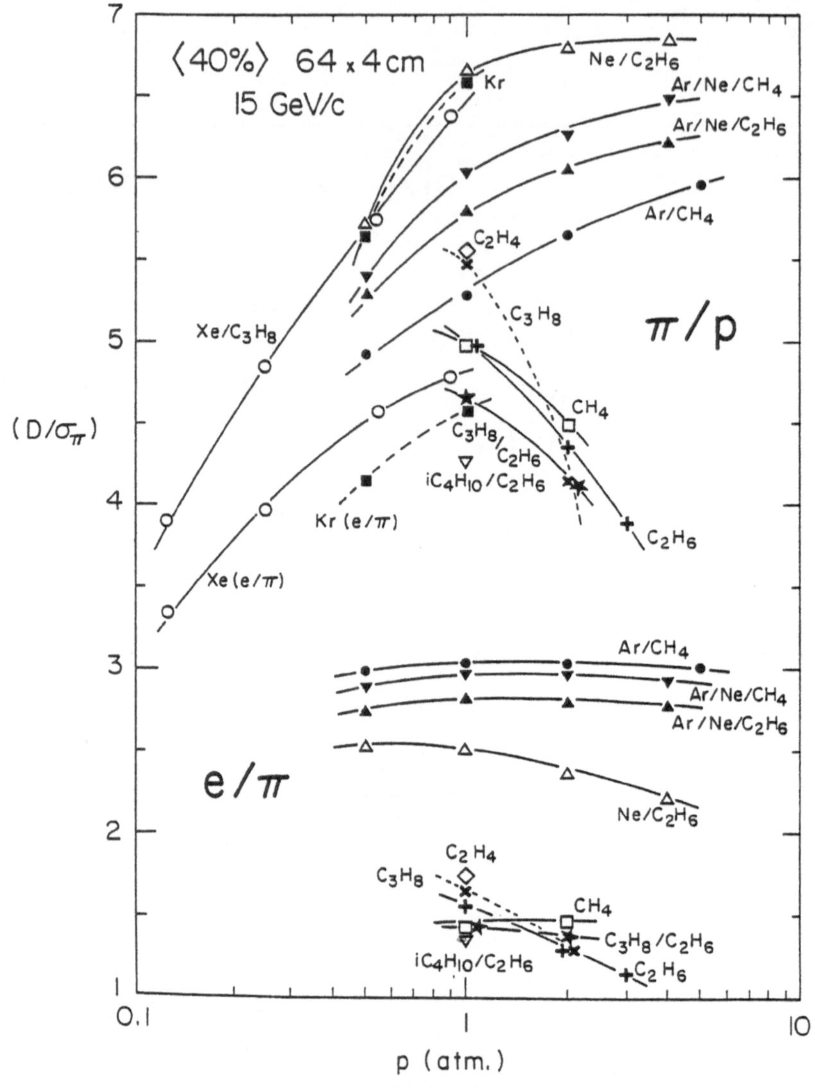

Figure 29

different (see Fig. 31), from Benjamin, Kemshall and Redfearn, 1968). Above a certain value of total charge, about 10^6 electrons detected, the resolution quickly deteriorates: this is understood as being due to space charge gain modification when the total charge exceeds a critical value. In ionization sampling detectors one tends therefore to work at the lowest possible proportional gain, depending on sample size and electronics noise: typically between 10^3 and 10^4. What happens is that when electrons from a track

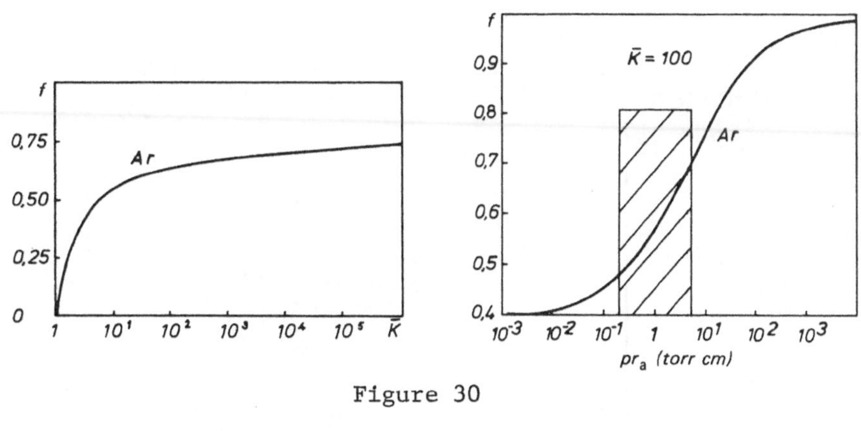

Figure 30

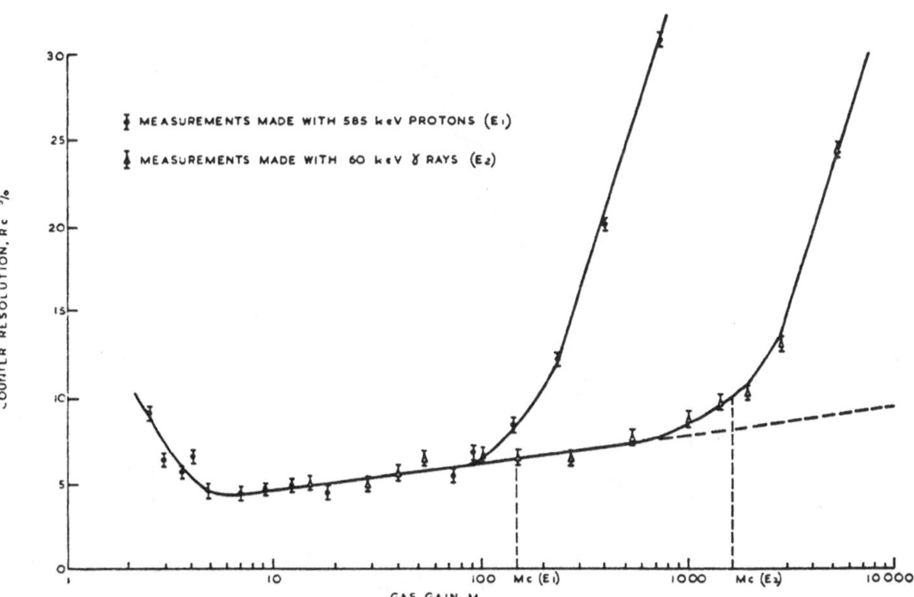

Figure 31

segment reach the anode in a localized spot but dispersed in time, the positive ions produced in the early avalanches decrease the electric field, and therefore the gain, for the following electrons. As the time structure of the detected electrons varies, a large fluctuation in the final detected size results. The effect is particularly relevant when particles traverse the detector at different angles: Figure 32 (Frehse, Lapique, Panter and Piuz, 1978) shows the variation in normalized detected charge as a function of gain for various angles of incidence (0° means a track perpendicular to the anode wire). At gains around 10^4 the variation is already 30%

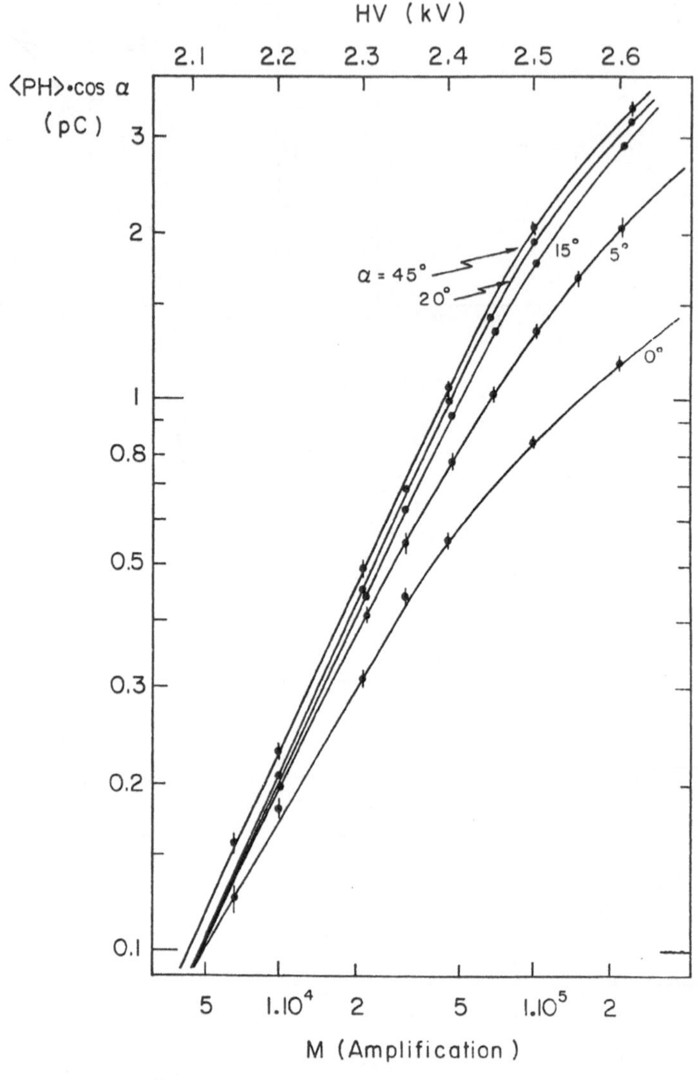

Figure 32

between tracks at 0° and 45°. Taking into account such geometric effects in the real data is very tedius and may be impossible if the data do not contain clean enough samples of events (such as cosmic muons).

Non-trivial electronics effects can also contribute to the loss of energy resolution. Since most of the amplitude measuring channels are a.c. coupled, a baseline shift can occur even at moderate rates. Moreover, the physical process of signal creation produces itself a tail extending to very long times (corresponding to the slow motion of ions); this has to be removed by suitable choice of the amplifiers' time constant. The resolution can be spoiled also by cross-talk, either internal (i.e. due to the charge induced mutually by positive ions in neighbouring channels) or external, in the chamber or in the electronics due to capacitive couplings. For an extended discussion on these points, see Allison, 1982.

Many of the above mentioned problems may be solved by replacing the global energy loss measurement by a measurement of the number of primary ionizing encounters (cluster counting). As indicated before, the origin of the large fluctuations of energy loss is the ejection of energetic electrons with their peculiar statistics. The number n of primary interactions, on the other hand, follows a Poisson distribution with a dispersion simply proportional to $\sqrt{n}$. In 1 cm of argon, while the dE/dx resolution exceeds 100% (see Fig. 21), one expects $n \simeq 30$ with a FWHM of $2.34 \times \sqrt{n}$ or about 40%.

Counting the number of clusters in a gas detector is, however, not straightforward. Indeed, the average distance between primary interactions, ($\sim$ 300 μm in argon at NTP) is close to the typical values of dispersion due to electron diffusion (see Fig. 5) and clusters quickly mix during the drift. Moreover, the minimum width of a proportional pulse (20-30 ns) results in substantial dead time losses. In the so-called Time Expansion Chamber (Walenta, 1979) good cluster separation is achieved by operating the detector at very low drift fields so as to spread out the time distance between clusters, and using thin anode wires and fast electronics to reduce dead times. A typical result is shown in Fig. 33, (Rehak and Walenta, 1980), measured in a 10 mm thick chamber filled with P-10 (10% CH_4 in argon). The distribution is indeed Poisson-like, but the average number of clusters disagrees with the expected value (around 25 cm^{-1}, see, for example, Korff, 1955) indicating that clusters largely overlap at the detection. Properties and limitations of cluster counting have been statistically analysed by Lapique and Piuz (1980) introducing realistic values for the diffusion coefficient and double-cluster resolution. The general outcome of this study is that with existing detectors the losses are too important and overcome the gain due to the favourable primary statistics.

While one can hope to improve the cluster counting technique for better resolution (see below), there is a more fundamental limitation in what concerns particle identification. It appears indeed that a good part of the relativistic rise in energy loss is due not to an increase in the number of clusters, but in the amount of energy imparted to inner shell δ electrons. Figure 34, from Lapique and Piuz (1980) shows indeed the result of the photoabsorption

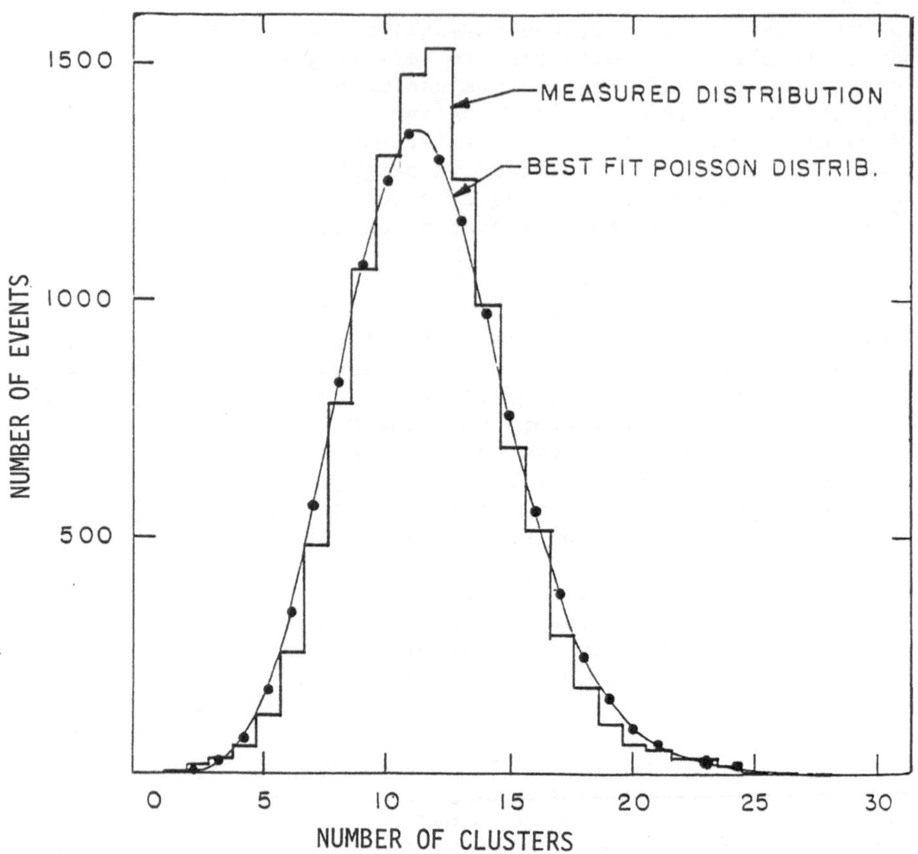

Figure 33

model calculation, described previously, for argon at NTP; the relative relativistic rise is computed as a function of γ using a truncated mean on the energy loss (curves c), total electron counting (b), and cluster counting (a); the dots are experimental points. The number of primary clusters saturates at lower momenta, a fact which is experimentally verified, see Fig. 35 (Rehak and Walenta, 1980).

A considerable advance in the obtainable resolution can be foreseen using the single electron counting technique developed to improve the energy resolution for soft X-rays (Siegmund et al., 1982). Coupling a low field drift space to a region of charge multiplication, and looking with a photomultiplier at the flashes of light emitted in the avalanches, these authors have optimized the gas mixture to obtain very short (2-3 ns) pulses for single

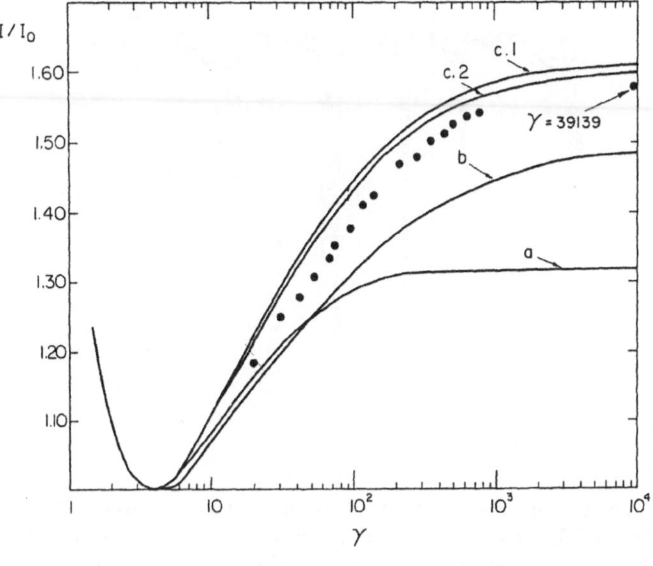

Figure 34

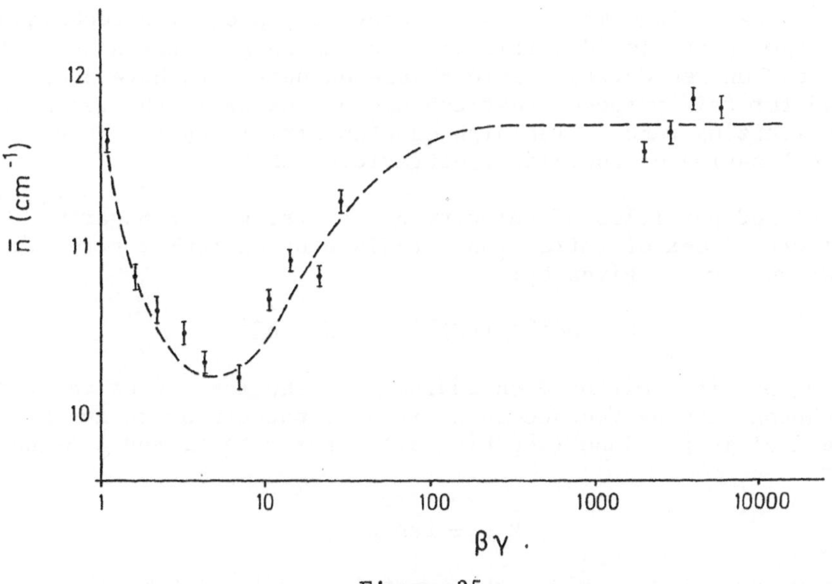

Figure 35

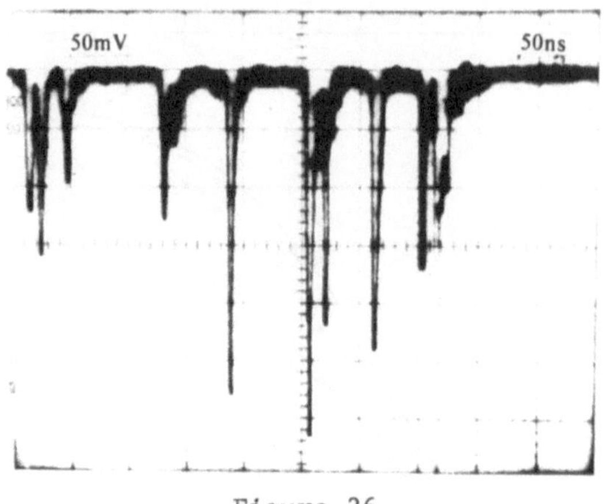

Figure 36

electrons; Figure 36 shows the photomultiplier output in the detection of 300 eV X-rays (producing a cluster of about 10 electrons). Individual electrons can be counted, with a time resolving power which is an order of magnitude better than in analogous proportional counters. It seems very interesting to apply this technique to cluster counting in the relativistic rise region.

4. PARTICLE IDENTIFICATION THROUGH CHERENKOV RING IMAGING

Cherenkov ring imaging is a recent very promising technique for charged particle identification in the range between several and a few hundred GeV/c. Various gaseous detectors have been developed for this purpose, that can operate using in the gas mixture a photoionizing vapour with high quantum efficiency for detection and localization of individual ultraviolet photons.

Charged particles of velocity β traversing a transparent medium with index of refraction n radiate by Cherenkov effect photons at an angle θ given by:

$$\theta = \cos^{-1} (n\beta)^{-1} , \quad \beta \geq n^{-1}$$

using a spherical mirror with radius r at the far end of the radiator, photons can be focused on a circular pattern of radius R on a spherical image plane (see Fig. 37). For r >> R, and a plane detector:

$$R = \frac{r}{2} \tan \theta$$

a measurement of R provides the particle's velocity and therefore its mass if the momentum is known. The number of detected photons is given by $N = N_0 \, L \sin^2 \theta$, where L is the radiator length, and

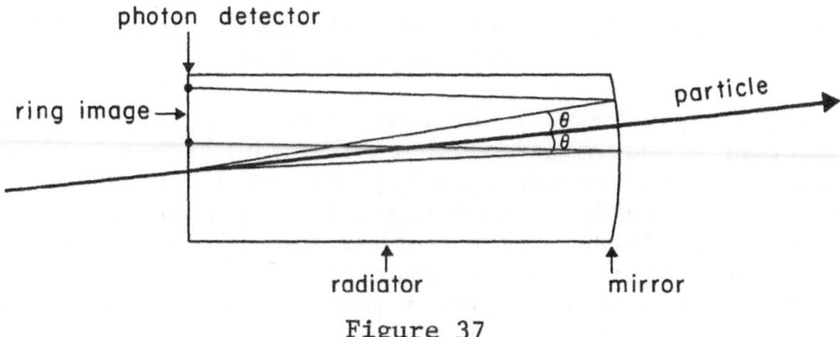

Figure 37

N_0 a constant whose value depends on the detected wavelength domain and on the various instrumental efficiencies. Typically, for the devices to be described, N_0 is around 70 cm^{-1}.

Photons are emitted all over the wavelength region for which the medium is transparent; an extended region of sensitivity results in more detected photons. On the other hand, since the index of refraction depends on wavelength, a larger dispersion (chromatic aberration) will be obtained. Such dispersive effect is particularly important at short wavelengths.

Cherenkov ring imaging has been demonstrated using electrostatic and channel plate image intensifiers at visible wavelengths (see, for example, Roberts, 1960; Robinson, 1980); this approach has, however, not been used much because of limited acceptance. It was suggested several years ago to shift the detection to ultraviolet wavelengths, using a photosensitive vapour in multiwire proportional chambers that allow in principle the covering of large sensitive areas (Séguinot and Ypsilantis, 1977).

The early works on the subject were realized using acetone or benzene as photoionizing vapours (ionization potentials 9.7 and 9.2 eV, respectively). This demands the use of lithium fluoride crystals as a window separating radiator and detector; LiF is expensive and highly hygroscopic. A considerable improvement was allowed by the introduction of triethylamine (TEA) as photoionizing vapour; with its 7.5 eV threshold, TEA allows the use of CaF_2 and MgF_2 crystals that are cheaper and stable in performances (Charpak et al., 1979 and 1981; Séguinot, Tocqueville and Ypsilantis, 1980; Comby et al., 1980; Williams et al., 1980).

More recently, a product having the extremely low ionization potential of 5.4 eV, tetrakis(dimethylamino)ethylene (TMAE) has been found, with considerable advantages because of the possibility of using fused silica as window (Anderson, 1981; Barrelet et al., 1982).

Figure 38 summarizes the quantum efficiency of some vapours and the cut-off of windows as a function of wavelength.

Not only a vapour with lower ionization potential allows the use of a simpler window material, but also the chromatic dispersions are reduced at the longer wavelengths; Figure 39 shows how the index of refraction depends on wavelength for argon and helium. In the region of sensitivity of TEA, with a CaF_2 window (1300 to 1600 Å) the expected aberration is about three times larger than in the region of sensitivity of TMAE with a fused silica window (1600 to 2200 Å). However, a substantial difference appears because of the vapour pressure of the two products at room temperature: 55 Torr for TEA, 0.35 Torr for TMAE. This means that with TEA full photon absorption can be obtained in 1-2 mm of gas while using TMAE even at high temperatures (35-40 °C), the mean path of absorption is 20-30 mm. This has implications both on the time resolution of detectors (at least an order of magnitude worse for TMAE than using TEA) and on the general background due to direct ionization in the gas by charged particles.

A detailed discussion of the particle identification properties of Cherenkov ring imaging devices, taking into account the various sources of dispersion, has been given by Ypsilantis (1981).

Let me here first consider the process of amplification of single electrons in proportional counters, extensively studied because of its relevance in determining the avalanche dispersion as indicated in the previous section.

Under simple assumptions, it can be shown that the avalanche size distribution when multiplying a single electron is an exponential of the kind:

$$P(n) = \frac{1}{\bar{n}} e^{-n/\bar{n}}$$

where $\bar{n}$ is the average avalanche size. Such pulse height distribution is indeed observed at small field values. At increasing fields, however, the distribution develops a peak with a reduction of the variance, as shown in Fig. 40 (Schlumbohm, 1958). Several theoretical models have been proposed to explain the observed avalanche size distribution (for a summary, see Alkhazov, 1970). They suggest a functional form of the Polya kind:

$$P(n) = \frac{b^b}{\Gamma(b)} \left(\frac{n}{\bar{n}}\right)^{b-1} e^{-b(n/\bar{n})}$$

with $b \simeq 2$. The practical implications of a peaked distribution are obvious: in the case of simple signal discrimination and

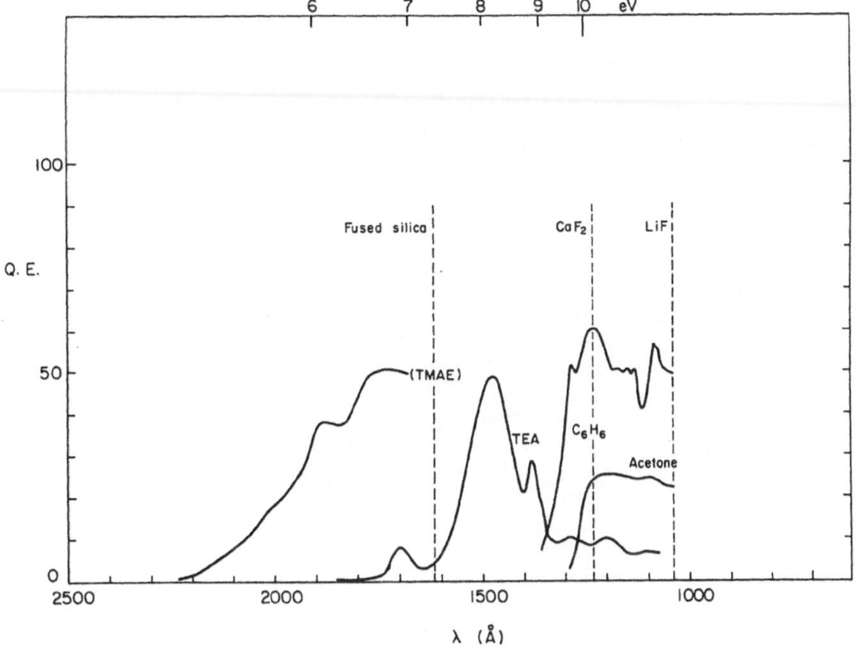

Figure 38

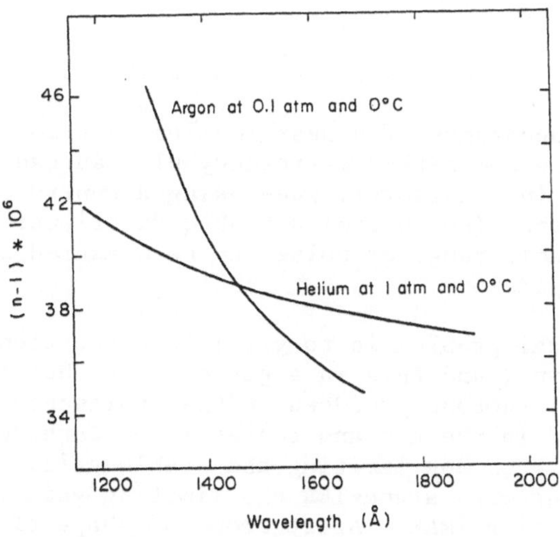

Figure 39

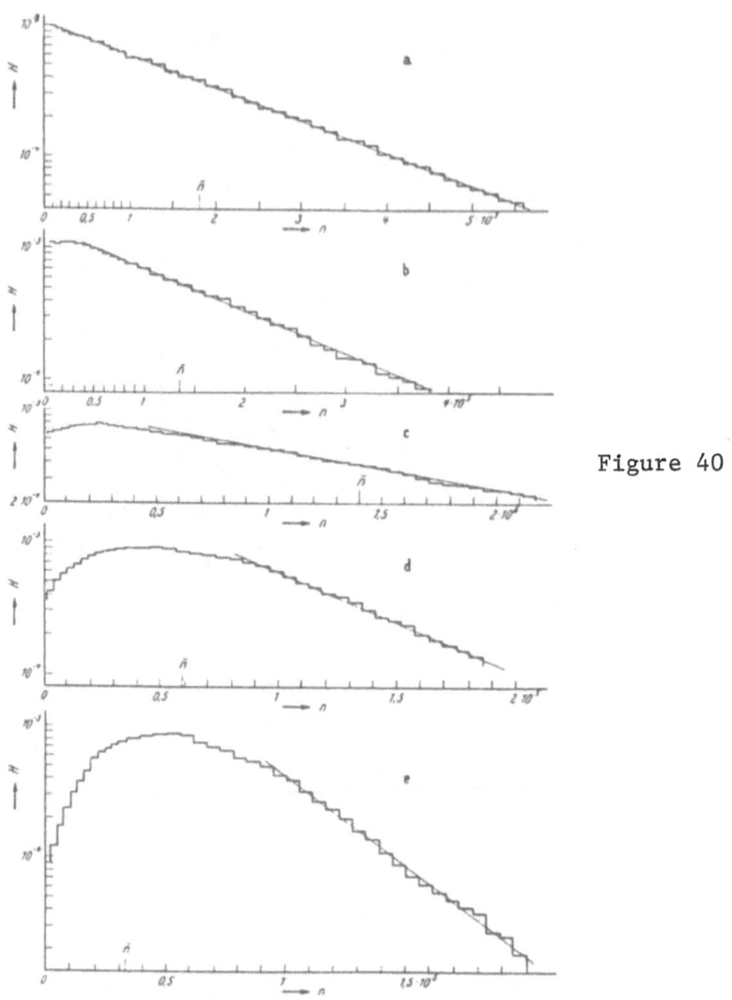

Figure 40

counting, the appearance of a peak implies that for low enough threshold setting a constant efficiency plateau can be obtained above a given gain. Moreover, when using a centre-of-gravity localization method (see below) a peaked distribution largely decreases the dynamic range of pulses to be measured and reduces electronics dispersions.

The practical problem is to get in a proportional chamber large enough gains, and this in a gas mixture that is by definition photosensitive. Photons produced in the avalanches themselves tend indeed to spread in the gas and reconvert on cathodes, inducing secondary discharges and limiting the stable gain. In a normal counter operating with argon-TEA the limiting gain is around 10^5, and even lower using TMAE. Apart from the shape of the distribution, 10^{-2} pC is a small charge to detect.

Larger gains, in excess of 10^6, have been obtained in two kinds of detectors developed for Cherenkov ring imaging: the multi-needle counter and the multistep proportional chamber.

The first device consists in a matrix of closely peaked needles, each acting as an individual Geiger counter, see Fig. 41 (Comby et al., 1980). The typical cell size is around 3 mm, which constitutes a natural limit in the localization accuracy. This kind of device has the advantage of being able to detect a large number of simultaneous photons; on the other hand, the complexity of construction and the number of electronics channels involved restricts its use in small size detectors. Moreover, the very long dead time per element typical of a Geiger discharge (several ms) limits the accepted rate, especially when charged particles traverse the detector. The device has been extensively tested in a beam using as

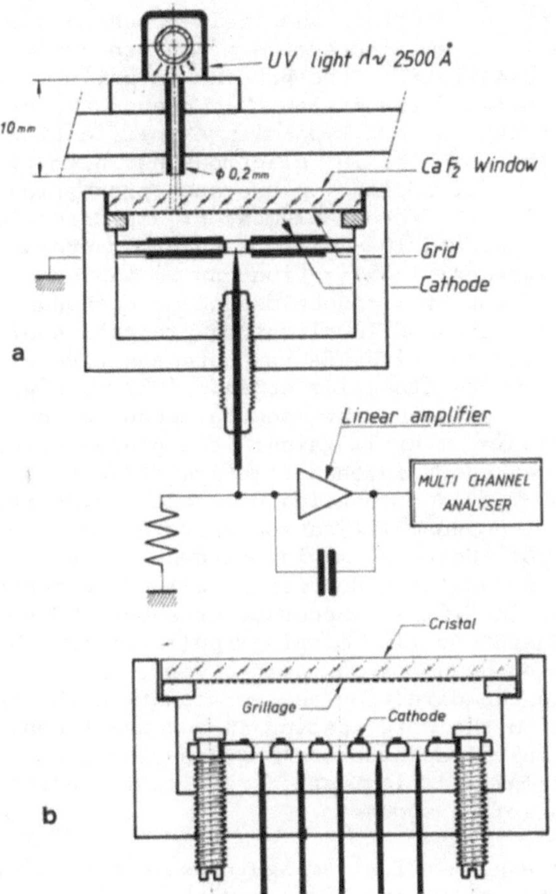

Figure 41

radiator a CaF_2 crystal, producing a large number of photons. Figure 42 shows an example of Cherenkov ring recorded for 1.2 GeV/c protons, and having a radius of around 30 mm.

A second approach to the detection problem is based on the multistep proportional chamber, a device originally developed to allow a gated operation of proportional chambers (Charpak and Sauli, 1978). By separating the overall gain in two amplification elements, each below the critical value for secondary effects to appear, this detector allows gains close to 10^7 to be obtained in a photosensitive gas mixture. The operation is as follows (Fig. 43). A photoelectron produced in the upper region, the conversion space, drifts into a region of very high field, the preamplification region. Here avalanche multiplication occurs, and in some gas mixtures (a noble gas and a photosensitivie vapour) the transverse avalanche growth is very large, 1 mm or so. Electrons in the avalanche produced within the field tubes connecting to the lower transfer region are injected there, in proportion to the total that equals the fields ratio: the process is uniform (independent on the original centre of the avalanche) because the spread exceeds the grid wire spacing. With a typical preamplification of 10^4 and a transfer efficiency of 10%, 10^3 electrons per primary charge are injected in the transfer region. A two-step proportional chamber is shown in Fig. 44: the second element of amplification, that brings the overall gain to 10^6-10^7 is a thin-gap proportional chamber, where bi-dimensional localization can be achieved for example using the centre-of-gravity readout method on cathodes. The efficiency, time resultion and localization properties of such a device in detecting photons with TEA as photosensitive vapour have been extensively studied (Bouclier et al., 1982). Figure 45 (Cattai, 1981) shows pulse-height spectra measured for single photoelectrons at increasing overall gains: the peaked structure is apparent. Figure 46 shows instead relative efficiency plateaus measured in helium-TEA as a function of anodic potential, for several values of the preamplification voltage and a detection threshold of 0.1 pC. Using 2 mm wide cathode strips readout, the centre-of-gravity method (see Section 2) allow the reconstruction of the photon position with an accuracy of about 400 μm r.m.s., suited for most Cherenkov ring imaging applications (the typical chromatic aberration being of the same order). Notice that localization accuracy in the direction perpendicular to the anode wires, bound to be equal to the wire spacing in a conventional chamber, is better in the multistep detector because the large avalanche spread in the preamplification region results in charge sharing and interpolation at the anodes.

To resolve the geometrical ambiguities in case of several simultaneous points, the anode wires can be mounted at 45° with the orthogonal cathodes thus providing three independent projections

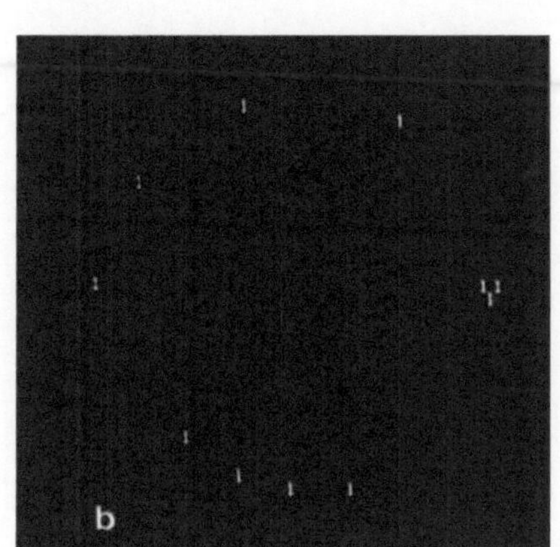

Figure 42

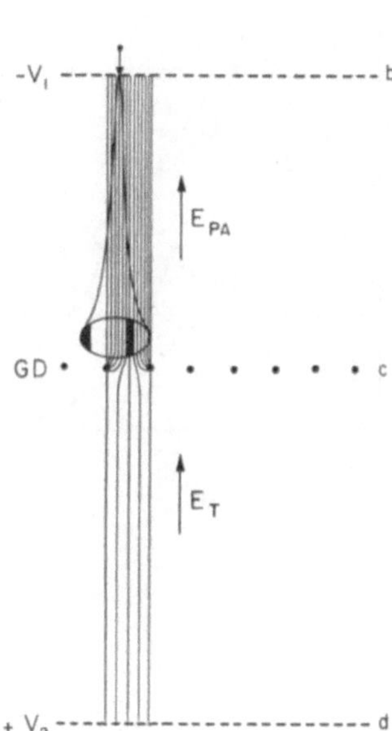

Figure 43

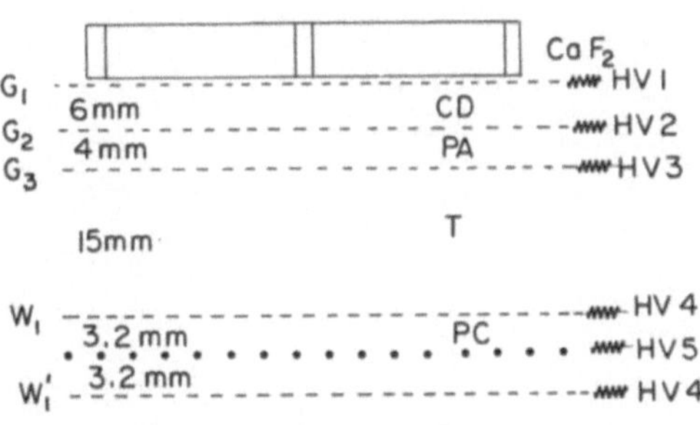

Figure 44

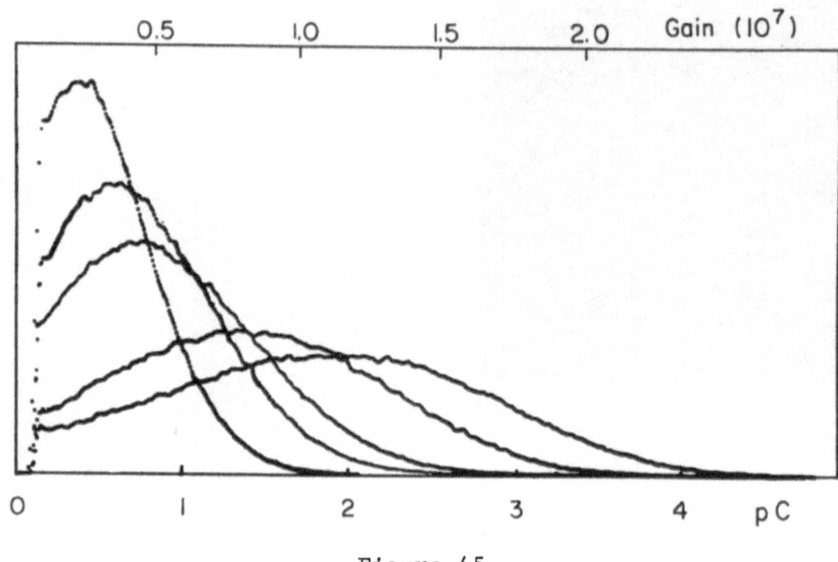

Figure 45

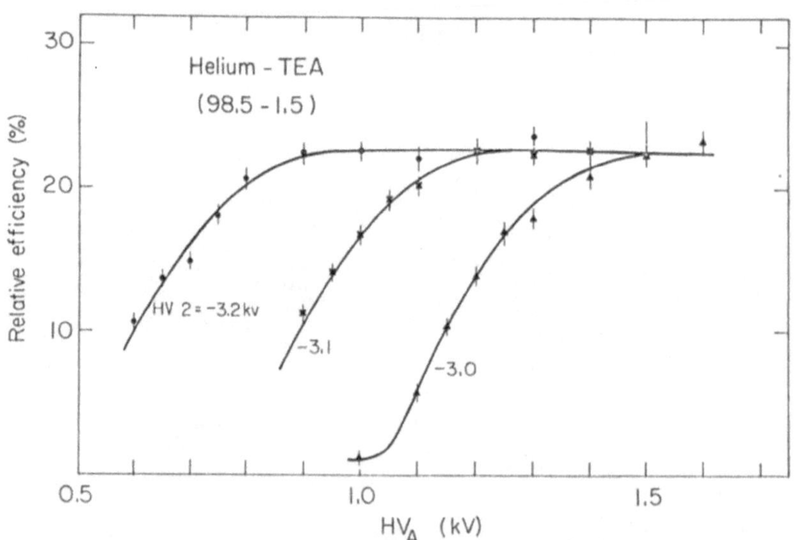

Figure 46

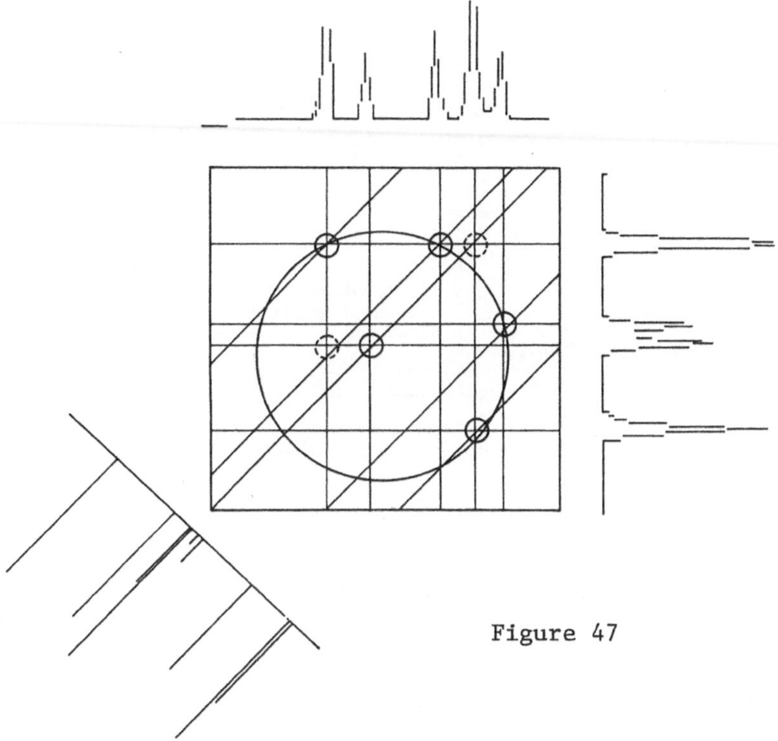

Figure 47

of the event. Figure 47 shows an example of ring reconstruction (Coutrakon et al., 1982). A detector of this kind is almost completed at Fermilab, and will be used for hadron identification up to 400 GeV/c. It consists of a 16 m long radiator, 3 × 3 m^2 in section, with two multistep photon detectors on the two sides where images are focused by the rows of mirrors (Fig. 48). Figure 49 shows one composite CaF_2 window, 40 × 80 cm^2, separating the detector from the radiator gas (helium at atmospheric pressure).

With a smaller size prototype detector of similar design, the results shown in Figs. 50 and 51 have been obtained in a test run at 200 GeV/c. The picture shows an overlap of ∿ 100 events, with the typical ring pattern, about 70 mm in radius, while Fig. 51 gives the reconstructed radius distribution, in logarithmic scale, showing the good πK separation obtained, around six standard deviations (Coutrakon et al., 1982). From the known detectors parameters and extrapolating the experimental results, one can compute the expected particle identification properties of the full device given in Fig. 52 (Peisert, 1982).

The multistep proportional chamber photon detector has the advantage of a good time resolution, 40-50 ns using TEA, but is

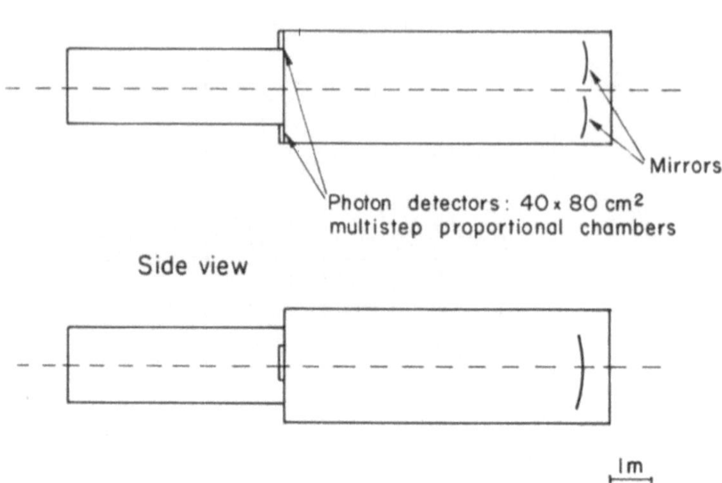

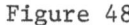

Figure 48

Figure 49

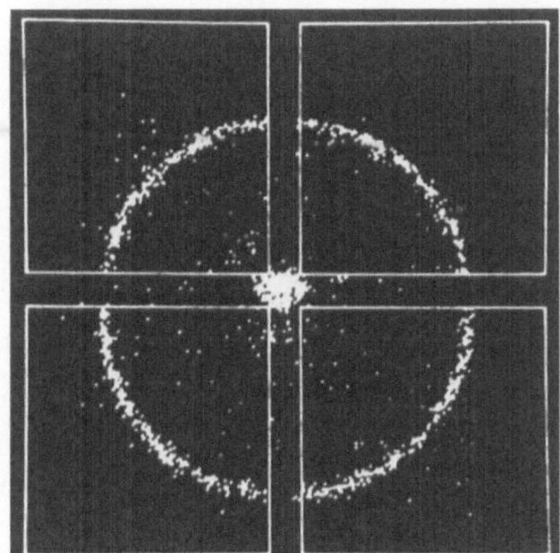

Figure 50

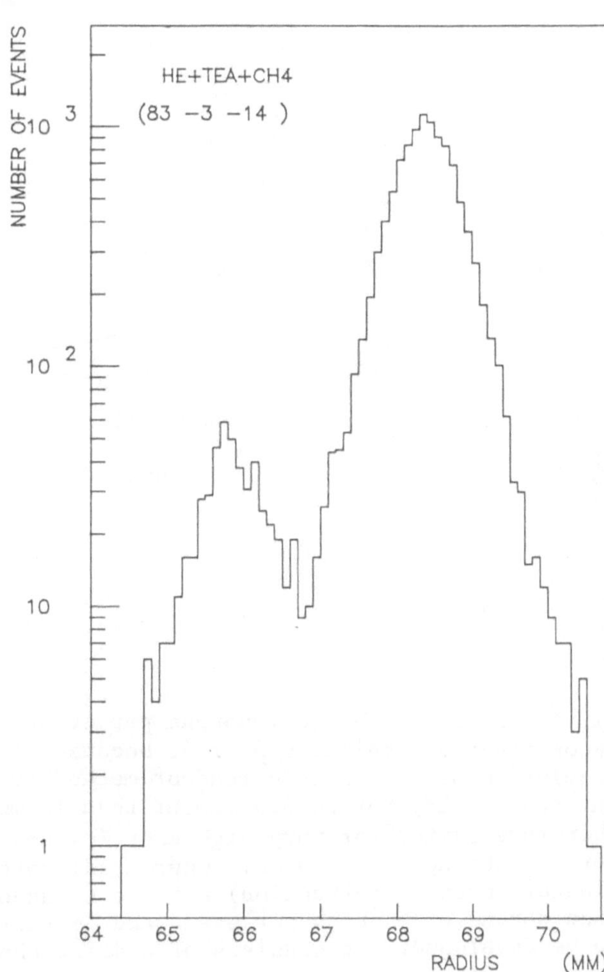

Figure 51

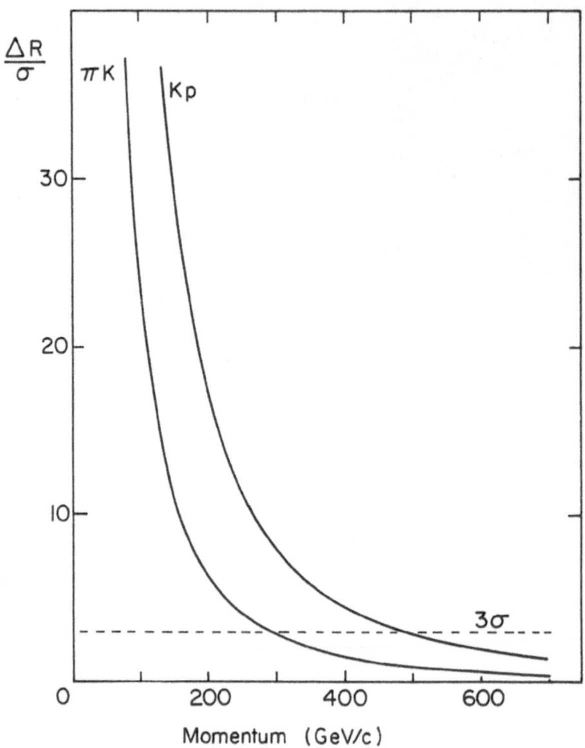

Figure 52

limited in the number of photons per event that can be unambiguously reconstructed, typically 3 or 4, because of the modest double point resolution of the cathode readout method (induced signals distributions overlap for points closer than 10 mm in projection). It is therefore suited for very high energies and high flux experiments. For momenta up to 60-70 GeV, where radiators with lower threshold (larger index of refraction) have to be used, one can easily produce dozens of photons per event and another kind of detector had to be developed. It consists of a drift chamber of special design,

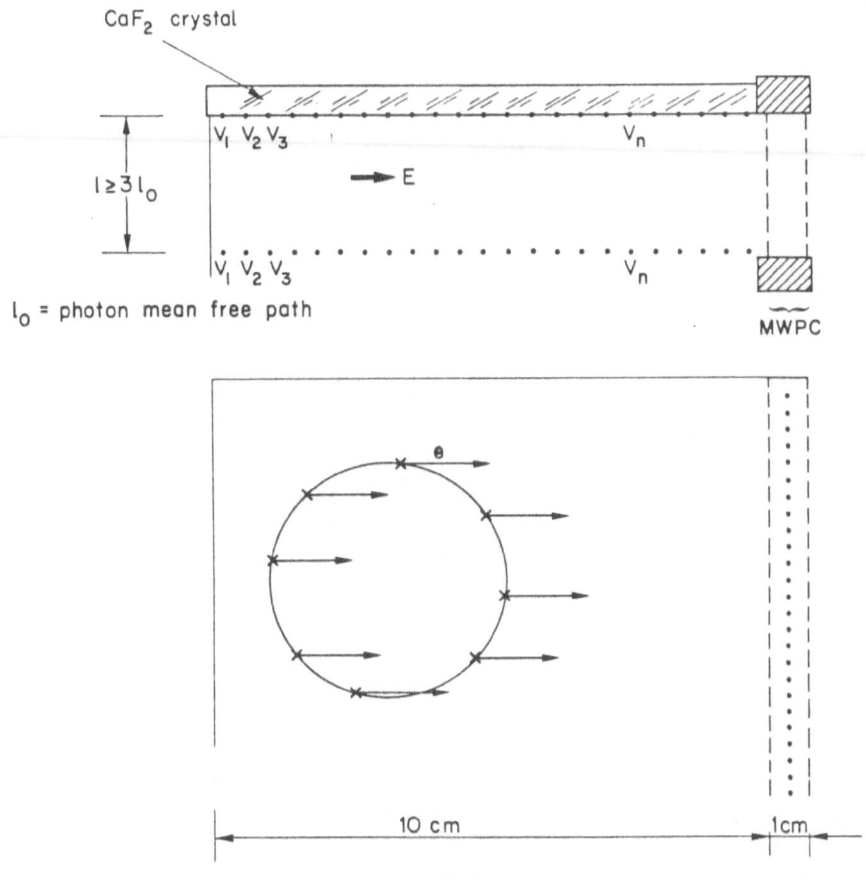

Figure 53

originally proposed by Séguinot and Ypsilantis (1977), see Fig. 53. Photons convert in the drift space, and the ring pattern is reconstructed measuring the drift time on a set of proportional wires as shown in the figure. The time resolution of the device corresponds obviously to the maximum drift length, around 1 μs or so, depending on the size.

Drifting electrons close to an insulator (the window) is, however, not straightforward, because of distortions in the electric field. The problem has been solved by doubling the drift field shaping electrodes on the outside of the window (Barrelet et al., 1982). Moreover, the diffusion of electrons during drift is an intrinsic limitation to the localization accuracy. Operating in pure methane with TEA and TMAE, the quoted authors have measured typical diffusion rates of about 250 $\mu m/cm^{1/2}$; for a 10 cm long drift, this implies a localization accuracy for one photon of around 800 μm.

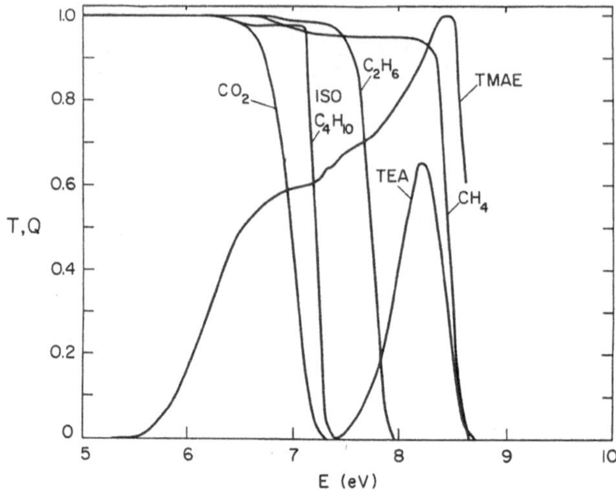

Figure 54

For the particular needs of DELPHI, a detector proposed for the LEP e^+e^- storage rings, a large amount of work is being done on the use of TMAE as photoionizing vapour. Apart from the already mentioned advantages of reduced chromatic aberrations and of using quartz windows, photon detection at longer wavelengths opens up the possibility of using radiating media other than pure noble gases. Isobutane, for example, has a transmission cut-off almost coinciding with the one of fused silica (Fig. 54, from Barrelet et al., 1982). Liquid freon can also be used as radiator for the lower momenta. Figure 55 shows the expected πK and Kp separations computed by the authors of the proposed detector in a combined liquid and gas Ring Imaging Cherenkov device (RICH). Figure 56 shows a simulated event with about 20 charged particles as seen by the detector.

I should mention in conclusion that operating with TMAE is very attractive but rather difficult because of its low vapour pressure. Photons emitted in the avalanches tend to spread far from the wires and reconvert in the gas producing spurious hits; a mechanical construction that prevents each wire from seeing the next one partly soves the problem, but is rather difficult to implement. Finding a low ionization potential vapour with decent vapour pressure would certainly constitute a considerable advance in the field.

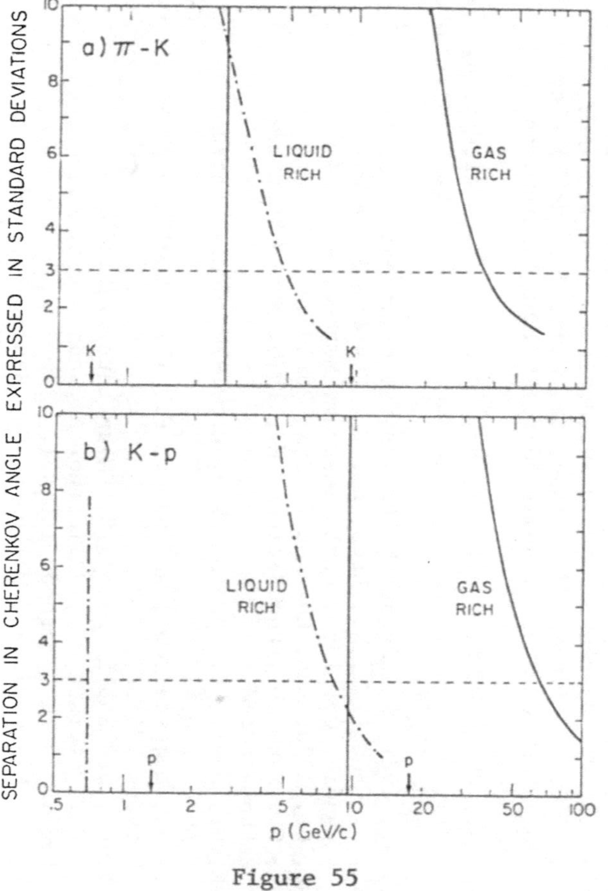

Figure 55

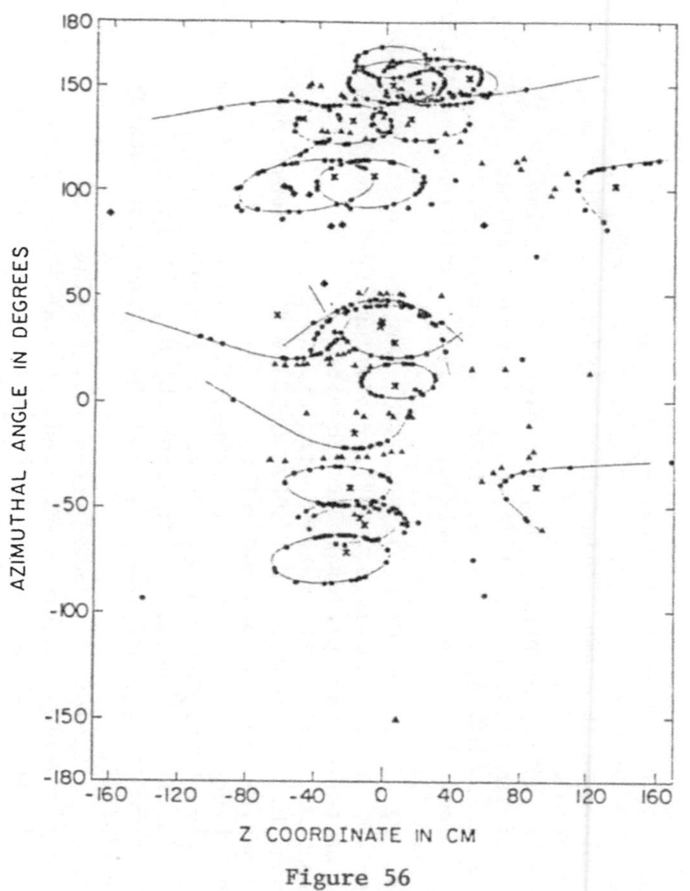

Figure 56

REFERENCES

Alkhazov, G. D., 1970, Nucl. Instrum. Methods 89:155.
Allison, W. W. M., Brooks, C. B., Bunch, J. N., Cobb, J. H., Lloyd, J. L., and Pleming, R. W., 1974, Nucl. Instrum. Methods 119:499.
Allison, W. W. M., Brooks, C. B., Bunch, J. N., Pleming, R. W., and Yamamoto, R. K., 1976, Nucl. Instrum. Methods 133:325.
Allison, W. W. M., Brooks, C. B., Lyons, L. Romaya, A. M., Shield, P. D., and McPherson, A., 1979, Nucl. Instrum. Methods 163:331.
Allison, W. W. M., and Cobb, J. H., 1980, Ann. Rev. Nucl. Sci. 30:253.
Allison, W. W. M., (Stanford, 1982), Proc. Int. Conf. on Instrumentation for Colliding Beam Physics.
Anderson, D., 1981, IEEE Trans. Nucl. Sci. NS-28:842.
Anderson, D., and Charpak, G., 1982, CERN-EP/82-52, submitted to Nuclear Instruments and Methods.
Barranco Luque, M., Calvetti, M., Dumps, L., Girard, C., Hoffmann, H., Maurin, G., Naumann, L., Perez, J., Piano Mortari, G., Placci, A., Queru, P., Rijssenbeek, M., Rubbia, C., Sadoulet, B., and Tao, C., 1980, Nucl. Instrum. Methods 176:175.
Barrelet, E., Ekelöf, T., Lund-Jensen, B., Séguinot, J., Tocqueville, J., Urban, M., and Ypsilantis, T., 1982, CERN-EP/82-09 (submitted to Nuclear Instruments and Methods).
Benjamin, P. W., Kemshall, C. D., and Redfearn, J., 1968, Nucl. Instrum. Methods 59:77.
Bouclier, R., Charpak, G., Million, J., Peisert, A., Santiard, J.C., Sauli, F., Coutrakon, G., Hubbard, J. R., Mangeot, Ph., Mullié, J., Tichit, J., Glass, H., Kirz, J., and McCarty, R., 1982, CERN-EP/82-83 (submitted to Nuclear Instruments and Methods).
Breskin, A., Charpak, G., Majewski, S., Melchart, G., Peisert, A., Sauli, F., Mathy, F., and Petersen, G., 1980, Nucl. Instrum. Methods 178:11.
Cattai, A., 1981, Thesis at the University of Trieste.
Charpak, G., Bouclier, R., Bressani, T., Favier, J., and Zupančič, Č., 1968, Nucl. Instrum. Methods 62:235.
Charpak, G., 1970, Ann. Rev. Nucl. Sci. 20:195.
Charpak, G., Rahm, D., and Steiner, H., 1970, Nucl. Instrum. Methods 80:13.
Charpak, G., Sauli, F., and Duinker, W., 1977, Nucl. Instrum. Methods 108:413.
Charpak, G., Majewski, S., and Sauli, F., 1975, Nucl. Instrum. Methods 126:381.
Charpak, G., and Sauli, F., 1978, Phys. Lett. 78B:523.
Charpak, G., Petersen, G., Policarpo, A., and Sauli, F., 1978, Nucl. Instrum. Methods 148:471.
Charpak, G., Majewski, S., Melchart, G., Sauli, F., and Ypsilantis, T., 1979, Nucl. Instrum. Methods 164:419.

Charpak, G., Melchart, G., Petersen, G., and Sauli, F., 1979, Nucl. Instrum. Methods 167:455.

Charpak, G., Peisert, A., Sauli, F., Cavestro, A., Vascon, M., and Zanella, G., 1981, Nucl. Instrum. Methods 180:387.

Clark, A. R., et al., (PEP-4, 1976), Proposal for a PEP facility based on the Time Projection Chamber.

Cobb, J. H., Allison, W. W. M., and Bunch, J. N., 1976, Nucl. Instrum. Methods 133:315.

Comby, G., Mangeot, Ph., Augères, J.-L., Chaudet, Y., Chalot, J. F., Tichit, J., de Lignères, H., and Zadra, A., 1980, Nucl. Instrum. Methods 174:93.

Coutrakon, G., Cribier, M., Hubbard, J. R., Mangeot, Ph., Mullié, J., Tichit, J., Bouclier, R., Breskin, A., Charpak, G., Million, J., Peisert, A., Santiard, J. C., Sauli, F., Brown, C. N., Finley, D., Glass, H., Kirz, J., and McCarty, R., 1982, IEEE Trans. Nucl. Sci. NS-29:323.

Drumm, H., Granz, B., Heintze, J., Heinzelmann, G., Heuer, R. D., van Krogh, J., Lennert, P., Nozaki, T., Rieselberg, H., Wagner, A., Eichler, R., Olsson, J., Steffen, P., Goddard, M. C., Pearce, G. F., and Warming, P., 1980, DESY 80/38.

Fabjan, C. W., and Fischer, H. G., 1980, Rep. Prog. Phys. 43:1003.

Farr, W., Heintze, J., Hellenbrand, K. H., and Walenta, M., 1978, Nucl. Instrum. Methods 156:283.

Filatova, N. A., Nigmanov, T. S., Pugachevich, V. P., Riabtsov, V. D., Shafranov, M. D., Tsyganov, E. N., Uralsky, D. V., Vodopianov, A. S., Sauli, F., and Atac, M., 1977, Nucl. Instrum. Methods 143:17.

Frehse, H., Lapique, F., Panter, M., and Piuz, F., 1978, Nucl. Instrum. Methods 156:87.

Friedrich, D., Melchart, G., Sadoulet, B., and Sauli, F., 1979, Nucl. Instrum. Methods 158:81.

Jean-Marie, B., Lepeltier, V., and L'Hote, D., 1979, Nucl. Instrum. Methods 159:213.

Korff, K. S. A., 1955, Electrons and nuclear counters (Van Nostrand, New York, 1955).

Lapique, F., and Piuz, F., 1980, Nucl. Instrum. Methods 175:247.

Lazeyras, P., Lehraus, I., Matthewson, R., and Tejessy, W., 1979, IEEE Trans. Nucl. Sci. NS-26:89.

Lehraus, I., Matthewson, R., and Tejessy, W., 1981, CERN-EF/81-14 (submitted to Nuclear Instruments and Methods).

Lehraus, I., Matthewson, R., and Tejessy, W., 1982, CERN-EF/82-1 (submitted to Nuclear Instruments and Methods).

Nygren, D., and Marx, J., 1978, Physics Today, Vol. 31, No. 10.

Palladino, V., and Sadoulet, B., 1975, Nucl. Instrum. Methods 128:323.

Peisert, A., 1982, Thesis at the University of Geneva.

Piuz, F., Roosen, R., and Timmermans, J., 1982, Nucl. Instrum. Methods 196:451.

Policarpo, A., 1981, Phys. Scripta 23:539.

Ramanantsizehena, P., 1979, CRN/HE 79-13.
Rehak, P., and Walenta, A. M., 1980, IEEE Trans. Nucl. Sci. NS-27:53.
Rice-Evans, P., 1974, Spark, streamer, proportional and drift chambers (Richelieu, London, 1974).
Ritson, D. M., 1961, Techniques of High Energy Physics (Interscience, New York, 1961).
Roberts, A., 1960, Nucl. Instrum. Methods 9:55.
Robinson, B., 1981, Phys. Scripta 23:716.
Salete Leite, M., 1980, Portugal Phys. 11:53.
Sauli, F., 1977, CERN 77-09.
Schlumbohm, H., 1958, Z. Phys. 151:563.
Séguinot, J., and Ypsilantis, T., 1977, Nucl. Instrum. Methods 142:377.
Séguinot, J., Tocqueville, J., and Ypsilantis, T., 1980, Nucl. Instrum. Methods 173:283.
Shultz, G., and Gresser, J., 1978, Nucl. Instrum. Methods 151:413.
Siegmund, D. H., Culhane, J. L., Mason, I. M., and Sanford, P. W., 1982, Nature 295:678.
Talman, R., 1979, Nucl. Instrum. Methods 159:189.
Walenta, A. H., Heintze, J., and Schürlein, B., 1971, Nucl. Instrum. Methods 92:373.
Walenta, A. H., 1979, IEEE Trans. Nucl. Sci. NS-26:73.
Walenta, A. H., Fisher, J., Okuno, H., and Wang, C. L., 1979, Nucl. Instrum. Methods 161:45.
Walenta, A. H., 1981, Phys. Scripta 23:353.
Williams, S. H., Leith, D. W., Poppe, M., and Ypsilantis, T., 1980, IEEE Trans. Nucl. Sci. NS-27:91.
Ypsilantis, T., 1981, Phys. Scripta 23:371.

LAND OF QCD

(To be sung to tune of Yellow Submarine.)

In the land, of QCD,
Lived a man, who probed the sea,
The quarks they played, the major role,
Till he found, the monopole.

Refrain:

> We all live in the land of QCD, Land of QCD, land of QCD,
> We all live in the land of QCD, land of QCD, land of QCD.

And the quarks, cannot be seen,
Some are yellow, some are green,
Some are blue, and some are red,
All this stuff is, above my head.

(Refrain)

And the group, is SU(3),
And the quarks, they are not free,
But they live, inside a bag,
And they don't come out, what a drag.

(Refrain)

Local group, the magic phrase,
And it disconnects, the points in space,
To fix it up, a B is used,
And it leaves us, all confused.

(Refrain)

A lattice is, the latest trick,
Folks are working hard, to make it stick,
They use a lot, computer time,
On just six points, it works quite fine.

(Refrain)

The only thing, that is still free,
In the land, of QCD,
Lambda bar, is a big mist,
Because it creeps, to higher twist.

(Refrain)

And we like, the monopole,
It's our friend, so we are told,
But one event, it's hard to see,
How its existence, will ever be.

(Refrain)

Unify, to SU(5),
Just to make, our theories jive,
SU(2), crossed with U(1),
Crossed with SU(3), and then it's done.

(Refrain)

SU(5), sounds very nice,
But we skate, on bumpy ice,
Lots of Higgs, we're in a mess,
Free parameters, not any less.

(Refrain)

But we like, our SU(5),
It's a group, that might survive,
But we don't, know just right now,
Will it make it, or take a bow.

(Refrain)

And this verse, it is our last,
'Cause the vacuum sucked it up.

Participants at the ASI. From left to right: G. Baranko, S. Kunori, H. J. Meyer, C. Castoldi, G. H. Wu, S. McHugh, F. Lomanno, P. Lebrun, T. Frank, C. Seez, J. Frank, M. Afdal, L. Gatignon, Y. Kubota, G. Taylor, A. Cattai, R. Wilson, R. Brock, A. Meneguzzo, P. Perez, J. D. Bjorken, P. Rankin, D. Caroumbalis, P. Auchincloss, L. Gladney, H. van Hecke, H. Weerts, N. Giokaris, M. Pimenta, G. Hogan, J. Sedgbeer, J. Fehlmann, M. Ruiz, D. Pandoulas, M. Ferguson, S. Beingessner, L. Olsen, F. Sauli, P. Ratoff, M. Diesburg, P. Kooijman, R. Steiner, C. Llewellyn Smith, S. Axelrod, D. Herrup, M. Bregman, J. Bofill, F. Dittus, A. Duncan, H. von Arb, D. Heyland, E. Lehmann, R. Nacasch, R. Horisberger, M. J. Yang, P. Soding, K. Kinoshita, C. Haber, R. Poling, F. Sciulli, T. Ferbel, T. Gentile, N. Cabibbo. Missing: F. James.

LECTURERS

J. D. Bjorken	Fermilab
N. Cabibbo	University of Rome
F. James	CERN
C. Llewellyn Smith	Oxford University
F. Sauli	CERN
F. Sciulli	Columbia University
P. Söding	DESY

SCIENTIFIC ADVISORY COMMITTEE

M. Derrick	Argonne National Laboratory
M. R. Jacob	CERN
D. W. G. S. Leith	Stanford Linear Accelerator Center
F. E. Low	Massachusetts Institute of Technology
S. Okubo	University of Rochester
C. Quigg	Fermilab
N. P. Samios	Brookhaven National Laboratory
J. Sandweiss	Yale University
R. R. Wilson	Columbia University
G. Wolf	DESY

SCIENTIFIC DIRECTOR

T. Ferbel	University of Rochester

INDEX

Avalanche dispersion, 334-336

Baryon decay, 27-33, 36-38
Beam-beam limit, 255-257
Beam dump experiments, 74-79
Beam optics, 257-269
Beta function, 258, 263, 265, 283
Betatron oscillations, 237-239, 257-259

Chebyshev fit, 221-215
Chromaticity, 268-269
Colored gluons, 164-168
Confidence intervals
  Bayesian theory, 203-204
  classical theory, 199-201
    binomial parameters, 201-202
  likelihood method, 205-207
Cosmology, 27-33

Design parameters for 500 TeV
  collision region, 283-286
  detection problems, 289-292
  injection, 286-287
  list, 296-299
  luminosity, 288-289
  RF system, 286

Diffusion coefficient, 306
Dirac monopoles, 59-60
Drift velocity, 303

Electron-positron collisions,
  and detectors, 126-130
  and photon-photon interactions, 168-177
    structure function of photon, 177-182
  and scalar particles, 140-143
  and scaling violations, 161-164
  and storage rings, 121-125
Electroweak interaction theory, 1-9, 138-140
  of leptons, 130-135
  of quarks, 135-138
Emittance, 264
Energy loss by ionization, 308-311

Fixed points, 278
Floquet transformation, 275
Focussing function, 258
FODO lattice, 264-266

Gauge theories, 2-9, 48-62
Gauss-Markov theorem, 209-210
Goodness of fit, 223-230
  Chi square test, 225-226, 228
  Kolmogorov test, 226-227, 228-229
  permutation tests, 229-230
  Smirnov-Cramer-Von Mises test, 227
Grand unified theories, 17-42
  SU(5), 19-33
GUTs (see Grand unified theories)

Hadronization, 159-161
Hill's equation, 257-259, 266, 275
Hypothesis testing, 216-222

Keil-Schnell criterion, 253-254

Least squares, 208-209
Longitudinal phase space, 243-248
Luminosity, 254-255, 257, 288-290

Momentum compaction, 244
Momentum dispersion, 241, 266-269
Monopoles, 57-62
Multistep proportional chamber, 338-344

Neutrino interactions, 79-93
  and deep-inelastic scattering, 82-90
  and scaling violations, 90-94
  and V-A theory, 79-82
  and Weinberg angle, 81-82
Neutrinos
  mass, 38-40, 66-74
  oscillations, 68-74
  Sources, 63-64
  types, 64-66
Nonlinear machine resonances, 269-282

Operating point, 282

Parameter estimation
  Gaussian measurements, 191-192
  least squares, 208-209
  likelihood method, 205-207
    maximum likelihood, 210-221, 221-222
  propagation of uncertainties, 193-198
  resolution function, 193
Particle identification
  through Cherenkov ring imaging, 332-247
    amplification, 334-338
    reconstruction, 338-341
Particle identification (continued)
  through energy loss, 316-332
    cluster counting, 329-332
    dispersion in gaseous counters, 323-328
    intrinsic resolution, 321-323
Phase function, 258-259
Plaquette, 53
Positive ion drift, 312-315
Probability theory, 189-190

QCD (see Quantum chromodynamics)
QED (see Quantum electrodynamics)
Quantum chromodynamics, 9-15
  and hadroproduction in e^+e^-, 112-118
    jet structure, 144-159
    scaling violation, 161-164
  and scaling violations 91-94
Quantum electrodynamics, 102-110
Quantum fluctuations for beams, 250-252
Quarkonium states, 118-121, 164-168

RF bucket, 246

Separatrix, 246
Single-beam instabilities, 252-254, 287
Supersymmetric models, 41
SUSY (see Supersymmetric models)
Synchrotron damping, 249-250
Synchrotron radiation, 242, 286

't Hooft-Polyakov monopole, 62
Topology, 55-59
Transition energy, 244
Transport matrix, 260
Tune, 259, 263
Tune shift, 256, 271-274